MANUAL OF
OSCILLOGRAPHY

MANUAL OF
OSCILLOGRAPHY

Howard W. Hoadley

THE FOCAL PRESS
LONDON and NEW YORK

Printed and bound in Hungary by Franklin Printing House, Budapest 1967

CONTENTS

X SYSTEMS AND PROCEDURES 327

PREFACE

The 300-year-old science of oscillography has recorded man's progress in nearly every metrological endeavor. Displaying in graphic form physical action against time, it has helped him to understand and solve his problems. In its simple form of stylus or pen recorder, the oscillograph itself was easy to comprehend. But today, oscillography is more difficult to understand because there are complex oscillographic recorders, many marking mediums, and extremely high-frequency data to be recorded. Photographic oscillography in particular has suffered from lack of knowledge of its potential and from poor co-ordination in achieving its capability.

It is the purpose of this book to acquaint the reader with the history of oscillography and its present state-of-the-art. In particular, photo-oscillography, as exemplified in light-beam and electron-beam recorders, is described from transducer to oscillogram.

The modern oscillographer, through training, experience and interest, is usually well versed in electronics and mechanics. He may be less cognizant of the photographic aspects of oscillography even though his success in light-beam and electron-beam recording is dependent upon photography. In other instances he will rely on photographic personnel who may be exceptionally adept at conventional photography, but who have no awareness of potential pitfalls that await them in this lesser known application. One purpose of this manual is to guide both individuals toward a better result—high-quality oscillograms. Therefore, practical information has been emphasized, with sufficient theoretical and reference data added for readers who may not have this background.

Oscillogram-record production in the space age is of such high volume that oscillography is often organized on a departmental basis. The final oscillogram may utilize the combined team talents of recording technicians, photographic technicians, data-reduction and data-analysis departments, and finally find its place in the oscillogram library. Therefore, this book has been written with sufficient coverage for large-scale operations. At the same time, the lower-volume user needs only to extract those data pertinent to his particular requirements or to scale down quantities to meet his own requisites.

The state-of-the-art survey covers recorders, transducers and accessories, materials, processors, readers, and support equipment. The descriptive product information is based on data furnished and subsequently reviewed by manufacturers. Although competitive products exist, the details of such items were not available during the preparation phase of this publication. The author expresses his appreciation to those firms who have co-operated on this project, but he has not evaluated the products and makes no guarantee of performance.

No attempt has been made to completely survey the numerous manufacturers of transducers because such an undertaking is beyond the scope of this single volume. Instead, transducer principles are discussed, and the examples of hardware are for the most part those supplied by oscillograph-recorder manufacturers.

There is every reason to expect that the growth and change in oscillography will occur at an ever increasing pace. In fact, technological progress has required revision to the manuscript even as it was being written. Therefore, the information contained in this manual reflects current models, materials, and methods, and future progress is expected to be reported in a revised edition.

Because of its widespread use, the oscilloscope and the recording of its CRT images has been well documented in electronic and photographic literature. Today, simple single-shot recording cameras are available to fit nearly every oscilloscope. It is easy to make and assess photo-oscillograms made on such equipment, for often Polaroid materials are used. However, continuous and drum recording of cathode-ray tube images, electron-beam recording, and light-beam recording are more involved. Light-beam oscillography in particular has suffered from the lack of printed information, not only concerning the recorder, but also the processing of the paper to an oscillogram. Technical representatives of equipment and materials manufacturers have long been aware of this problem. One representative, Harold I. Smith, recognizing the need for a manual on oscillography, generated with the author the basic concept for this book. Later, the efforts of John H. Jacobs brought about the fulfillment of the plan. The author expresses his appreciation for their consultation, zeal, and energy in shaping the contents of the manuscript.

Special thanks are also due to two other authorities. The author is grateful to Thurlow M. Morrow for his guidance on metrology, and to A. C. "Tex" Thomsen for his review of the photographic aspects of this book.

A 1960 issue of a Society of Photographic Scientists and Engineers periodical carried an editorial in which I wrote: "What will the '60's bring? Continued perfection and miniaturization of course. But this decade will be marked by the putting together of the known in exotic combinations to form new concepts, methods, and materials... These will emerge at such an accelerated change-rate that photography ten years from now may bear slight resemblance to photography today."

Progress in the field of photo-oscillography is being paced by advances made not only in photography, but by those made in electronics as well. Now that we have passed the halfway mark in this decade we can see that discoveries of these two sciences have combined to advance oscillography to new heights. The goals of compact equipment and improved records obtained under rapid access conditions have already been realized, and each advancement is being continually improved.

Howard W. Hoadley

ACKNOWLEDGEMENTS

The author sincerely appreciates the assistance and cooperation, and the information and illustrative material furnished by the following organizations:

Agfa-Gevaert

Analab Instrument Corporation

Anken Chemical & Film Corp.

Beattie-Coleman Products, Coleman Engineering Company, Inc.

Bell & Howell Company

Benson-Lehner Corporation

Brooks Instrument Company, Inc.

Brush Instruments

The Budd Company, Instrument Division

Century Electronics & Instruments

Chicago Aerial Industries, Inc.

Consolidated Electrodynamics Corporation

Cossor Instruments Limited

E. I. du Pont de Nemours & Company

Eastman Kodak Company

Edgerton, Germeshausen & Grier, Inc.

Electro-Medical Laboratory, Inc.

Endevco Corporation

Fairchild Industrial Products and Fairchild Instrumentation, Divisions of Fairchild Camera and Instrument Corporation

Oscar Fisher Company, Inc.

General Aniline & Film Corporation

General Atronics Corporation

The Geotechnical Corporation

The Gerber Scientific Instrument Co.

Granger Associates

Hartmann & Braun AG

Hathaway Denver, Division of Hathaway Instruments, Inc.

Hi-Speed Equipment Inc.

Honeywell Denver Division

Houston Fearless Corporation

Ilford Limited

Information International Inc.

Itek Business Products Division, Photostat Corporation

Kodak Limited

Lawler Automatic Controls, Inc.
Midwestern Instruments
Minnesota Mining & Manufacturing Company
The Morse Instrument Company
Offner Division of Beckman Instruments
Osram GmbH
N. V. Phillips' Gloeilampenfabrieken
PEK Labs Inc.
Photo-Control Corp.
Polaroid Corporation
Ramapo Instrument Company, Inc.
Sanborn Company
Savage and Parsons Limited
SE Laboratories (Engineering) Limited
Siemens & Halske AG
Southern Instruments Limited
Southern Railway System
Tektronix, Inc.
Thermal Instrument Company
Trüb, Taüber & Co., AG
UGC Instruments, Inc.
United States Radium Corporation
Vought Camera Division, Computer Equipment Corporation
Frederick P. Warrick Company, Electronics Division of McDonnell Aircraft Corporation
Western Electrodynamics, Inc.
Xerox Corporation
Yellow Springs Instrument Company

I INTRODUCTION TO OSCILLOGRAPHY

History

Wherever man travels today–whether in his quest for the distant planets or in seeking renewed life at the skilled hands of the surgeon–his adventure is chronicled by deft fingers writing each episode as it unfolds. With quiet, precise perception, sensed in chronological progression, the scene, its atmosphere, and its principals are jotted on the scroll. Its author is oscillography.

The first known recording instrument appeared about 300 years ago in Oxford, England, and was used to record varying temperatures against a time base. The early recording techniques included the use of a needle to scratch through a smoked surface previously applied to glass or paper.[1.1] During its evolution, numerous materials and marking methods were attempted as increased mechanization of the 19th and 20th centuries created the need for accurate recording media. This is exemplified in the following except from a Southern Instruments publication: "The earliest type of engine indicator was devised by James Watt during the early part of the last century for use on steam engines to determine the horse-power from the pressure-piston displacement relation. These small compact instruments had a paper chart on a drum and the pressure scale was obtained by the movement of a small spring-loaded piston that moved a pencil up and down the chart through a link mechanism. At the same time the engine piston displacement was obtained by rotating the drum and chart by a cord attached to the crosshead. The diagram or 'card' so obtained was informative and led to more improvements in the design of valve gear and compound engines. In due course the advent of the internal combustion engine necessitated similar cylinder pressure measurements and all kinds of slow and medium speed engines including diesel types and air-compressors have been and still are being indicated by these extremely simple and reliable mechanical devices.

"The mechanical indicator has, however, the inevitable disadvantage of inertia which reduces the speed with which it can respond. Even with the latest improved types there comes a point beyond which the accuracy of measurement decreases considerably as the engine speed increases. The need for instruments capable of faithfully indicating cylinder pressure fluctuations in high-speed internal-combustion engines has led to the development of a number of extremely useful and interesting types, some of which use electric spark systems to mark the chart. The best known and most versatile, however, use a cathode-ray tube as the display element".

The following paragraph, reprinted here from "Kodak Recording Materials", published by Kodak, Limited, places the first use of photographic recording about a century ago:

"The advantages of photographic recording were appreciated nearly a hundred years ago. In fact, a photothermograph and a photobarograph were developed and used by the Royal Meteorological Society in 1867 to satisfy the need for continuous records of temperature and pressure. Tribute must be paid to the ingenuity of the 19th-century scientists who first devised such equipment, which can be seen at the Science Museum, London, England."

The development of recorders, however, continued to be centered for many years around mechanical, ink, pressure, heat, and electric marking instruments. The principles of each are described in Chapter 2.

A review of mechanical recorders and early photo-oscillographs will reveal how inherent limitations of each system led to the development of modern rapid-access equipment. The mechanical-stylus recorder had a useful frequency range of only 0–100 cycles per second. In addition, mechanical restrictions of size and stylus swing limited the number of information channels and amplitude of the deflection that could be recorded on paper of a given width. The advent of photo-oscillography overcame the low frequency-response limitation, and made it possible to overlap traces so that as many as sixty channels might be recorded on 12-inch wide sensitized paper. However, the paper had to be chemically developed and this precluded immediate readout.

Since this book is primarily concerned with photographic oscillography, let us now review some of the milestones in the development of photographic recorders (see Fig. 1.1). The following descriptions of equipment avaliable at the close of World War II are reprinted with permission from the Eastman Kodak publication "Kodak Recording Materials".

"Trace-recording instruments in present use fall into two general types which differ in the source and characteristics of the illumination. Group one includes recorders using tungsten illumination where the position of the recording trace is controlled by a galvanometer or other indicating instrument. Group two includes instruments using cathode-ray tubes. Recorders of the first group are widely used for slow and medium speed phenomena, while cathode-ray instruments – though useful at lower speeds – are particularly valuable for high-speed work.

"In oscillographs and other recording instruments using tungsten illumination, either the String or Einthoven Galvanometer or the Moving Mirror Galvanometer measures the electrical impulses.

"A string galvanometer is highly sensitive to extremely weak electrical impulses, such as the electrical energy set free by the contraction of the heart muscles. Fundamentally, it consists of a string of suitable conducting material suspended under tension in a magnetic field, causing it to be displaced in proportion to the current passing through the string. The photographic record is produced by the movement of the shadow cast by the string. Since the displacement of the string is extremely small, an optical system is used to enlarge it, sometimes as much as a thousand times. Even this enormous magnification produces only a relatively small trace, which is in the form of an unexposed line on an exposed background.

"In the moving mirror galvanometer, a tiny mirror is mounted on the moving

Fig. 1.1. Typical photographic oscillographs. Produced in the 1940's. *Top*: William Miller Oscillograph.
Bottom: Early CEC Oscillograph

element, which may take any of several forms, depending on the type of instrument. Light from an illuminated slit is reflected by this mirror through a lens which forms an image of the slit on the surface of the recording paper or film. As the mirror is rocked by the galvanometer element, the spot of light is moved back and forth across the sensitive material, and after development the trace appears as a dark line on a light background (see Fig. 1.3)".

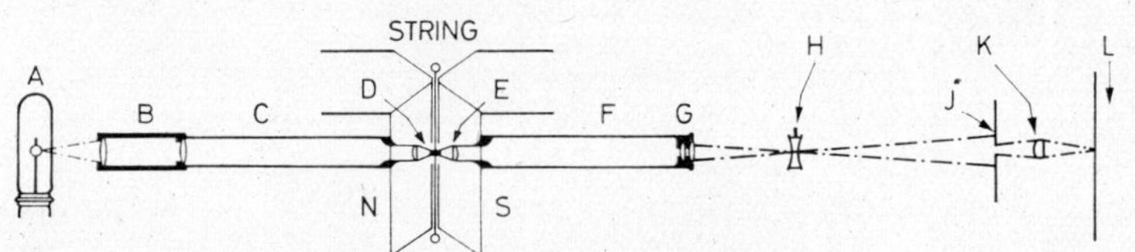

Fig. 1.2. Schematic diagram of string galvanometer type electrocardiograph. A — Light Source B, C, D, E, F, G, K, — Optical System H — Time Marker J — Sutter L — Recording Material N, S, — Magnetic Poles

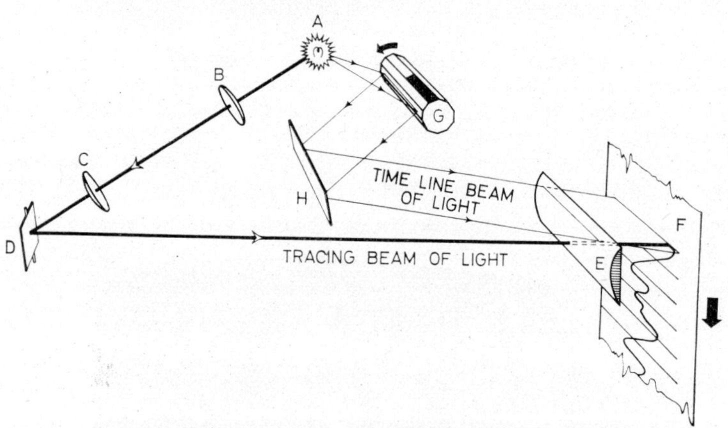

Fig. 1.3. Schematic diagram of moving-mirror type oscillograph. A — Light source; B, C, E — Optical system; D — Galvanometer mirror; G, H — Time-line reflecting mirrors; F — Recording material

PATENT HISTORY. Present-day photographic galvanometer recorders are the outgrowth of two areas of development: mechanical-recorder improvement early in this century, coupled to the photographic-recording medium prior to World War II. Poorly documented in literature, the historical facts about early photographic recorders require diligent search. A more fruitful survey can be obtained from early patents which eloquently convey many novel approaches.

Some of the patents described here represent truly significant work. Others in this cross section are an insight into the thinking that developed today's recorders. It is the author's intent to trace some of the thinking, rather than to present a cold, analytical patent search.

20

Extensive contributions to photo-oscillography were made by members of the seismology profession. A U. S. Patent, Number 1, 451, 080, entitled "Field Seismograph", was issued to Ludger Mintrop, of Hanover, Germany, on April 10, 1923. The patent claims include the use of a light source, preferably electric, lenses to collimate the light, and a mirror to reflect the beam onto photosensitive material.

The reduction of friction in stylus-type recorders received extensive consideration, and light was one way to eliminate it. A U. S. Patent, Number 1, 597, 487, issued August 24, 1926, to Byron W. St. Clair, proposed a stylus made of fused quartz to transmit radiant energy. The patent describes the transmission of both light and ultraviolet radiation from an electric lamp at one end of the quartz stylus, around its curved end, to the stylus tip. The tip had a lens configuration to concentrate the radiation on the sensitized material. Infrared radiation could also be used as a source for heat-sensitive paper recordings.

As early as 1921, the use of highly specialized electric arcs, involving the use of iron or other metallic terminals, or the use of such sources as vacuum tubes and mercury arcs was proposed as a source of radiant energy rich in extreme ultraviolet. A U. S. Patent, Number 1, 385, 657, was issued to Louis Bell and Norman Marshall. It discussed the use of radiant sources which could be focused by refraction or reflection, and then recorded by exposure of a suitable sensitized film.

SOUND-RECORDING PATENTS. In the mid and late 1920 period, the advent of sound for motion pictures greatly advanced the state-of-art of light-beam recording. Research and development were conducted on sensitized materials, chemistry, light sources, and optics. All facets of science related to sound, its transmission and amplification, the techniques for recording and reproducing it, were explored. Naturally, the fund of knowledge gained for this specific art spilled over into related fields, one of which was oscillography. Many patents were issued, a few of which are given in the list that follows: U. S. Patents, Number 1, 482, 119 to Lee de Forest, January 29, 1924; Number 1, 590, 185 to Josef Engl et al, June 29, 1926; and Number 2, 030, 760 to C. L. Oswald et al, February 11, 1936.

GEOPHYSICAL RECORDER PATENTS. The science of seismology, particularly the application of the seismic method to oil prospecting used by the oil industry, was responsible for many improvements of photographic oscillography in the United States. In addition, related instruments developed for well drilling made use of photographic methods. Among such instruments were well-logging devices and inclinometers. A U. S. Patent, Number 2, 072, 950, granted March 9, 1937, to F. W. Huber, disclosed a method of recording two curves on photosensitive material. The curves, recorded from electrodes lowered into the drill hole, represented vertical distance and resistivity of the formation adjacent to the hole (at the depth indicated on the first curve). Improvements and refinements to the Huber patent were disclosed in U. S. Patent, Number 2, 170, 857, issued to Raymond D. Elliott, in 1939.

In well-drilling operations, the drill bit is sometimes deflected from the desired course by inclined strata, or by other causes. It is, therefore, necessary that the driller check the inclination of the bore hole from time to time as drilling proceebs, to verify the true path of the drill. The deviation, known as "drift", is recorded dy a mechanical or a photographic recorder. In operation, the recording device is lowered to the bottom of the bore hole, the record is made, and the instrument is withdrawn. The chart indicates inclination and azimuth as determined by the position and location of the device during the recording period.

In early mechanical drift recorders, the recording was made by a pendulum or plumb bob, the point of which pierced the chart. The device was ruggedly constructed to withstand the shock of piercing the chart. This conflicted with the need for sufficient sensitivity to respond to position changes and resulted in inaccuralies of measurement. Ideas for better equipment created a flurry of patent applications being filed in the late 1930 and early 1940 period. The new devices were based on the principle of photographic recording.

A U. S. Patent, Number 2, 246, 519, issued to Philip Jones on June 24, 1941, describes a drift recorder which could use printing-out paper. The change from developing-out to printing-out material was dictated primarily by the recording operation. Most drift recorders had a built-in timing mechanism which was set before the recorder was lowered into the bore hole. In making the timer setting, the time necessary to position the recording device had to be calculated. Simultaneously, the start time had to be logged and monitored. When insufficient time was allowed, the chart would not give an indication of the drift at the bottom of the bore hole. When too much time was allowed, the drilling crew had to wait until the preset-time lag elapsed and the record was made, a costly and time consuming procedure. Therefore, the longer exposure, required to produce a drift indication on printing-out paper, allowed a greater tolerance in timing. By using printing-out paper, a visible trace could only occur when the light beam was stationary. Any motion occurring during descent would not be recorded if the exposure began before the bottom was reached. The use of printing-out paper also had the advantage that it required no darkroom, and did not have to be developed. Processing procedures were unsuited to field conditions and to personnel unskilled in photographic development.

Among other patents that also mention the use of printing-out material are U. S. Patents, Numbers 2, 255, 295, issued September 9, 1941, and 2, 327, 658, issued August 24, 1943, to Leonidas C. Miller; U. S. Patent, Number 2, 296, 996, issued September 29, 1942, to Philip Jones; and U. S. Patent, Number 2, 413, 005, issued December 24, 1946, to George A. Smith.

Improvements to methods for geophysical exploration were described by Carl A. Heiland in a 1932 publication of the American Institute of Mining and Metallurgical Engineers. The magnetic measurements were to be made from an aircraft. Further development of the aerial gravimetric-prospecting method resulted in a patent application being filed by Heiland in 1946. U. S. Patents, Numbers 2, 626, 525 and 2, 659, 859, were issued in 1953, the latter describing and illustrating a recording system for aeromagnetic measurements.

Heiland's patents covered many techniques and equipment, among them the application of the "hollow charge" explosive technique to beam seismic energy for mapping subsurface geological formations (U. S. 2, 601, 522). Other patents related to recording equipment: U. S. Patents, 2, 535, 065; 2, 677, 105, and 2, 678, 424 for galvanometers; and 2, 580, 427 for a recording system.

Fault-recorder Patents. In the 1920's, the increased use of electric power caused utility companies to install larger and larger power stations. This resulted in more economical power distribution, but the larger capacity, interconnected, and long distribution networks were more prone to line surges and transients. These line variations, sometimes evidencing themselves at points far distant from the point where the original fault had occurred, forced utility companies to make an extensive study of the problem. Recording instruments played an important part in such investigations, and improvements to the recording systems had to be developed. The fault-type recorder was, therefore, the byproduct of trouble shooting power distribution. Development of such special-purpose devices has continued at a steady pace to the present day.

Two patents for such instruments were issued to Chester I. Hall (U. S. 1, 713, 226, issued in 1929); and to Howard E. Dyche (U. S. 1, 824, 469, issued in 1931).

Medical-recorder Patents. The medical profession also adopted the use of photographic recording methods. In 1931, a U. S. Patent, Number 1, 794, 685, was granted to A. T. Hayman and Guy L. Campbell, for a cardiological instrument employing film as the recording medium.

Emil Blum was granted U. S. Patent, Number 1, 903, 890, on April 18, 1933, for a portable optical polygraph. The recording device featured spherical condensing lenses, and had a film magazine and transport section that could be removed and taken to the darkroom for development.

Two patents, U. S. Number 2, 099, 938, and U. S. Number 2, 298, 574, were issued to M. L. Lockhard for an electrostethograph. The apparatus used a microphone and amplifier to drive the galvanometer. Removable film magazines were employed, the takeup magazine having a self-threading feature and a cut-off device. The use of a filter to screen out infrared and to allow ultraviolet to pass was mentioned.

The application of photography to the medical field was not limited to sphygmomanometers, pulsimeters, and other similar recorders. A U. S. Patent, Number 2, 184, 131, was issued to E. A. Taylor in 1939, for an eye-treatment instrument. The equipment, used for recording simultaneous eye movements, consisted of a light source which was reflected from each eye and focused by a lens system onto the moving film. The use of materials sensitized to ultraviolet and infrared, as well as visible rays, was discussed.

Sensitized-material Patents. The foundations for direct-writing photographic recorders were laid in the early days of photography with the development of the positive printing process.[1,2]

In 1921, James Addison Johnson filed a patent application for an improvement of print-out papers. The basis of his claims, for an emulsion several hundred times faster than then existing print-out materials, was the use of silver iodide, rendered light sensitive by the action of sensitizers. The use of the paper for "self-recording meters" and "radium pencil" devices is mentioned, and the ability to shift the spectral sensitivity of the material by the use of sensitizing dyes is described. Johnson was granted a U. S. Patent, Number 1, 582, 050, on April 27, 1926.

Samuel E. Sheppard and Leon W. Eberlin were granted U. S. Patent, Number 1, 934, 451 in 1933 for the use of certain metals to sensitize paper to ultraviolet. Formulas using molybdenum, tellurium, and tungsten compounds as sensitizers for the 3, 500 to 4, 000 Angstrom-unit range are given.

In 1937, Sheppard and Waldemar Vanselow were granted a U. S. Patent, Number 2, 066, 582, for ultraviolet printing-out materials, using pyrazolone compounds as the UV sensitizer.

A series of patents for print-out emulsions was granted to George E. Fallesen in collaboration with others during the mid-1930 period. The patents, assigned to Eastman Kodak Company, achieved increased sensitivity by the addition of halogen absorbers and organic and inorganic alkalies. The U. S. Patent numbers are 2,030, 860, 2, 126, 318, 2, 126, 319, and 2, 129, 207, issued between 1936 and 1938.

The work of Dr. Edith Weyde led to significant developments in direct print-out recording. Dr. Weyde, with associates Drs. Gustav Schaum and Werner Stracke, disclosed in German Patents, Numbers 872, 155, and 899, 586, methods for producing improved print-out paper for photo-recording purposes.

Discussed in the patents is the use of a high-intensity, high-pressure mercury lamp for the original exposure, followed by additional exposure with light of lower intensity. The use of halogen acceptors, such as thiourea, and emulsions preferably composed of a mixture of silver thiocyanate and silver bromide, are among the claims. Image formation in the emulsion within approximately 1/10th second after exposure, is mentioned in one claim.

Related work published[1,3] by Dr. Weyde reviewed the development of the Agfa Type L print-out paper as used in the Hartmann & Braun Lumiscript recorder.

While some researchers were directing their efforts toward better emulsions, others were giving attention to the support. Present day transparentized paper stocks are the result of many years of development of base material, transparentizers, coatings, and rigid control of materials. A U. S. Patent, Number 1, 997, 745, issued to Max Renker on April 16, 1935, discussed the manufacture of translucent parchment-like papers for photographic use. Methods of applying cellulose esters and/or cellulose ether, together with other coatings, were described.

HIGH-INTENSITY LIGHT SOURCES. Some 60 years elapsed between the first practical incandescent lamp and the development of a point-source glow-discharge lamp, suitable for exposing print-out materials in a galvanometer recorder.

In 1879, Edison constructed the first feasible incandescent lamp. Three years later, Siemens and Halske established a lamp manufacturing plant. In 1884, the German Edison Company, and in 1901, the Auer Company, began manufacture of incandescent lamps. In 1918, these three firms merged, using the "Osram" trademark, already registered in 1906. The combined research facilities produced many lamp improvements, among them the high-pressure mercury and the xenon gaseous-discharge lamps. The point-source version of this quartz tube, when teamed with improved print-out papers, produced a breakthrough in oscillography.

Typical of the research on high-pressure, point source discharge lamps, was the work of Dr. Rompe, Kurt Ittig, and Wolfgang Thouret, who were granted German Patent, Number 767, 232. The patent discloses electrode clearances, voltages, pressures, and construction materials. The use of various gases, including argon, krypton, and xenon, is discussed and compared with mercury.

An Osram publication tells of research by P. Schulz in the early 1940's on the use of xenon in high pressure lamps. The xenon "brings about a spectrum closely resembling that of natural daylight." The first commercially available xenon lamps had a life of only 500 hours. In over a decade of development and production experience at Osram, the life was increased to 2, 000 hours for the XBO short arc series.

Print-out Recorder Patents. By the early 1940's, the state-of-the-arts of relatively highspeed print-out paper, and of a point-source electric lamp suitable for exposing it, had progressed sufficiently to produce a practical photographic print-out paper galvanometer recorder.

In German Patent, Number 879, 921, Heinrich Stabe disclosed the use of print-out paper and a high-intensity ultraviolet light source. Within a two-year period, German Patents, Numbers 880, 496, 900, 754, and 904, 474, were also issued to Stabe. Each described the application of the new method for various types of recorders.

Stabe also wrote a paper[1.4] in which he told of his experiments using a high pressure mercury lamp developed by Osram to expose a print-out paper produced by Agfa. His research was instrumental in the development of the print-out recorder, and its later production by Hartmann & Braun.

In the United States, Dr. Carl A. Heiland filed a patent application on August 11, 1944, for a direct print-out recording system. The claims included the use of a source of invisible radiations in conjunction with a specially prepared photo-sensitive material primarily responsive to those radiations. A U. S. Patent, Number 2, 580, 427, was granted to Heiland for this invention on January 1, 1952. A copending application covering an improved galvanometer for the recorder, filed on the same date, resulted in U. S. Patent, Number 2, 535, 065 being issued on December 26, 1950. In 1956, the Heiland Division of Minneapolis-Honeywell, (now Honeywell, Denver Division), Denver, Colorado, announced the Model 906 Visicorder, utilizing ultraviolet radiation to write on print-out paper. Usable records to 5, 000 cycles per second, and the convenience of the print-out method, caused immediate acceptance of the recorder. Improved, models, with more channels and added features soon appeared.

Patent applications were filed September 14, 1959, by R. P. Brown and J. H. Jacobs for a rapid latensification method. U. S. Patent Numbers 3, 143, 940 and 3,

144, 332 were issued April 11, 1964, assigned to Consolidated Electrodynamics Corporation. The claims covered heating the print-out emulsion while latensifying it with intense illumination from black light blue (BLB) lamps. This method is sold by CEC under the trade name "Dataflash."

INTEGRAL RECORDING AND DEVELOPMENT. A 35 mm processor, simple, compact, and clean enough in operation to be built into recording equipment was described by G. I. P. Levenson.[1,5] Following their application of March 9, 1948 for the hot-drum machine, British Patent No. 647, 922 was issued December 28, 1950, to A. Batley, E. R. Davies, and G. I. P. Levenson.

In 1953, Carl Heiland disclosed in a patent application, methods for pre-exposure chemical developer application on developout material to provide immediate readout at higher speeds. A U. S. Patent, Number 2, 902, 331, was issued to him for this invention on September 1, 1959.

Integral post-exposure high-temperature chemical development received attention from Dr. C. F. Robinson and J. H. Jacobs of Consolidated Electrodynamics, who filed a U. S. patent application in 1957, resulting in U. S. Patent number 3, 130, 000. They also filed a British application in 1958 on the CEC "Datarite" system, which resulted in a British Patent, Number 888, 700 being issued on January 31, 1962. Jacobs also received a U. S. Patent, Number 3, 088, 824 in 1963 for the specialized chemistry used in the process. In the same year, H. I. Smith received a U.S. Patent, Number 3, 094, 914, for improvements to the developer-application station of the Datarite equipment.

OSCILLOGRAM PROCESSORS. Processing of developing-out records was initially done by hand and with batch-type rewind processors ordinarily used to develop aerial roll films. For many years, the rewind-type machine was the principal method for processing oscillograms. While it was satisfactory for limited production, the increased use of oscillography created a demand for a rapid, reliable, continuous processor. Finally, compact, continuous stabilization machines were introduced in the mid 1950's, and produced a rapid and progressive advance from hand processing to continuous machine-processing methods.

Early stabilization processors were manufactured by several firms, among them Consolidated Electrodynamics, General Electric, and Southern Equipment, Limited. The compact size and convenience of such processors proved a valuable aid for handling the ever increasing volume of data required by the new-born space age. However, the print-out technique, which became available at about the same time, caused many users to change to that method. Between the shorter stabilization-processing time required for developing-out papers, and the use of print-out materials, the user was better able to cope with the deluge of records. The integral recorder-processor, the CEC Datarite, was also an important advancement toward quick access. By the late 1950's, recording and read-out, at close to real time, had become a goal sought by nearly every recording-equipment manufacturer.

As print-out methods became more commonplace, recorders and materials were improved. There was a consequent shift to the print-out technique for many fields of activity that previously had been the exclusive province of the develop-out method. Naturally this resulted in more and more pressure being brought to bear, to reduce even further, the time required in stabilization processing. General Electric and Southern Equipment discontinued their processors, leaving CEC to fill the market for stabilization processing equipment. The CEC processor was improved, and in 1961, a new model made its appearance.

During the 1950's, users, faced with the problem of processing large volumes of charts, purchased standard processors and modified them. Equipment of the type used by photofinishers was often reworked and used to process paper records at speeds in the range of 40 to 75 feet per minute. A few special machines were designed and installed on research contracts, but were not marketed.

In 1962, Eastman Kodak Company announced its Ektaline 200, Model 1M processor. This continuous stabilization machine has the capability of processing at speeds up to 200 feet per minute. A new paper and special process-chemistry had to be developed with characteristics suitable for use in the new machine.

COLOR OSCILLOGRAPHY. During World War II, aerial reconnaissance in color was practiced with integral-tripack color materials. These wide-film materials could be processed in field laboratories by armed services personnel.

It was a natural step for oscillographers to adapt such materials to record oscillogram traces in color instead of black. The method had the advantages of positive trace identification, and made reading easier and faster. The disadvantages were the high cost of color film and the complex and long processing procedure.

Among the materials used for color oscillography in the 1940's were Aerial Kodacolor and Anscocolor reversal films. These films had integral tripack emulsions (a single support with three light-recording layers on one side). Each layer was sensitive to a specific spectral range. The three layers were capable of recording the full visible spectrum. During processing, the reversed image was converted from silver to dye in the appropriate emulsion layer, resulting in a subtractive transparency.

Although color materials were improved in quality and speed during the 1950's, the costs were still prohibitively high and the process too critical for oscillography. One simplified process, Chromart, briefly created some interest as a means of recording in color.

In 1963 both Ansco and Eastman Kodak announced new papers suitable for photorecording in color. In these simplified processes, color is introduced by filters in the light beam of the galvanometer oscillograph. The exposed oscillogram can be processed in a four-tank stabilization processor with proper color processing chemicals in the tanks. Details on color oscillography are given in Chapter 5.

THE CATHODE-RAY TUBE. The discovery of the electron is said to date from 1897. In that year, Sir J. J. Thomson, at the Cavendish Laboratory in Cambridge, measured the ratio of the charge of the electron to its mass. For his historic experiments he used the cathode-ray tube, shown schematically in Fig. 1.4. It can be seen that the tube is remarkably similar to present-day devices, not only in configuration, but also its methods of electrostatic and electromagnetic deflection. Although H. A. Lorentz and P. Zeeman, using a different method, had measured the charge-mass ratio during the previous year, Thomson is usually credited with the discovery of the electron.

Professor F. Braun first applied the cathode-ray tube to industrial testing (also in 1897). Although his instrument was more rudimentary than Thomson's, it nevertheless had the shape of present-day tubes, which increased the possibility for accurate readings.

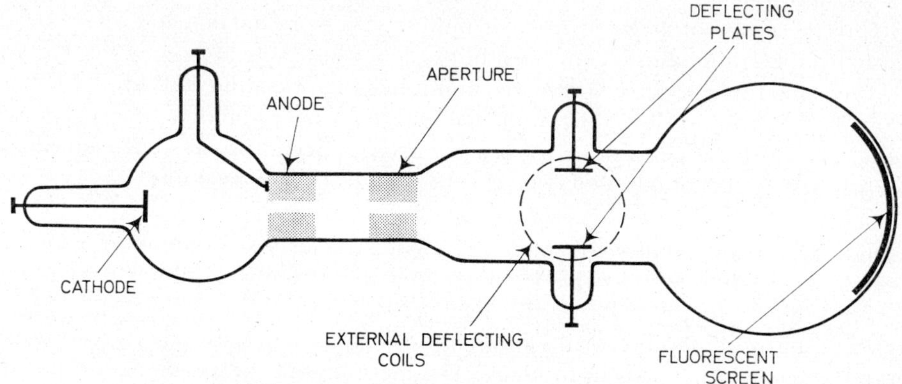

Fig. 1.4. Diagram of cathode-ray tube used by Sir J. J. Thomson

MAGNETIC-TAPE DATA RECORDING. In recent years, data recording via the magnetic tape method has gained prominence. This medium, with its broad frequency range capability, and its associated integration with computers, was at one time expected to make the oscillograph obsolete. However, the volume of oscillogram data continues to increase as new recorders, direct-readout, and rapid processing systems appear on the market.

Better equipment and materials are not the only reasons for using oscillography. The following quotation from an article in "CEC Recordings", first quarter, 1963, by R. J. Horak, gives these explanations for its popularity:

"The primary virtue of the oscillograph is its presentation of data in graphical form. Records permit the engineer to first, "intuitively" examine trends and slopes; second, determine correlation between different variables; and finally extract quantitative precision data for more exacting calculations or performance determination. Ordinarily, these latter calculations need only to be performed at a few selected times during a given test. Thus, the oscillographic record permits the engineer to do only those calculations he "sees" as necessary. This "selectable data reduction" feature, coupled with the low original cost per channel, offers both the lowest cost and fastest retrieval time per USED data point.

"The second major virtue of the oscillograph is its ability to provide good dynamic response as well as precision d-c response. The instrumentation engineer does not necessarily have to know before the test the probability of test-item-instability.

"The selection of any oscillograph must be determined by the nature of the data and its ultimate use. In a production test for performance, for example, the time of the critical data is known beforehand. In this case, digital processing may prove superior to oscillography since only useful data need be handled (through the use of test preprogramming). Consequently, digital techniques will offer the lowest cost per HANDLED data point. However, if dynamic data is necessary, as well as discrete time performance data, the oscillograph may be needed as an adjunct or may be the best overall single-system compromise."

In many cases, oscillography is the quick look, and magnetic tape recording the backup and the means of direct computer handling. There are also many instances in which magnetic tape is played back and re-recorded photographically in galvanometer oscillograph recorders.

PHOTOCOPY HISTORY. Generally, there is a need to make copies of an oscillogram or of some portion of it that contains critical data. Today, diazo and numerous other photographic processes are available to provide many types of duplicates; these are discussed in Chapter 11.

Modern photocopy methods were popularized by a special camera, capable of making readable, negative-photographic-paper prints in a single operation. The principle is as follows: The camera is equipped with a 45-degree prism in front of the lens. The prism transposes the image laterally so that a readable image is obtained on the emulsion side of the sensitized material. This permits one-step prints to be made on photographic paper.

The technique has been in general use since close to the beginning of this century. Rene Graffin, a professor at the Institute Catholique de Paris, was awarded a silver medal for the prism photocopy camera which he entered in the International Exposition at Paris in 1900.

In the United States, the first Photostat camera, shown in Fig. 1.5, made its appearance in 1909. It was preceeded by a competitor, the Rectigraph, which was built in Rochester, New York, by George C. Beidler. Both the Photostat and the Rectigraph cameras are still used for making copies of all kinds of documents, letters, office material, printed matter, and records. Photostat Corporation is now the Itek Business Products Division of Itek Corporation.

The Rectigraph is manufactured by Xerox, (the firm that also markets xerographic copiers described in Chapter 11). The Commercial Rectigraph Model 3 is shown in Fig. 1.5.

In Europe the prism-camera method has been utilized in models similar to the Photostat and Rectigraph. These are known by such trade names as the Alos Super-Repro, the Fotocopiest, the Kontophot, the Lucigraph, the Omniphot, the Technophot, and others.

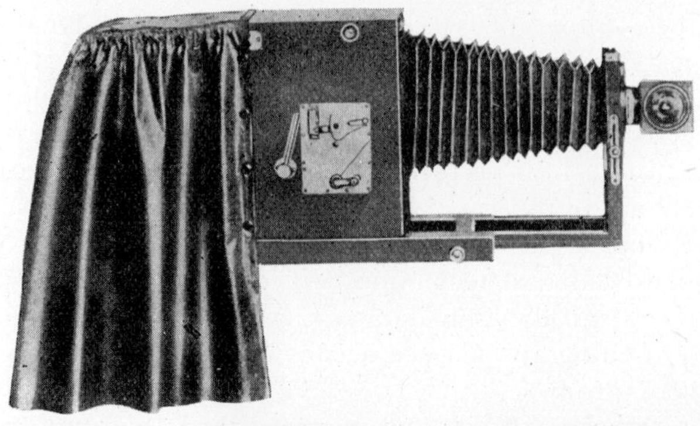

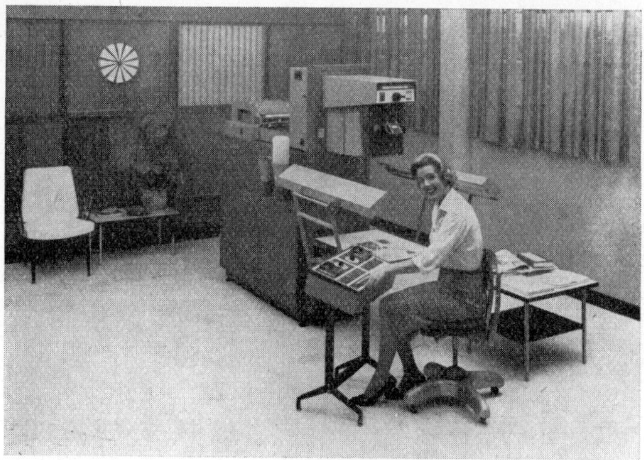

Fig. 1.5. Prism-type copy cameras. *Top*: The original model Photostat camera is compared here with the Photostat 10.14 Photocopier. *Bottom*: The Commercial Rectigraph Model 3.

GEOPHYSICAL OSCILLOGRAPHY. Seismography, the science of recording natural and man-made earth motions, is used to detect and to evaluate earthquakes and explosions. Such data are necessary in the fields of geology, geomagnetism, meteorology, and oceanography. Practical applications include oil exploration and geological surveys for civil engineering purposes.

The following description, reprinted from a publication of Century Electronics and Instruments, illustrates how the seismic refraction method is used:

"The problem of determining the thickness of overburden and depth to bedrock has always been of a pertinent nature in the civil engineering, construction and mining fields. This problem has been solved by numerous means, including the core drill, resistivity and, more recently, the seismic method.

"With the proper use of the seismic refraction method and equipment, the depth to media of certain velocity differentials can be accurately determined. When these data are plotted, a graphic cross-section of the earth (up to certain limits, dependent upon the application) will be determined. This makes it possible to conduct a survey over any kind of terrain and under conditions such as river crossings, metropolitan areas, etc.

"The seismic method for determining thickness of overburden and depth of bedrocks is based on the measurement of the velocity of a seismic wave front passing through the earth. Ordinarily such a wave front is generated by exploding a small dynamite charge.

"After the explosion, the arrival of a refracted wave front is picked up by the seismometer. The mechanical earth-motion is changed to an electrical impulse or signal, which is transmitted by means of a cable to an amplifier. The signal is amplified and transmitted to the galvanometer, where it is recorded.

"Figs. 1.6 and 1.7 illustrate a typical seismograph set-up or 'spread', showing the seismometers placed in progression at predetermined intervals away from the point of explosion, or shot-point. The recording apparatus is usually set up near the shot-point. A small charge of dynamite is placed in a shallow bored hole and tamped with dirt. When the charge is exploded, the exact instant of the explosion is recorded on the oscillogram (Fig. 1.7(b)) as 'shot time' or 'time break'. The energy from the explosion radiates in all directions. Figure 1–7(c) shows the direct and refracted paths of the energy or wave front. The direct or horizontal wave reaches the seismometers nearest the shotpoint immediately through the low velocity overburden because of the short travel path. As the wave front progresses downward, it soon reaches bedrock which is much more dense in its composition and has a much higher velocity. This change in media causes some of the energy to be refracted along and through the bedrock. Because of the higher velocity the travel time is greatly accelerated and, consequently, the refracted wave will soon overtake and pass the horizontal wave. It may be noted in Fig. 1.7(c) that this occurs at Station 4. Until this time, all that has been recorded is the travel time of the wave through the overburden. Now that the refracted wave has passed the horizontal

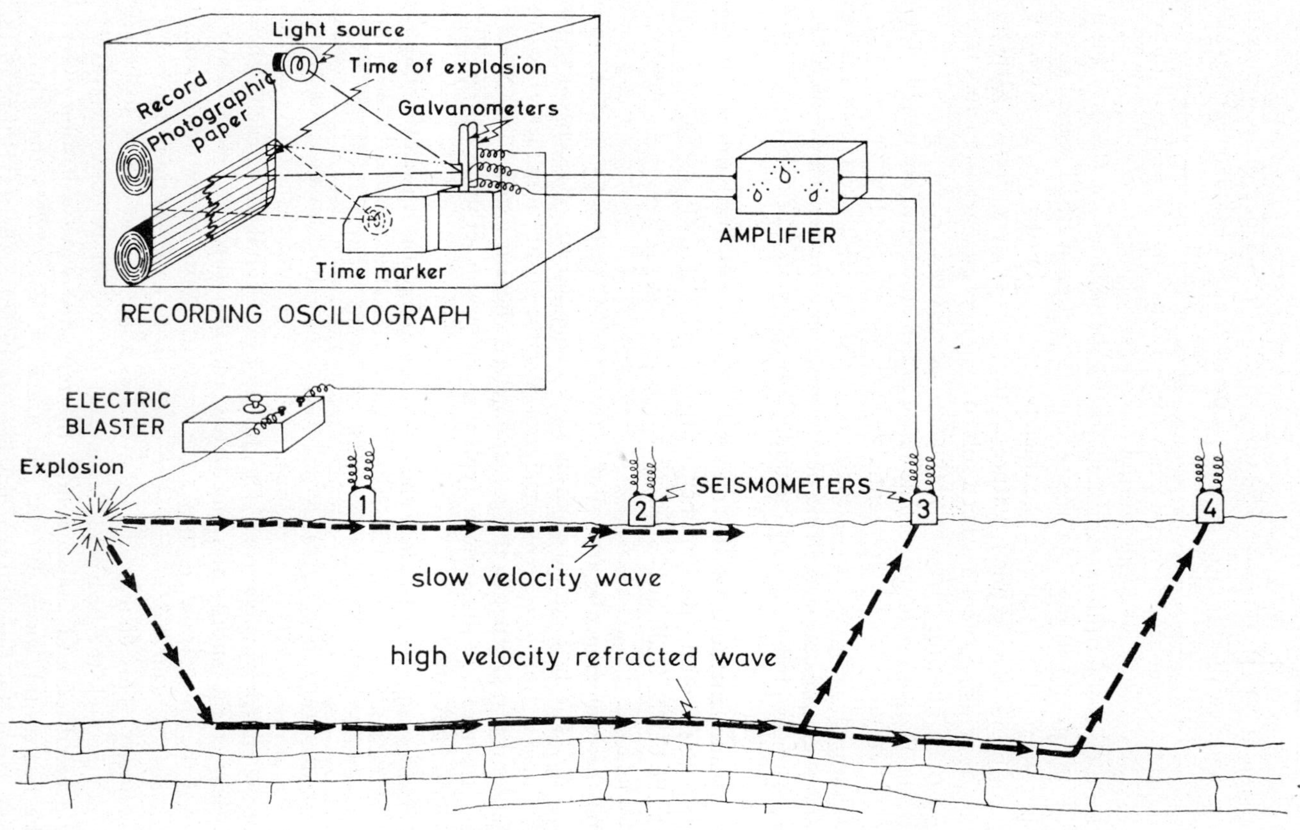

Fig. 1.6. Typical set-up for refraction seismographic method

wave, Stations 5 through 12 will record only the refracted wave from the bedrock.

"Fig. 1.7(b) represents an actual oscillogram, or record. The vertical lines are for timing and each division is 0.01 second. The heavy horizontal lines are the galvanometer traces. When the wave front arrives at the various stations, it is recorded instantaneously and shown as a break in the continuity of the galvanometer trace.

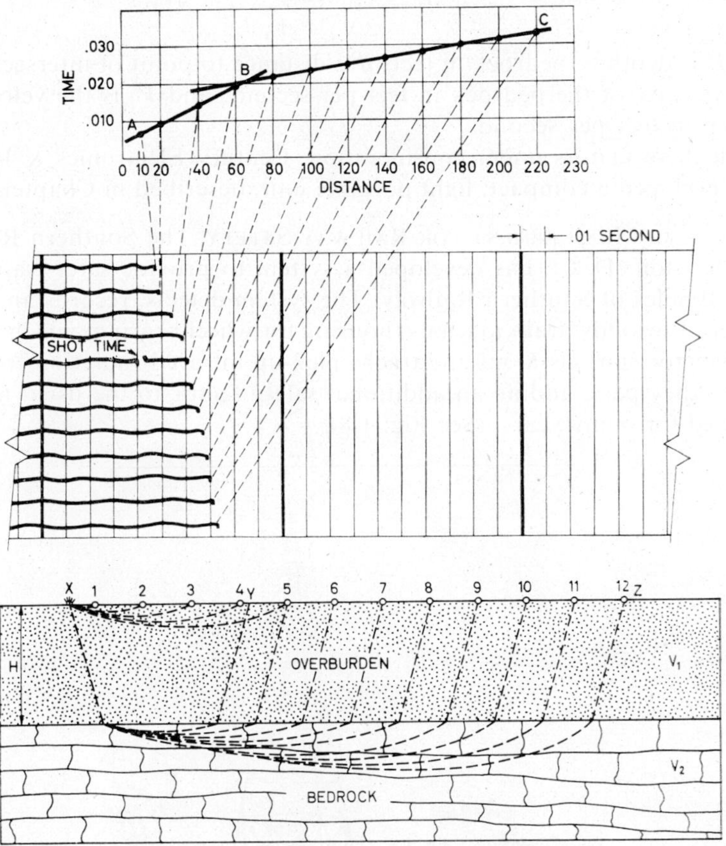

Fig. 1.7. Seismic refraction method details

The traces still continue, but they are of such amplitude that they cannot be seen on the record for approximately 0.5 second. This amplitude is the result of the arrival of secondary waves. The first arrival information is all that is necessary for the computation of bedrock depth.

"The record is developed in a few seconds after the recording, and is available for immediate study. The "first arrivals" are measured in time from the "shot time". These times are then plotted on graph paper to form a Time-Distance Curve, as illustrated in Fig. 1.7(a). With this curve, the velocities can be accurately determined by d/t, with d as the distance, and t as the time. Also, the exact point where the refracted wave overtakes the horizontal wave may easily be measured. This is commonly referred to as the "point of intersection", and is measured in feet.

"When these factors are known, the thickness of the overburden or depth to bedrock may be determined by the formula:

$$H = \frac{d}{2} \sqrt{\frac{V_2 - V_1}{V_2 V_1}}$$

in which H is depth to bedrock in feet; d is distance to point of intersection in feet; V_2 is the velocity of the bedrock in feet per second; and V_1 is the velocity of the overburden in feet per second."

With all these factors under consideration, Century Electronics & Instruments, Inc. has developed a compact, light portable unit, described in Chapter 3.

TEMPERATURE MEASUREMENT FOR RAILWAY SAFETY. The Southern Railway System, Washington, D. C., has developed a system to provide safer train operation along 8,300 miles of Southern Railway. The system detects, records and locates all in a matter of seconds, train journals (housing for wheel bearing) which are running "hot". Detector stations along the tracks pick-up infrared radiation from all train journals as they pass, adding an additional safety factor to the usual hand testing method used for many years. (See Fig. 1.8).

Fig. 1.8. Typical detector location. The infrared detectors and train-radio VHF communication tower and station are shown here

Fig. 1.9. Southern Railway detector system center at Atlanta, Georgia

As a train enters the operational zone of a detector its wheel flanges roll across a magnetic transducer located on the inside of one rail. A few feet further are two track-level infrared detectors, one on each side of the track. A shutter in front of the lens system in the detector is opened by the transducer allowing the detector to pick up infrared radiation from the hot journal. The relative heat measurements of each journal are transmitted via an FM Carrier over wire lines to Atlanta, Georgia, the hot-box detection center. At the center, one of the 53 Sanborn 150 Series recorders begins to chart the condition of each journal in the train. As the events are recorded, they are read by an officer who is trained to spot any variation in normal readings. When a serious variation is detected, the recording is annoted with vital information as to train identification, detecting station, speed, weather conditions, etc. (See Fig. 1.9.)

If Atlanta detects the existence of a hot box, the train is immediately told to stop via a VHF radio. Proper interpretation of the record permits the man on duty at the Atlanta center to tell the train crew the exact location of the hot journal in the train.

MEDICAL INSTRUMENTATION. The following information concerning medical applications for oscillography has been furnished by Honeywell, Consolidated Electrodynamics, and Itek Business Products.

Possibly the most important industrial dynamic measurement instrumentation used by the medical profession is the recording oscillograph. The instrument has made possible many advances through its probings into the heart, the brain, the nervous system, and the muscular and respiratory systems.

In spite of the limitations at low-frequency, few channels, small peak-to-peak deflection, and low sensitivities which necessitate preamplification, the direct-writing oscillograph is the most common type in use in the medical field today.

The most popular application of the direct-writing oscillograph is in making electrocardiograph records for human heart studies. Here voltages indicative of cardio-muscular action are produced from electrodes taped to various portions of the body. Although this type of record has normally been made on a single-channel oscillograph, particularly in clinical diagnosis, there is a growing trend toward simultaneous multichannel recording from a number of electrodes so that the pattern of potentials can be studied.

Photographic-type oscillographs are frequently used for multi-channel electroencephalographic studies in which the electrical activity of various nerve centers in the brain are recorded. Electrodes attached to the scalp pick up the brain potentials, which are then analyzed by comparing amplitudes and phases of simultaneous vibrations. Electroneurology is another closely related field where spontaneous nerve action and reaction to external stimuli are recorded. In phonocardiography, a crystal microphone is used to pick up heart sounds.

It has been common practice to record simultaneously on the same chart such variables as oxygen in the blood, respiration, heart rate, blood pressure, temperatures, and pulse rate. It is for such applications that the photographic oscillograph holds the greatest promise.

A Consolidated oscillograph, for example, has been used with pressure pickups for measuring blood and spinal fluid pressures at the University of California Medical Center in San Francisco.

To study heart action, electrocardiograms are made on mechanical recorders, or on light beam or CRT photographic oscillographs. Connections to the patient's body transmit electrical variations to the recorder. The variations are created by various phases of heart action.

After reading the resulting oscillograms, selected areas of the records that show representative patterns of heart behavior are clipped by the cardiology laboratory

technicians. These clippings are mounted on a standard form that shows the name
of the patient and other case history data. A photocopy of the group of recordings
becomes a part of the patient's chart. A typical electrocardiogram record, reproduced on a Photostat 10.14 Photocopier, is shown in Fig. 1.10.

Cardiac System Measurements. The critical mysteries of heart action as related
to intra-cardiac blood pressures and respiration are being charted by a Model

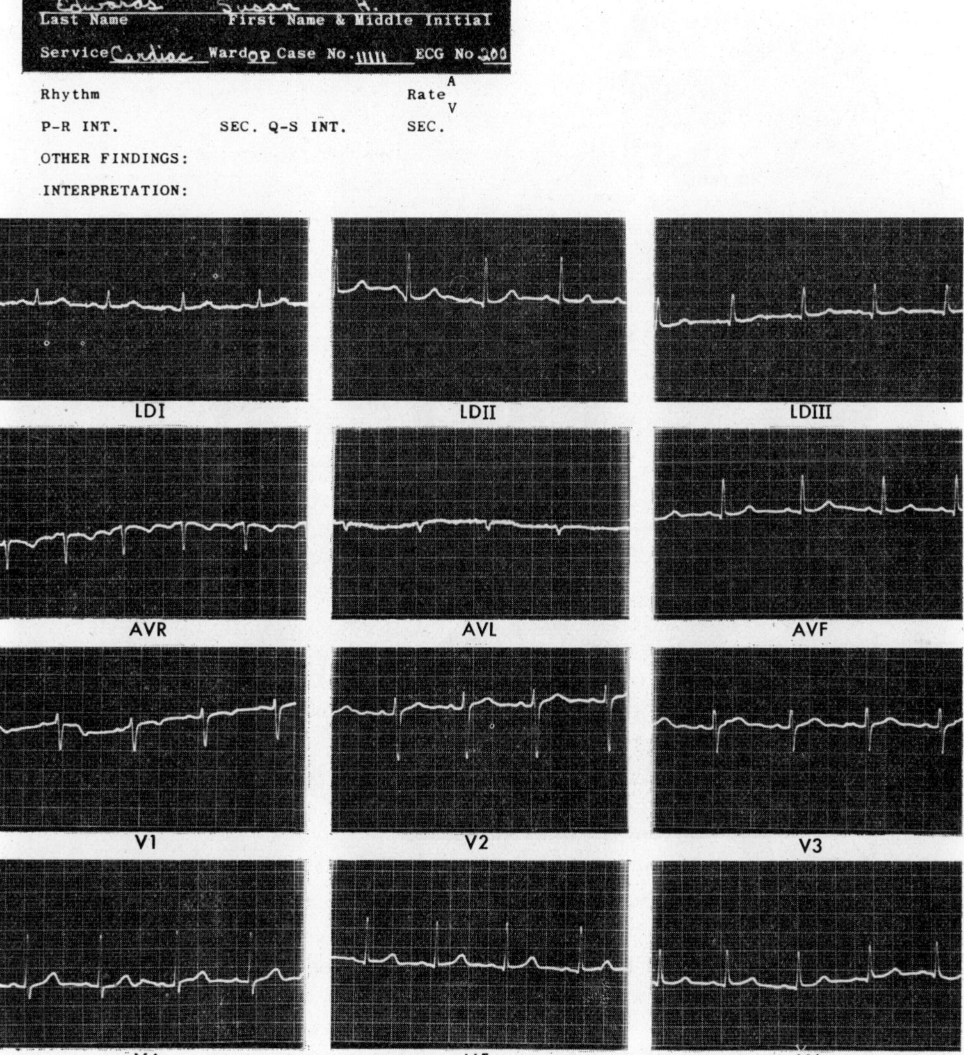

Fig. 1.10. Typical electrocardiogram record

906A2 Visicorder at the Children's Hospital in Fresno, California. (Fig. 1.11).
The accompanying record (Fig. 1.12) shows the use of the Visicorder as a cardiac
catheterization recorder, marking an event.... the introduction of a catheter into
a heart chamber (see Marks 1A, 1B, and 1C on the record). The equipment used

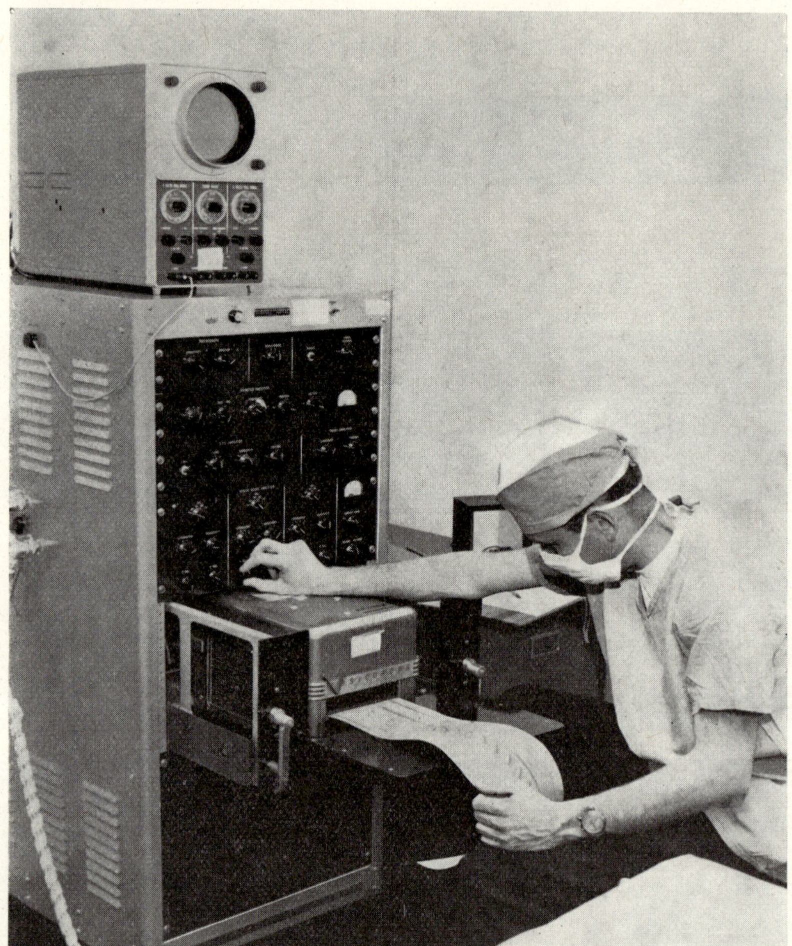

Fig. 1.11. Dr. Roger Larson monitors the catheterization of a heart chamber on the Visicorder
recorder

in these tests was developed by Electro-Medical Engineering Co., Burbank, California. The timing trace appearing on the record shows timing spikes spaced 1/10
second apart.

A pneumograph trace (marked No. 2 on the record) registers the respiration rate
of the patient. This trace was activated by a Baldwin SR–4 strain gauge bonded

to an aluminum strip strapped around the patient's chest, passed through an E-M-E
respiration bridge, and recorded by a Honeywell V–40–350 galvanometer.

Trace No. 3 is an EKG channel at idle, not being used in this test. No. 4 is a
phonocardiography trace actuated by a stetho-microphone strapped to the patient's
chest, amplified, and recorded by a Honeywell V–100–350B galvanometer.

Traces 5, 7 and 8 show intra-cardiac pressures in the chambers of the heart.
These traces were activated by miniature Statham Model P23G strain gauges
(range: 0–750 mm Hg), amplified by a three-channel strain-gauge amplifier and
recorded by Honeywell V–450–55B galvanometers. Traces 6 and 9 are static reference traces.

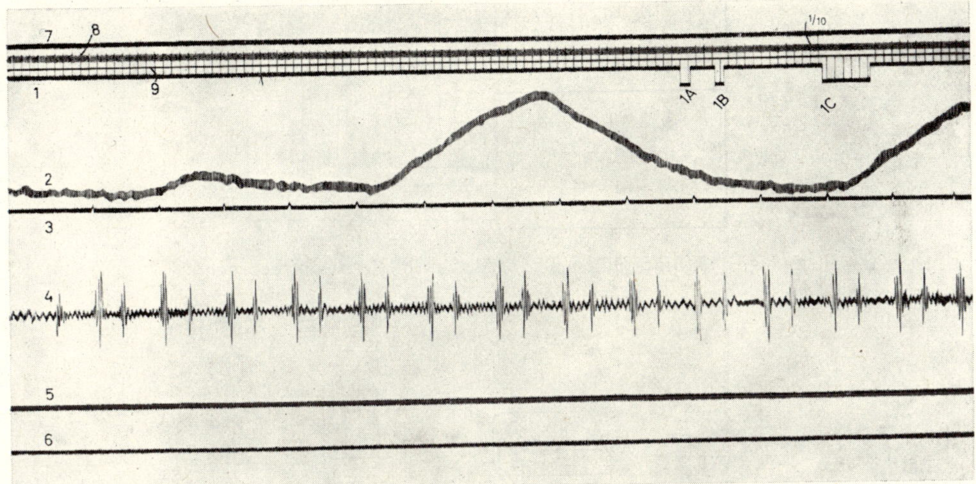

Fig. 1.12. Visicorder record of heart catheterization

Also included in the convenient, rack-mounted instrumentation system is a
provision for recording an oximeter trace (through an E-M-E oximeter amplifier)
in place of one of the intra-cardiac pressure points. A Waters Corp. photoelectric
pickup originates the oximetry trace.

The equipment, designed and manufactured by Electro-Medical Engineering
Company, consists of a 3-channel strain gauge amplifier for Statham gauges, an
electrocardiographic amplifier, a phonocardiographic amplifier, a respiration bridge,
an oximeter amplifier, a timing generator, and an event marker. (See Fig. 1.13).

Restoring Heart Action. Electronic instrumentation is also solving the problem
of cardiac arrest–the stopping of the heart during surgery. Usually the surgeon
can only resort to thoractomy (cutting the chest) to get at the heart and manually
squeeze the heart back into action. An electronic "external pacemaker" and a new
technique for external electric stimulation have been developed at Harvard Medical

School. The pacemaker may have great usefulness since there are about 5,000 cardiac arrests a year in operating rooms in the United States.

Alveolar Ventilation Measurements. At the U. S. Public Health Service in Cincinnati, Ohio, one of the most interesting physiological studies is the determination of

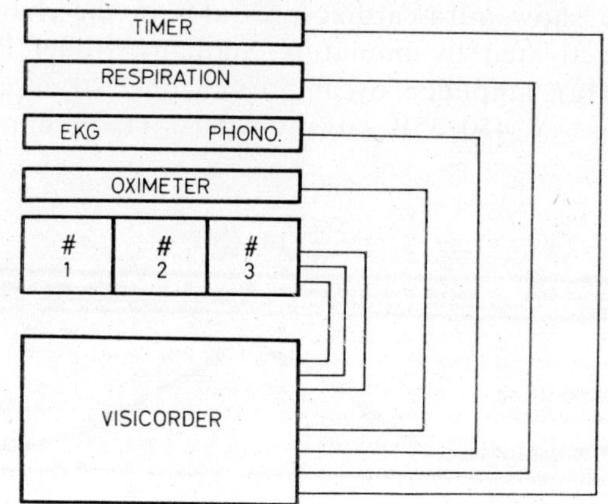

Fig. 1.13. Block diagram of cardiac catheterization recorder

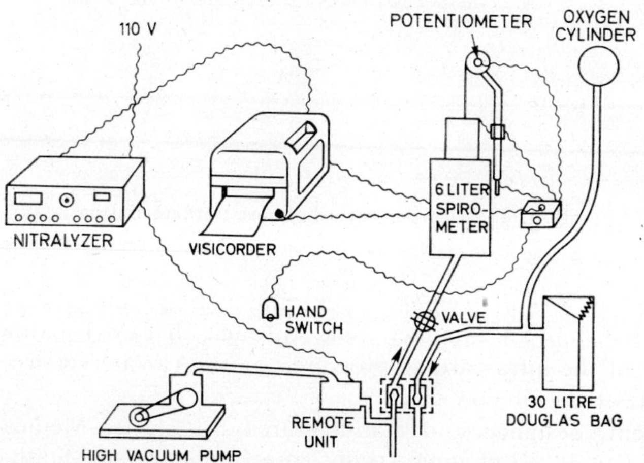

Fig. 1.14. Schematic wiring diagram for determination of uneven alveolar ventilation

lung functions through comparisons of lung volumes, respiratory dead space, and abnormalities in distribution of inspired gas during a breath.

In the lung function tests, the subject inhales a single maximal breath of oxygen and then exhales the breath maximally while a nitrogen meter continuously analyzes the nitrogen. The Visicorder (valuable in this use because of its high frequency response) continuously records the nitrogen concentration during the exhalation.

40

If the oxygen is distributed uniformly throughout all the lung, it will wash the alveolar nitrogen out of all parts of the lung uniformly.

Equipment used (see Figs. 1.14 and 1.15) is a Nitralizer nitrogen meter (Custom Engineering & Development Co., St. Louis), a Welch 1406 H high vacuum pump, a 30-liter Douglas bag, the Visicorder, a nose clip, sterile disposable paper mouthpieces, and a six-liter spirometer (Vitalometer) with a cut-off switch.

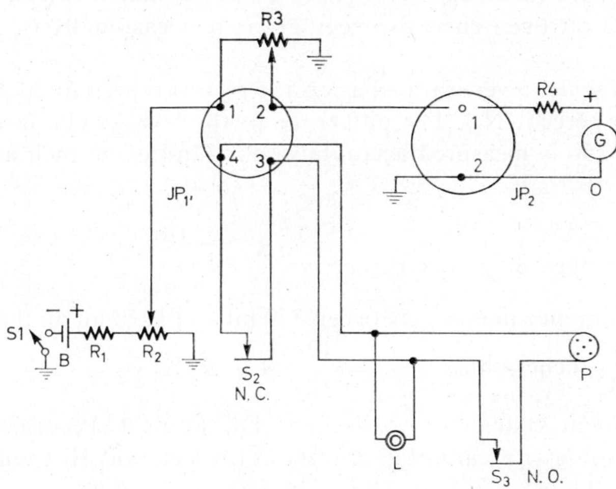

Fig. 1.15. Sensing system for expired gas volume and controls for recording-paper motor drive

A special pulley wheel with a 100-ohm Helipot is used for sensing the volume flow. Volume flow markers are set for 0, 750, and 1250 milliliters. An automatic cut-off switch is attached to the supporting post in a position corresponding to about 1500 ml. As the spirometer bell passes the switch, it automatically stops the Visicorder.

Static reference traces on the Visicorder are set to indicate 0 to 80 per cent N_2 for a spread of 5 inches on the paper. The expanded scales 0–20, 20–40, etc., fall on the same static indicator lines 5 inches apart. Three other indicator lines are set as volume indicators to correspond to 0, 750, 1250 ml. Zero in this group is the same as the zero-percent nitrogen line. Paper speed on the Visicorder is set at 5 inches per second.

The test begins with the Douglas bag filled about half with oxygen, and the subject seated or reclining comfortably at the mouthpieces, noseclip in place. The subject exhales maximally and then inhales oxygen maximally through the right-hand mouthpiece. He holds his breath for a moment while the oscillograph is started, and then exhales through the left-hand mouthpiece at a moderately fast and even rate (at least 60 liters per minute). When the spirometer bell passes the 1500 milliliter mark, the remote switch turns off the Visicorder. The direct-recording feature of the Visicorder reveals in seconds whether another test is necessary.

41

Re-recording may be necessary if the operator selects an improper scale (0–20, 20–40, etc.) for the subject under test. Generally, the 0–20 scale is satisfactory for recording the test for normal young persons. For older persons and for patients with certain pulmonary diseases, the 20–40 scale may be more proper.

For interpolation, lines perpendicular to the paper edge are drawn on the record through the 750 and 1250 intersections where the volume line crosses.

Next, a line is drawn along the nitrogen curve so that it intersects the 750 and 1250 lines. If the nitrogen curve is irregular, as it occasionally is, a best-fit line is approximated.

With the 0–20 scale, 5 vertical inches on the record represent 20 percent nitrogen (1 inch equals 4 percent N_2). The difference in the level of the nitrogen curve between 750 and 1250 is measured accurately to a 32nd of an inch and expressed in the same units.

Formula:

Equation: Nitrogen difference percent $= \dfrac{x}{32} \times 4$

Example: Difference in inches between 750 ml and 1250 ml on the nitrogen curve is $\dfrac{12}{32}$ inches, thus: $\%N_2 = \dfrac{12}{32} \times 4 = 1.5\%$

This measurement system was devised by Dr. Louis J. Pecora, Director of the Pulmonary Physiology Research Laboratory at the Veterans Hospital in Cincinnati.

Monitoring Patient Condition during Operations. As the modern surgeon operates with procedures more dramatic and more radical than those in use only a short time ago, he needs more precise, more rapid, and broader information about the condition of his patient as the operation progresses. ENSCO, Salt Lake City, has made the Visicorder a component of its medical instrumentation system to give operating-room personnel a continuous, directly-recorded encephalogram, electrocardiogram, blood oxygen content analysis and blood pressure traces.

To accomplish the EKG measurements, transducers used are, of course, the standard EKG electrodes found in any hospital. Blood pressure transducers are Statham pickups, Model P23D.

Amplifiers are: EEG Amplifier, Model EEG–1; EKG, Model EKG–1; Galvo Adjustment Panel, Model GP–4; Oximeter Amplifier, Model OSA–2B; and Pressure Amplifier, Model S6A–3. All are manufactured by ENSCO.

The resistor network contains the 350-ohm shunt, 100,000-ohm series. Galvanometers are Honeywell Sub-miniature Model M40–350, and static reference traces. (See Fig. 1.16).

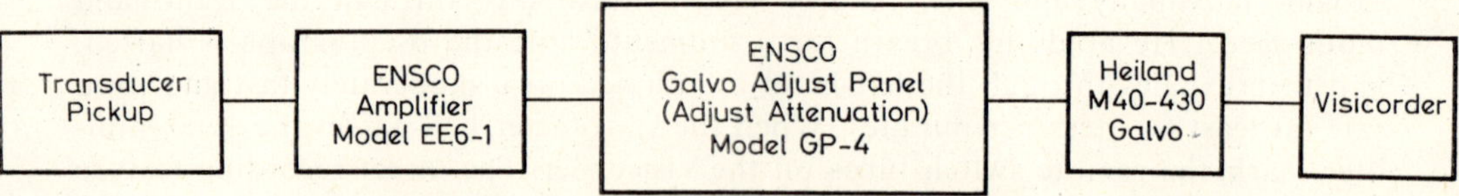

Fig. 1.16. Block diagram for one channel (EEG)

Electronic Respirator for Infants. Another interesting development is an electronic respirator which helps prevent suffocation either because of lung defects or by a general weakness which prevents normal breathing. A machine developed at the Royal Maternity and Women's Hospital, Glasgow, Scotland, provides electronic controls which govern the rate and frequency at which air is delivered, overcoming many of the problems encountered with earlier devices.

Electronic Cancer Detection. A newly developed automatic optical electronic machine, the Cytoanalyzer, may greatly speed detection of cancer of the uterus. The machine consists of a scanning microscope, computer and analyzer, and recorder. The computer measures the nuclear size and nuclear optical density of the cells, and distinguishes between signals arising from normal and suspicious cells.

A nuclear measurement graph plots each accepted measurement so that the cells can be rated as normal, suspicious, or deficient in required information. A high-intensity cathode-ray tube and an oscillograph camera similar to CEC's Type 5–140 Cathode Ray Recording Oscillograph make recordings of the nuclear measurement graph. All computations are made as the slide is being scanned in less than one fifth of a millisecond. The Cytoanalyzer was developed by the Airborne Instruments Laboratory, Inc., of Mineola, New York.

Specialized recorders for medical use are described in Chapters 3 and 4.

LITERATURE REFERENCES

[1.1] ERICKSON, R. D., The Changing Role of Oscillographs, *Industrial Research*, Vol. 5, No. 6, 32, 1963.

[1.2] Neblette, C. B., *Photography, Its Materials and Processes*, p 1, D. Van Nostrand Company, Inc. (New York 18, New York), 1962.

[1.3] WEYDE, DR. EDITH, Upon a Practical Application of the Clayden Effect. *Zeits. Wiss. Phot.*, 48, 1–3, (1953).

[1.4] STABE, H., Contribution to the Development of a New Photographic Writing Method with Immediately Visible and Durable Light Traces. *Zeits. Wiss. Phot.*, 48, 1–3, (1953).

[1.5] LEVENSON, G. I. P., A Machine for Rapidly Processing Photographic Trace Recordings. *J. Sci. Instrum.*, 27, 170, June 1950.

II RECORDING SYSTEMS

Indicating and Recording Methods

There are two basic classifications of measuring devices: 1) indicating instruments; and 2) recording instruments. Indicators, such as pressure gages, manometers and voltmeters, are necessary to monitor test operations and are well suited to static tests in which a specific condition is reached, measured, and noted. Indicators are also suitable for quasi-static tests and/or tests that may be repeated.

Two limitations[2.1] imposed by the use of indicating instruments (for recording purposes, such as a photopanel installation) are: 1) the response time of the instrument, i.e., its mechanical limitations; and, 2) the response time of the operator, who must visually observe, mentally react, and then physically record the reading of the meter. If many channels of data are involved, a number of operators are necessary to monitor the data, and all conditions must have a steady-state life sufficiently long to permit manual recording. If there are numerous variables which change rapidly, the engineer must change from manual to simultaneous (and automatic) recording. Often, instruments are mounted on a common photopanel, which is illuminated and photographed with an automatic camera. After the test is completed, the photopanel data is processed, and the readings of the dials are read and logged. A timer on the photopanel permits the indicated data to be plotted against time. Within certain tolerances, it is possible to use indicating instruments (pointer-types and similar devices) for both static and dynamic tests. Manual recording may be used for steady-state conditions, and cameras may be used for rapidly changing conditions. However, the photopanel method is limited by the following factors:

1) The speed of response of the indicating instrument is limited by the mass of its components, and by friction.

2) The speed of response may vary from one instrument to another, so that transient-condition readings would produce nebulous information among the various parameters.

3) The limitations of the camera system such as time between records, blurred pointers (caused by pointer movement during exposure) and the physical space required for such an installation.

4) The lag time between the test and the availability of the processed film, precluding quick access.

5) The difficulty of making precise readings from very small images containing distortion, parallax, and blurred pointers.

The search for better instrumentation has led to the development of techniques that would circumvent the disadvantages of photopanel and other bulky, unmatched systems. It is apparent, that to improve the system performance, the basic limitations of the instrument itself and its lag time must be reduced as much as possible. This means that for data having high frequencies, friction must be eliminated, mass must be reduced as much as possible, and sensitivity should be the best obtainable. Although mechanical recorders are well suited to their purposes and electro-mechanical pen writers are adequate for frequencies up to several hundred cycles, other methods must be used if high frequencies are to be accurately recorded. High frequency requirements dictate that the moving component must accelerate to its top

TABLE 2.1

**FREQUENCY-RANGE CAPABILITIES OF
VARIOUS RECORDING METHODS**

Type of Instrument	Frequency in Cycles per Second 1 10 100 10^3 10^4 10^5 10^6 10^7 10^8	Type of Recording
Capillary tube, mechanical liquid-level pointer	One cycle per minute	Human monitor
Meter or gage		Camera monitor
Drum or circular chart recorder		Mechanical-linkage pen, ink on paper
Potentiometric recorder		Pen record on chart
Pen, electric, or heatsensitive paper recorder		d'Arsonval movement; ink, conductive, or wax paper
Ink jet recorder		Jet of liquid on paper
Light-beam oscillograph		Light beam on photosensitive paper
Electron-beam recorder		Cathode-ray tube, photosensitive material

46

speed, decelerate, stop, and accelerate in the opposite direction, all in a time interval ranging from 1/100th to 1/10,000 of a second. Oscillograph systems, employing either electron beams or mirror-galvanometers for writing, meet these requirements.

Six common categories of marking are ink, heat, electric discharge, pressure, light and electrons, and are discussed in greater detail by Keinath, Nolte and Owens[2.2, 2.3, 2.4].

Table 2.1 lists the frequency spectrum of various recording methods; starting with manual observation and logging of the indication and ascending to high-frequency electron-beam analog recorders.

Low Speed Recorders

Low speed recorders include x-y plotters,[2.5] event recorders, and analog recorders[2.6, 2.7, 2.8] using ink, heat, electric discharge, and pressure as the marking method. Because such systems are inherently simple, have been in use for many years, and have had extensive development, they will be discussed only briefly in the following paragraphs. Some low-speed recorders have interchangeable marking systems so that the recording medium may be selected to meet the requirements of the application. Converting from one medium to another can be accomplished in a few minutes. Table 2.2 presents data to aid in the selection of the optimum recording medium, and may be used for recorders having interchangeable marking systems, or for separate recorders as available.

INK RECORDERS. Ink is the most popular marking method, and is applied to the chart by means of capillary pens, ballpoint pens, and jet sprays. Ink marking recorders have had years of development of the instruments themselves, the ink, the paper and the marking components.

The chief advantages of pen systems are simplicity, low initial cost, and low operating cost. Chart papers for ink recorders are the lowest priced papers of the various types used in the several recording systems.

Ink pens are designed to be light weight and rigid, so they will not whip or throw ink at high frequencies. A typical pen design, used in Brush recorders, is illustrated in Fig. 2.1.

Ink recorders require proper maintenance: cleaning of the marking element by flushing, changing it periodically, and filling the ink reservoir. Many of the problems associated with drying and clogging of capillary feeds have been reduced by the use of pressurized-ink feed systems.

Even though the chart paper is much less costly than the papers used in other marking systems, its desirable characteristics include good dimensional stability, an accurately printed grid, and correct ink absorption. A reproduction of a typical ink paper record is shown in Fig. 2.2. The curved "time" lines of the grid match the arc of the pen radius, and the paper is called curvilinear for this reason. Multiple channels are printed side by side. The thin, high-tensile-strength paper base is transucent and permits reproduction by conventional diazo machines.

Ink recorders must be mounted correctly to avoid ink spillage. Recorders with ballpoint markers eliminate this problem, but greater pen pressure is necessary; consequently response is adequate only for slow speed recording, and galvanometers must have a high torque.

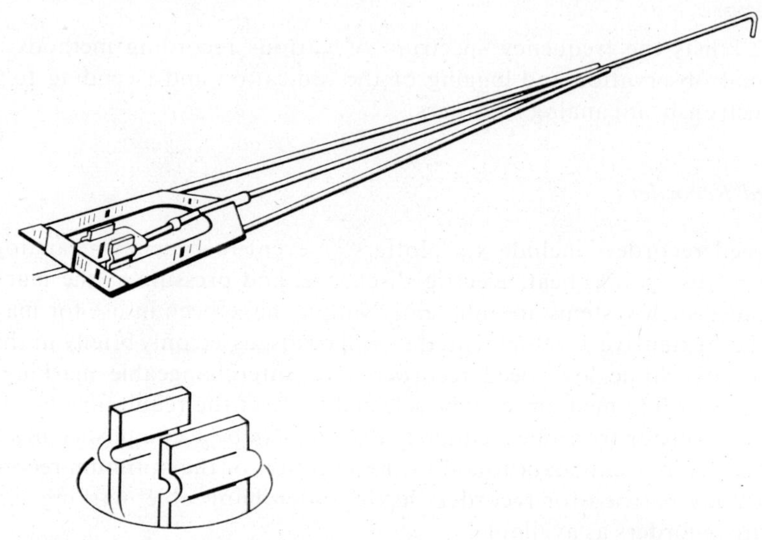

Fig. 2.1. Brush Instruments ink-recording pen with pen mounting shown inset

Among the various writing methods used in ink oscillographs is an ink-jet recording system. A recorder of this type, the Oscillomink, is illustrated in Fig. 2.3. The instrument will record frequencies within the range of 0 to approximately 1,000 cycles per second. This frequency maximum is well above that of conventional mechanical recorders using pens or stylus type marking methods. For this reason, the Oscillomink fills a need for an intermediate frequency-range instrument.

The design of the transducer element of the jet oscillograph is similar to that of a galvanometer. Instead of a mirror or a pointer, however, a glass capillary tube approximately as thick as a horsehair is set in motion. One end of the tube is tapered forming a nozzle of 10 microns in diameter, bent at a right-angle to the rotation axis.

The liquid-jet recording system of the Oscillomink is fed by means of an electric vibrator pump which produces a pressure continuously adjustable from 15 to 45 kilogrammes per square centimeter (210 to 640 pounds per square inch). In this way, the required quantity of recording liquid may be adjusted with respect to the set paper speed and the writing speed to be expected. The liquid jet does not exert any reaction on the measuring element which has a very low inertia.

The individual writing jets do not produce any mutual disturbance, even if the curves are traced one upon another. The natural frequency of the measuring element

48

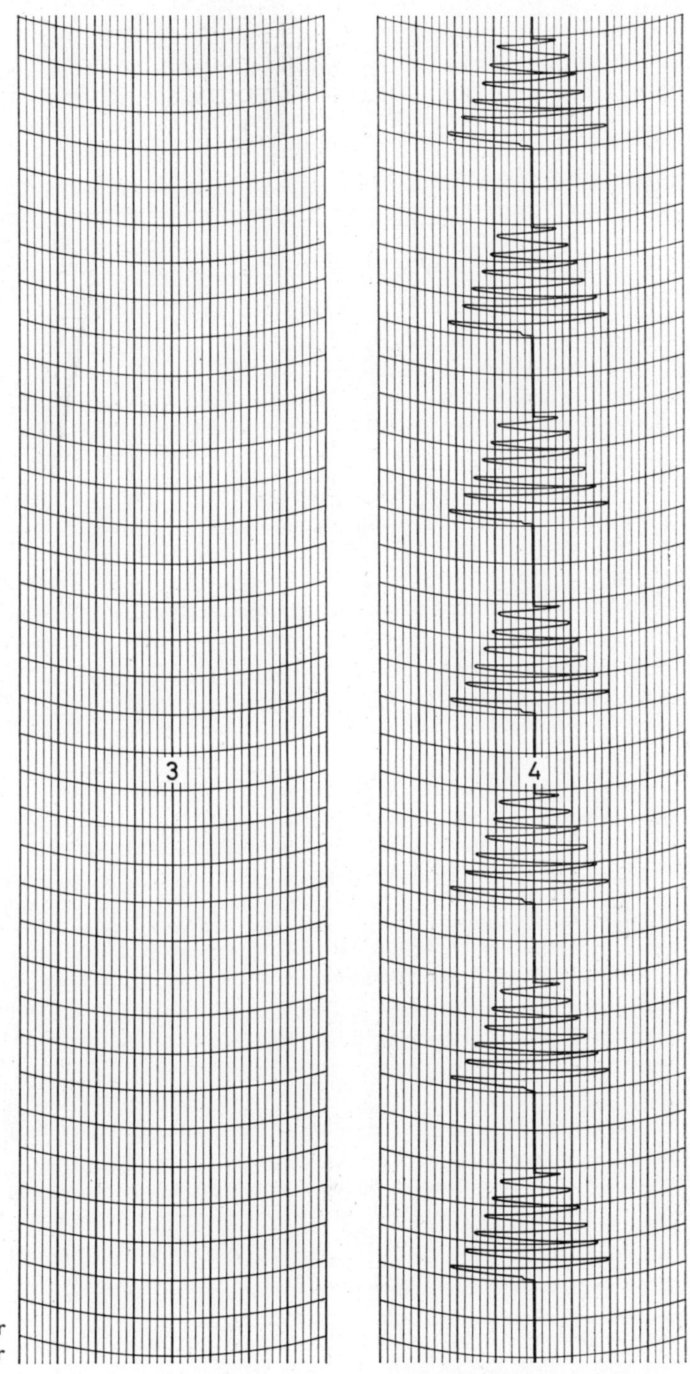

Fig. 2.2. Typical curvilinear ink chart paper

TABLE 2.2

RECORDING MEDIUM SELECTION CHART

(Reprinted through the courtesy of the Offner Division of Beckman
Instruments, Inc.)

Medium	*Application*	*Advantages*	*Disadvantages*
Ink	General Purpose	Wide range of recording speeds without readjustments. Low stylus friction gives accurate records. Economical. Records reproducible by ozalid and similar processes.	Curvilinear records. Ink may spatter at high frequencies and amplitudes. Ink may spill under severe vibration (mobile use), and freeze at low temperatures.
Electric Curvilinear	Where ink is not usable.	Avoids ink spatter, spilling or freezing. High resolution with fine recorded line.	Curvilinear records. Relatively high cost recording medium. Recording current must be changed when going from very slow to fast phenomena. More stylus friction than ink. Reproduction requires photographic processes.
Heat Rectilinear	Where rectilinear recording required, and only relatively slow phenomena to be recorded.	Rectilinear record avoids curvilinear distortions. At slow speeds gives high resolution, well defined line, with no attention required.	Cannot record rapid stylus motions, limiting use to relatively low frequencies and/or amplitudes. Relatively high stylus friction results in errors in low amplitude signals. Stylus temperature must be readjusted with changes in signal frequency and amplitude. Relatively high cost recording medium. Reproduction requires photograph processes, unless special paper employed.
Electric Rectilinear	Where rectilinear recording required, and wide range of recording speed necessary.	Rectilinear record avoids curvilinear distortions. Will record wide range of speeds without readjustment. Stylus friction no higher than with ink recording.	Stylus must be cleaned occasionally. Relatively high cost recording medium. Reproduction requires photographic process.

amounts to approximately 650 cycles per second. The linear response up to, 1,000 cps is obtained by electric compensation.

A blotting roller, automatically running on the paper strip, blots the wet traces which occur frequently when slowly changing phenomena is recorded at low paper speeds. The blotting roller can be cleaned with pure water and used several times.

It is also possible to compensate for the surplus of recording liquid, appearing at low recording speeds, by superposing a wobbling current or adjustable amplitude to the measuring current. This current, which causes the moving system of the meas-

Fig. 2.3. Siemens and Halske Oscillomink ink-jet recorder

uring element to vibrate and thereby to write a broader line, is generated by a built-in 2,000 cps oscillator and may also be used as a time-marking system when recording at high speeds.

The paper feed has six speeds ranging from 0.5 to 20 cm per second (0.2 to 7.9 in per second). Speeds can be adjusted during operation due to an overrunning clutch of the synchronous motor drive. Higher paper speeds up to 2 m per second (79 in. per second) can be adjusted by means of an external high-speed attachment.

THERMAL-WRITING RECORDERS. Another type of low-speed recorder employs electrically-heated styli to produce visible traces on heat-sensitive paper. Heat from the styli melts the wax or plastic coating, revealing the dark base of the recording paper. The paper may be obtained in both opaque and reproducible types, and costs from two to five times as much as ink-recorder paper. Most thermal papers may be easily scratched and must be handled with more care than ink-marked charts. Improved papers are Mylar-coated to minimize friction and prevent the formation of deposits on the stylus. Thermal recorders are best suited for relatively slow speed recording, and because of high stylus friction, errors may occur in low-amplitude signals. Most thermal recorders have a rheostat-type control to adjust the stylus-tip heat, so that the width and density of the recorded line may be adjusted for easier reading.

ELECTRIC-WRITING RECORDERS. Electric-discharge recorders produce charts that have excellent resolution under a wide range of ambient and operating conditions. Typical trace widths of 0.005 inch may be achieved without the dangers common in ink recorders: ink splatter, spilling, and freezing at low temperatures. Electric recorders may be operated without attention over prolonged periods, and require only occasional cleaning of carbon from the styli tips. In some electric event recorders, styli tips are so oriented that the light contact with the paper surface itself removes the carbon and renews the tip. These advantages make electric-writing recorders ideal for unattended operation.

The principle of the electric method is to break down and burn away the gray-colored, insulating-surface coating of electrically conductive black-carbon chart paper. The back, or electrically conductive surface, travels over a grounded curved platen and the electric writing styli contact the top or visible surface. A high d-c voltage at the styli tips, usually 150 to 300 volts, causes a high electric field to develop. This breaks down the gray, insulating layer and allows the black carbon backing to show through as a trace.

Electric-writing recorders are available for both curvilinear and rectilinear recording. The paper is opaque and cannot be duplicated by diazo and similar reproduction processes that print by transmitted light. Recording materials are relatively expensive when compared with the cost of ink-recording papers.

Sensing Devices

Physical qualities such as pressure, force, strain, vibration, and acceleration can be measured by converting the energy into an electric signal. The following discussion contains excerpts from an article by L. Fleming[2.9], and describes the use of transducers to make this conversion.

"Most present-day measuring, control and communication systems are concerned with mechanical, acoustical, thermal or similar inputs, and with equally non-electrical final outputs. The electrical signals involved are only an intermediate product, cast in that form for convenience in the transmission and processing of data. Therefore, it is reasonable to say that today nearly all measuring and control systems begin and end with transducers. However, since output transducers in the military and industrial field are commonly in the form of indicators, servo valves, recorders, or motors, they are not considered here. Only sensors or input transducers in the industrial and military context are discussed.

"Transducers are of two basic types – active and passive. An active transducer can be defined as a calibrated device for converting a small sample of some form of mechanical energy to analogous electrical energy. Examples are piezo-electric and magnetic transducers for sensing vibration and dynamic pressure. Generally speaking, active transducers cannot handle d-c or steady mechanical inputs. A thermocouple is an example of a different type of active transducer.

"Passive transducers act as valves. A mechanical input, such as displacement, controls the flow of electrical energy from an outside source. Strain-gage instruments are an example. In general, all transducers which "go down to d-c", i.e., deliver a continuous electrical signal in response to a steady input condition, must be passive.

"Table 2.3 presents a qualitative comparison of the more common types of input transducers used in the aerospace industry for telemetry and ground testing. The prime characteristic of transducers in this class is the ability to operate in adverse environments of temperature and vibration. Accuracy is normally in the 0.5 to 1 percent range.

Some of the more recent transducer types in limited use are magnetoelastic, capacitive, and ionization. Photoconductive, magnetoresistive, and Hall-effect devices are beginning to appear. The pyroelectric effect has been used to some extent in heat-transfer gages; the piezocapacitive effect has been investigated but found impractical, at least with presently available materials; thermionic diodes with a movable element are troubled by excessive drift." References[2.10, 2.11] contain further discussions of sensing devices.

"Southern Instruments, Limited has published the following statement concerning capacitive type transducers: "With a radio frequency carrier system the condenser pickup is an extremely sensitive device that gives static calibration conditions and faithfully records low frequency vibrations. It can be adapted to widely different forms. Because of its high sensitivity the moving element can be very stiff giving a mechanical system with a high natural frequency, and it can be easily made to withstand very severe overloads. The disadvantages are complexity of construction, poor temperature stability and the necessity for shielded coaxial cable connections."

ACCURACY OF SENSORS. Southern Instruments has published the following statements concerning the accuracy of sensing devices: "The user should realize that these various electrical gages and pickups are not inherently devices of great precision. They have to be calibrated by other standards or sub-standards. Condenser or resistance pressure gages, for example, have to be calibrated against Bourdon-tube dial gages or dead-load test pumps, (called dead-weight gage testers in the United States). Movement gages have to be calibrated against micrometers. Unfortunately, as they do not hold their calibrations indefinitely, rechecks are necessary. Hence the accuracy of the electrical gages cannot exceed their mechanical counterparts. The justification for the use of electrical gages is their rapid response, their small size, their flexibility of control, the ease with which several can be used at the same time, the ease with which they can give remote indication, and last but

TABLE 2.3

COMPARISON OF SEVERAL TYPES OF MECHANICAL-INPUT TRANSDUCERS
(CEC Table)

Principle	Note	External Excitation	Electrical Output		Response to steady Mech. Input	Freq. Range	Life	Mech. Rug-gedness	Environmental Temp. Range
			Level	Impedance					
Piezoelectric	1,2	None	Med. Low	Very High	No	Widest of all	Longest	Best of all	Moderate
Magnetic Induction	1,3	None	Med.	Med. Low	No	Few kc	Long	Good	Fairly High
Potentiometer	4	D-C	High	Med.	Yes	Low	Limited	Fair	Moderate
Strain gage (wire or foil)	5	D-C	Med. Low	Low	Yes	To a Few kc	Very Long	Very Good	Moderate
Strain gage (Semiconductor)	5, 6	D-C	Med. High	Low	Yes	To a Few kc	Very Long	Very Good	Moderate
Variable Inductance (Short stroke)	7	A-C	Med High	Low	Yes	Low	Long	Good	Moderate
Variable Inductance (Long stroke, e.g. differential trans-former)	7	A-C	Med. High	Low	Yes	Low	Long	Good	Potentially High

1. Active transducer, for dynamic inputs only
2. Cable and preamplifier requirements are peculiar to this type.
3. Widely used as a vibrometer. Output voltage proportional to rate of change of position of moving element, i.e., velocity.
4. Requires less external electronics than other types.
5. May use A-C excitation but D-C is now more common.
6. Present types have a higher temperature coefficient of sensitivity than wire gages.
7. Output is A-C unless demodulator is packaged in transducer case: presents cross-talk and balance difficulties in multi-channel systems with long cables.

not least, the fact that the electrical signal from them can be amplified and applied to a cathode-ray tube. They do not replace the long established mechanical measuring instruments, but are supplementary to them."

Re-calibration against known measurement standards is necessary to certify that the gages are within specified tolerances of measurement capability. Today, it is necessary to separate the capability of measurement tests (calibration) from the capability of operation tests (functional testing).

Gage application is also important. It is true that there are many instances where the long-established mechanical measuring instruments have been supplemented by the transducer-type sensing and measuring devices. There are also as many cases where phenomena are being measured by the transducer-type gage that could not have been measured through any known fully-mechanical means. Thus the transducer, in these instances, is an entirely new application, and is an extension of capability and new application rather than a supplement to the mechanical type gages.

Further, there are many cases where the mechanical type gages cannot be exchanged for the transducer; the characteristics of the phenomena are such that measurement by transducers is neither applicable nor practical. Consequently, the old fully-mechanical gages continue to be used.

Types of Measurements

FORCE MEASUREMENTS. Measurements of such forces as thrust and torque are measured by force rings and load cells, and strain is measured by strain gages. Each is a passive sensor. Strain gages are often used on force rings and load cells. The strain gage has a varying resistance, i.e., the resistance changes in proportion to the applied strain. When attached to the specimen undergoing test, it experiences the same strain as the specimen, and converts this into a measurable output.

A Southern Instruments publication makes the following statements about variable-resistance measurements: "Either as a bonded or unbonded strain gage, the variable-resistance method can be applied to all kinds of measurements. Its sensitivity is less than the condenser gage by an appreciable factor. The small physical size of the resistance wire winding, its linearity of response, freedom from hysteresis and the possibility of temperature compensation are its outstanding advantages. It is applicable to static and dynamic measurements, but the former demands a stable high-gain direct-coupled amplifier or a low-frequency carrier system if high frequency response is not required." (An amplifier is not required when coupling is to a light-beam oscillograph.)

The following excerpts are taken from the CEC publication "Strain Gages": (Reference [2.12]). "The usual form of strain gage consists of a short length of small (approximately 0.001 inch) diameter wire of high electrical resistance. To keep the gage length short, the wire is formed into a grid. To simplify its mounting and protect it, the wire is cemented between two thin pieces of paper. To apply, the gage is cemented to the member to be tested. Flat or curved shapes are easily accommodated, no extensive preparation is required, no gage lines need be scribed.

"When the test member is strained, so is the bonded gage. As the wire is strained its electrical resistance changes. This resistance change is directly proportional to the strain in the wire, and the strain in the wire is directly proportional to the strain in the member. A number of simple techniques are available to give accurate measurement of the resistance change; the proportionate strain is quickly calculated.

"To make a selection of the best type of gage for a specific application, let us consider the constituents of the gage likely to be affected by the test conditions.

"TYPE OF STRAIN: *Area of Measurement.* In use, each infinitesimal portion of the strain gage is intimately bonded to the member being tested and follows its movements in both tension and compression. It measures the average of the strains along the whole gage length. If strain in a localized area is needed, a short gage length must be chosen.

"*Amplitude.* Gage factor and the expected amplitude of the stress and consequent strain, determine the gage output at maximum excitation power. The amplitude (and frequency) of the output will determine whether sensitive galvanometers (such as CEC 7–318) may be used directly, or whether amplification is required. Gages with high gage factors may be obtained, but the range of ready availability for such gages is limited.

"*Complexity of Stress.* Single-grid strain gages are designed to measure only those strains parallel to the strain axis. Transverse-axis sensitivity is generally less than 2 percent of the major axis sensitivity. When strain directions are unknown, it is necessary to use a multiple array of gages disposed to permit later calculations of the magnitude and direction of the principal strains. To aid in this type of measurement, special multiple arrays called Rosette gages have been developed by the manufacturers and are available as standard catalog items. Some of these are shown in Fig. 2.4.

"*Frequency and Duration of Strain.* In general, two classes of strain work are encountered. The first is long-term examination of static or slowly varying forces, requiring a gage which gives consistent output under varying environmental conditions. While all strain-gage wire is specially selected and drawn to maintain uniform diameter, ductility, temperature coefficient and resistance, static or "quasi-static" work requires a special wire with an extremely low temperature coefficient (1.0×10^{-5} ohm/ohms/°C). Such a wire has been developed under the trade name Constantan. It is a copper-nickel alloy (60–40 percent) and has a gage factor of approximately 2.0.

"The other class of work deals with high frequencies, often at low amplitudes. This work requires wire of good fatigue resistance and gages of high output, even at the sacrifice of long-term stability. For this, Iso-elastic wire was developed. It is a complex alloy of iron (52 percent), nickel (36 percent), chromium (8 percent) manganese, silicon, molybdenum (4 percent), carbon, and vanadium (trace). It has

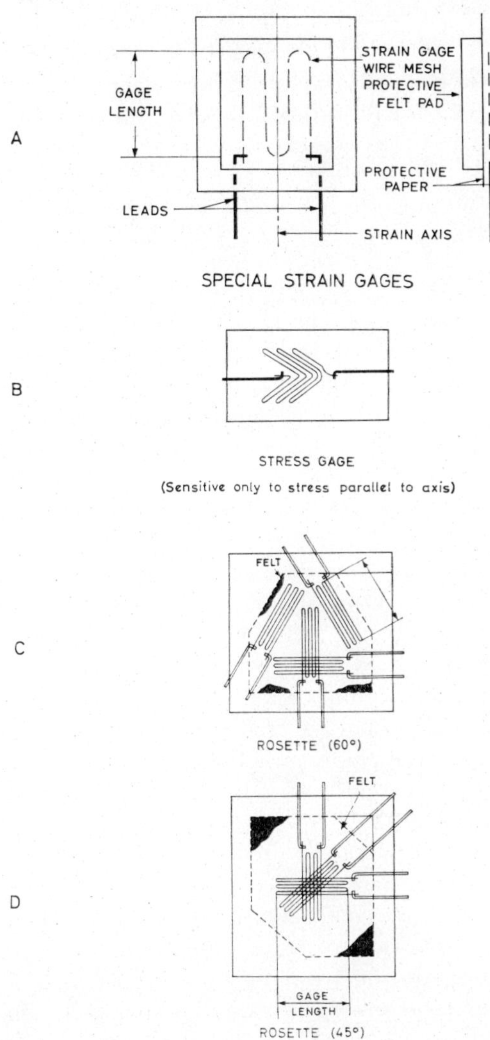

Fig. 2.4. Strain gages: a) typical construction b, c, d) special types

a gage factor of approximately 3.5. Its temperature coefficient is 47×10^{-5} ohm/ohms/°C. When fatigue studies are made, the strain-gage leads must be also specially manufactured. For fatigue work, "dual-lead" gages are available with long-flex-life, low-hysteresis leads.

STRAIN GAGE APPLICATION. "Like most engineering problems, strain-gage technique breaks down into a logical series of steps: –

1) Determine where and how the gage should be mounted on the test member.
2) Select the type of gage best fitted to give the information needed under the test conditions.
3) Prepare the surface and mount the gages.
4) Select the power supply and measuring instrumentation.
5) Determine technique for calibration and select the necessary components.
6) Arrange for test conditions matched to use conditions to be imposed upon members.
7) Prepare the necessary log sheets–brief the test personnel.
8) Run the test.
9) Analyze the results.

"The place for measurement and the measurement axis may be cearly shown by design calculations, engineering judgment, or previous tests. Sometimes, however, information from such studies is insufficient. When areas so identified are too general, more specific information can be gained by running preliminary tests using Stresscoat. This is a brittle lacquer which cracks when strained. The pattern of cracks indicates direction and order of magnitude of subsequent strain.

"In deciding how the gages are to be mounted it must always be kept in mind that strain gages are just what their name implies. They measure strain only. The relationship between the indicated strain and consequent stress must be decided by calculations, tempered by good engineering judgment.

"With the initial work completed, all that remains is to prepare for a test run matching the conditions of use to be investigated. The test equipment should be neatly set up. Transducer leads should be well shielded and kept away from spurious voltage pickup. Belden 8424 rubber-covered cable using diagonally opposite wires as pairs has given good service. Suprenant Electrical Manufacturing Co., Clinton, Massachusetts, supplies cable that has individually shielded and twisted pairs, which is good for long runs.

"When the test equipment is ready, brief the test personnel throughly on their respective duties. Prepare log or data sheets which will ensure test data are adequately correlated, the sequence of events established, channels and instruments identified. Go through a dry run to prove out instrumentation, calibration, and personnel. With good preparation, running the test becomes almost a matter of routine—apart from inevitable emergencies and last minute changes. All that remains is to analyze the results of this work."

The CEC publication from which the above excerpts have been taken contains many more details on strain gages and their use. Additional information may be obtained from Hathaway, [2.12, 2.13].

FOIL STRAIN GAGES. Foil-type strain gages are produced by photo-etching from a variety of specially controlled alloy foils. Foil thicknesses range from one hundred-to two hundred-millionths of an inch. Close control of the alloy and manufacturing operations results in self temperature compensation, a high degree of repeatability, and uniformity in the finished gages. Some advantages of foil-type strain gages are precision pattern orientation for highest accuracy, temperature compensation for a wide range of materials, large power-handling capability, and extended fatigue life.

BUDD METALFILM STRAIN GAGES. The Budd Company, Instruments Division, P. O. Box 245, Phoenixville, Pennsylvania, supplies MetalFilm foil-type strain gages and accessories for their installation and use. Active area dimensions range from

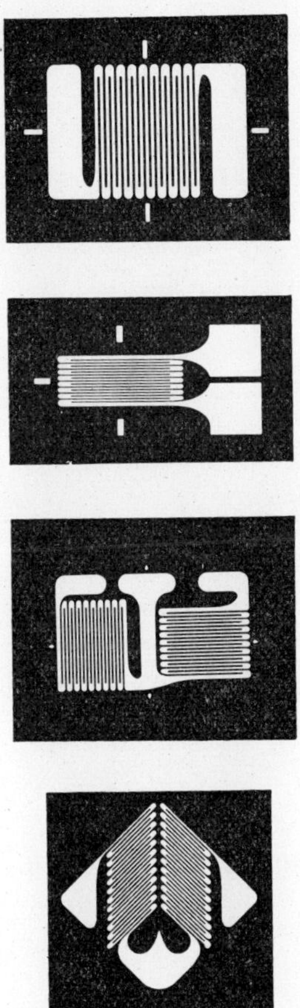

Fig. 2.5. Typical Budd foil-type strain gage patterns

0.015 inch in the microminiature series to 5.000 inches in several standard series. See Fig. 2.5.

Budd also supplies the Flexagage (TM), a double-deck strain gage that can separate and identify direct stresses (also called tensile, membrane, or hoop stresses) and stresses produced by bending moments acting in a plane perpendicular to the surface of the structure. See Fig. 2.6.

Table 2.4 provides useful information for strain-gage applications.

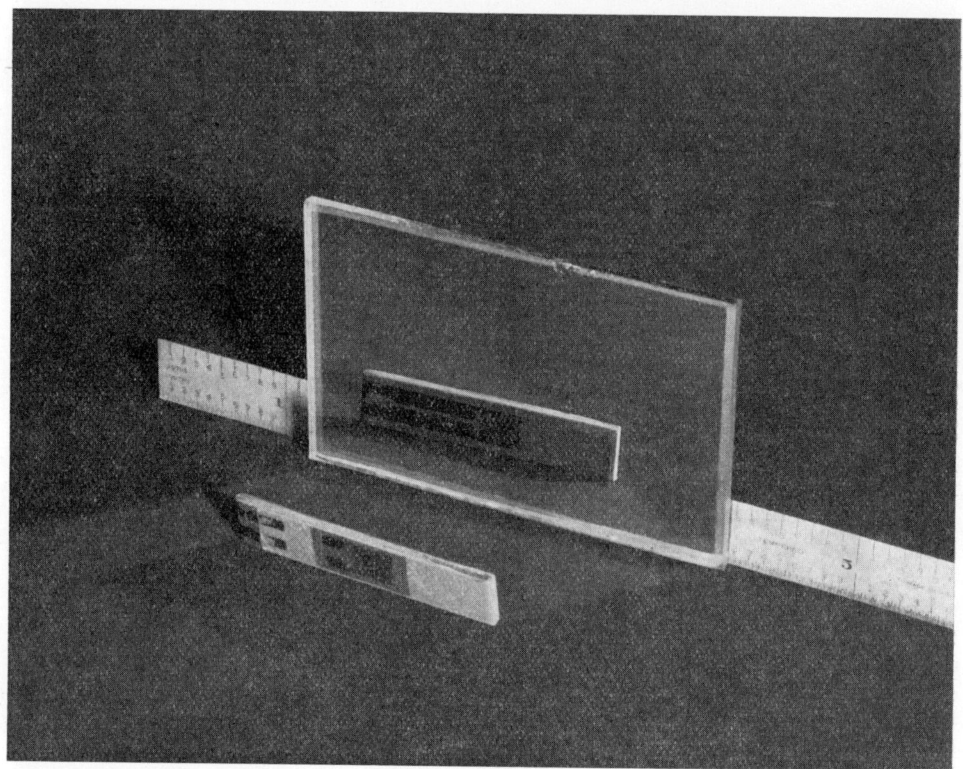

Fig. 2.6. Budd Flexagage

DISPLACEMENT MEASUREMENTS. Differential transformers or linear-motion potentiometers may be used to record physical displacement.

Differential Transformer. Operating on the principle of mutual inductance, the transformer, illustrated in Fig. 2.7, has three coils: two primaries connected in series and one secondary. A metallic movable armature attached to the test item moves in relation to the core and changes the inductance between the primary and the secondary. The variation in a–c voltage can be recorded when connected to an oscillograph in a manner similar to that used with pressure transducers. The characteristics of differential transformers in tabulated form were previously shown in Table 2.3. Rotational displacement transformers employ the same principle as the linear-displacement units.

60

TABLE 2.4

STRAIN GAGE APPLICATION DATA

USEFUL DATA FOR STRAIN GAGE WORK

(Courtesy Instruments Division, The Budd Company)

Budd Metal Film strain gages

For equal-arm strain gage bridge:

$$E_0 = K\varepsilon V$$

E_0 = output voltage in microvolts
K = gage factor

$$\varepsilon = \frac{\varepsilon_1 - \varepsilon_2 + \varepsilon_3 - \varepsilon_4}{4} = \text{average strain in each arm in microinches/inch}$$

V = bridge excitation voltage in volts

Note: use (+) sign for tensile strain, (−) for compression

Shunt calibration:

$$R_c = \frac{R_g}{\varepsilon_s K} - R_g \quad \text{or} \quad R_c = \frac{R_g}{\varepsilon_s K} \quad \text{if } R_c \gg R_g$$

$$\varepsilon_s = \frac{R_g}{K(R_c + R_g)} \quad \text{or} \quad \varepsilon_s = \frac{R_g}{KR_c} \quad \text{if } R_c \gg R_g$$

R_c = shunt cal resistor in megohms
R_g = gage resistance in ohms
K = gage factor
ε_s = Simulated strain in microinches/inch

Table For Annealed Copper

AWG Wire Size	Res. in ohms per 100 Ft.
10	.100′
12	.159
14	.253
16	.402
18	.639
20	1.02
22	1.61
24	2.57
26	4.08
28	6.49
30	10.3
32	16.4
34	26.1
36	41.5
38	66.0
40	105.

Strain Gage Lead Wire Consideration

Desensitization of active strain gage, of resistance R_g, in series with lead resistance R_L, is given by:

$$\frac{R_g}{R_g + R_L} \times 100 \text{ per cent}$$

For 2 wire lead circuit, R_L is total resistance of both wires. For 3 wire lead circuit, R_L is resistance of *one lead only*.

This desensitization may be compensated by setting a new gage factor, K', on the strain indicator.

$$K' = \frac{R_g}{R_L + R_g} K$$

where K is original gage factor of the active gage

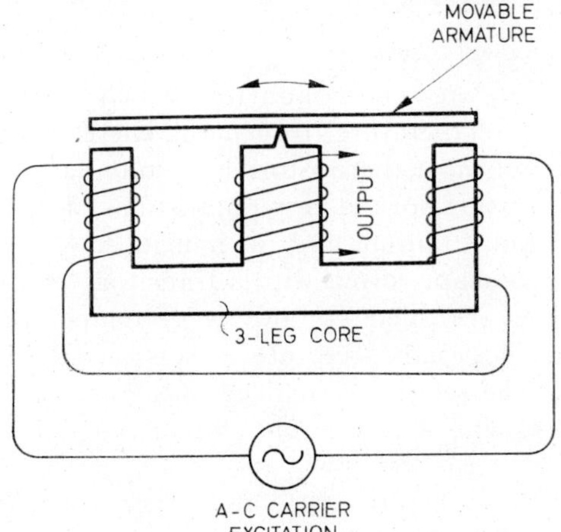

Fig. 2.7. Differential transformer. One form of short-stroke inductance-type transducer

Linear-Motion Potentiometer. The resistance of this device is changed in proportion to the movement of the test item which is mechanically linked to the wiper arm of the potentiometer.

TEMPERATURE MEASUREMENT. The interrelation between thermal and electrical forms of energy was discovered more than a century ago. The phenomenon, investigated by Seebeck, Peltier, Thomson and others, subsequently was developed into principles that later were applied to electrical temperature-measurement equipment. Thermocouples, resistance thermometers, and thermistors are commonly used today as sensors to measure and/or record temperatures in a wide range of applications.

Thermocouples. Wires of two dissimilar metals are electrically connected in series to make a thermocouple. When the temperature at this probe is varied, the thermal energy change is converted into a corresponding electrical change. The device can be connected to an indicating or recording instrument and calibrated to measure temperature.

Thermocouples produce voltages in the order of millivolts; typical ranges are 30 mv at temperatures around 100 °F to 50 mv at above 2000 °F. Thermocouples are available in a variety of different metal combinations, each of which is best suited for a particular temperature range. For example, a copper-constantan combination may be used for a –100° to 600 °F range, an iron-constantan combination for a 0° to 1400 °F range, and a chromel-alumel combination for a 1000° to 2200 °F range.

Resistance Thermometers. Electrical resistance of metals increases as temperature increases, and this principle is used in resistance thermometers. Because it is a passive transducer, a power source must be used with it. Resistance thermometers have quick response, high accuracy, and have excellent remote wiring characteristics.

Thermistors. Thermistors are semi-conductors which exhibit an extremely large change in resistance with temperature. In general, there are negative and positive coefficient thermistors, but most thermistors now made are negative. Thermistors are made from specific mixtures of the very pure oxides of nickel, manganese, iron, cobalt, copper, magnesium, titanium and other metals. A mixture of these oxides is formed into the desired shape, often with a transitory binder mixture, and then sintered under carefully controlled conditions of temperature and atmosphere. The sintered piece either has leads fired into it or is carefully silvered on two sides and leads are attached. The relatively small physical size of the thermistor can be visualized from Fig. 2.8. This photograph compares a typical thermistor with a postage stamp.

Normally thermistors vary greatly as to resistance at a particular temperature and as to rate of change. The usual tolerance for resistance at a standardizing point is ±10 to 20 percent. Because of the variation of rate of change, even thermistors selected to have the same resistance at a standardizing point may well develop a substantial resistance difference at temperatures removed from the standardizing point. These characteristics previously made the use of thermistors in many situations difficult or even impractical. The necessity to treat each thermistor individually in precision and semi-precision applications has greatly limited common use of thermistors in those application areas.

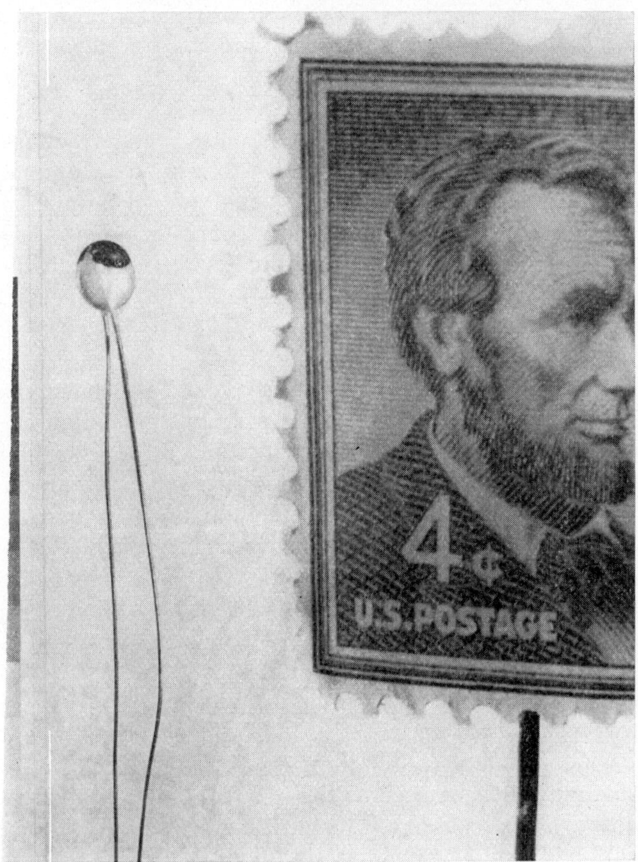

Fig. 2.8. YSI thermistor compared in size with a postage stamp

The Yellow Spring Instrument Company (YSI), Yellow Springs, Ohio, has developed a method (U. S. Patent No. 2,970,411) for producing interchangeable or precision thermistors. These thermistors are within ± 0.5 per cent of each other at a standardizing point and stay within a narrow tolerance over a wide temperature range. This thermistor allowed multiple thermistor probe inputs to a single circuit, simple exchange of sensing probes for different measuring or control jobs, easy replacement of damaged probes from the shelf. A family of thermistors is available which follows the same resistance-temperature curve from -80 to $+150$ °C to tight tolerances. This family covers the range of 100 to 1,000,000 ohms at 25 °C in nine steps.

All YSI thermistors are the same general size, 0.09 inches in a diameter by 0.05 inches thick. They are furnished with standard #32 tinned solid copper parallel leads three inches long. Each thermistor is supplied with temperature-resistance tables covering the useful range within $-80°$ to $+150$ °C and with application notes including power dissipation capability and simple instructions for handling. YSI precise thermistors are epoxy encapsulated for ruggedness and increased life expectancy.

LIGHT MEASUREMENTS. Light-sensitive cells have been in common use, particularly by photographers, since the early 1930's, to measure illumination. Photometers typically contain a light-sensitive cell, usually a selenium-coated steel plate. The selenium surface is electrically connected to one terminal of a microammeter, and the steel plate is connected to the opposite terminal. Light, falling on the cell, generates a current in microamperes which is registered on the meter. The meter scale is suitably calibrated so that a direct reading in footcandles (or in other units) can be made.

The type of photocell used in most photometers can be considered an active transducer because it generates a current. Such cells are available in numerous configurations and with various characteristics, and can be used for recording as well as for indicating purposes. For example, solar cells are made of cadmium sulphide, or cadmium selenide, or silicon, usually in a rectangular or circular disc configuration.

Passive photocells are also obtainable. The resistance of the cell is lowered when the cell is exposed to light so that an external source of current flows through the cell. Thus the cell functions as a valve.

Photocells measure variations in light intensity, but they can be used to measure some other parameter using light as the medium.

An application of a photocell to oscillography is Southern Instruments Type M. 738 Photocell Sweep Unit, shown in Fig. 2.9. The purpose of the unit is to provide a signal voltage which has a direct relationship to the rotation of a shaft. The signal, suitably amplified, may then be used as the X-sweep on a cathode-ray oscillograph tube so that another signal voltage, representing, say, a variable such as cylinder pressure, when applied to the Y-plates, is plotted as a function of shaft angle. This is especially useful for engine indicating. It is a great advantage to have the diagram continuously presented on a crank-angle base irrespective of the engine speed. The diagram then stays in position even if the engine speed is varying. This photocell sweep is therefore preferable to the more conventional electronic time-base.

The M. 738 photocell pick-up is a cast aluminium box, 13 inches by 7 1/2 inches by 5 inches overall, fitted with lugs for bolting to an engine bedplate or rigid bracket. One end houses a projecting shaft 3/8 inch diameter mounted on ball races, which is to be coupled to the engine under test. This shaft drives a helical shutter which controls the amount of light from a small lamp falling on a photocell. The shutter is shaped so that, in conjunction with a mask fitted to the photocell, the cell gives an electrical output proportional to crank angle over the major portion of the 360 degrees rotation. A screened connector on the unit provides the termination for a cable which supplies power to the lamp and photocell and carries the output signal from the unit.

Fig. 2.9. Photocell Sweep Unit (Southern Instruments Type M. 738)

The Southern Instruments Type M. 738 Photocell Sweep Unit produces spot movement that is directly related to crank rotation at all times. The spot position and amplitude of sweep are entirely independent of engine speed. A calibrated control is provided on the M. 738 unit to phase the lamp and photocell assembly with relation to the shutter so that the sweep can be made to start at any point in the engine cycle. This unit is complete with degree marker disc and pick-up.

Flow Measurements. Fluid (liquid or gas) flow is measured by a class of instruments usually referred to as flowmeters. There are many types of flowmeters in use today, but those discussed are frequently used with the oscillograph. Among these are turbine meters, strain-gage flow transducers, and thermal-conductivity-bridge flowmeters.

Turbine Flow Transducers The Hydropoise Turbine Meter, manufactured by Brooks Instrument Company, Inc., Hatfield, Pennsylvania, is pictured in Figs. 2.10

and 2.11. It is an in-line, fluid metering device in which a multi-bladed turbine rotates in precise proportion to the velocity of the fluid through the meter. An electrical pick-off coil, located adjacent to the turbine, generates electrical pulses absolutely proportional to the flow. Since each pulse represents a discrete volume of

Fig. 2.10. Brooks turbine flow transducer

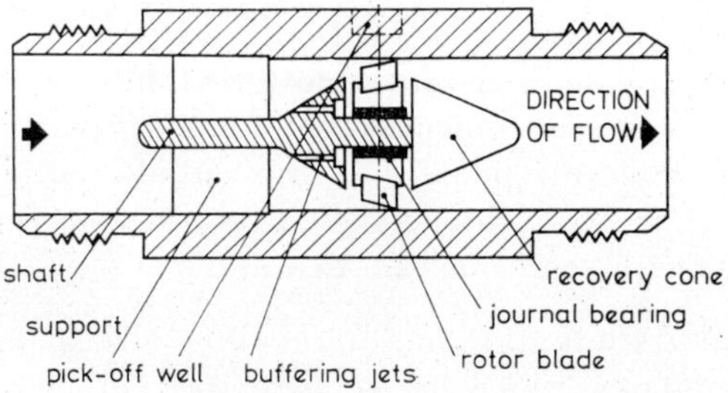

Fig. 2.11. Cross-section view of a turbine transducer

fluid, it can be fed directly into digital indicators and totalizers; or, through a frequency-to-d-c converter, into potentiometric devices for indicating, recording, and controlling. The transducer can be used under a wide range of flows, and at various temperatures and pressures.

66

A turbine flowmeter manufactured by another company is described in reference[2.15].

Strain-gage Flow Transducers. The strain-gage flow transducer senses the rate of fluid flow as a product of the dynamic forces acting upon a fixed body immersed in the flow stream.

This may be expressed as:

$$\text{Force} = C_d \, A p \, \frac{V^2}{2g}$$

Where:

C_d = Drag Coefficient
A = Area of Sensor,
p = Fluid Density,
$\dfrac{V^2}{2g}$ = Velocity Head

An example of this type of flow transducer is the Mark V, manufactured by the Ramapo Instrument Company, Inc., Bloomingdale, New Jersey. An outline diagram of the Mark V, together with a graph of its operating characteristics is shown in Fig. 2.12.

The transducer operates as follows: Bonded strain gages in a four active arm bridge circuit, completely isolated from the fluid by a wall of stainless steel, faithfully translate this physical force into an electrical output which is proportional to the original flow rate squared.

Bi-directional flow metering, and the indication of exact flow direction by the polarity of the output signal can be obtained with this type of transducer. Several models are available for various tube or pipe sizes and for selection of the desired flow range.

Thermal-conductivity-bridge Flowmeters. An example of a flowmeter that utilizes the thermal-conductivity bridge principle is the Model 59 Electric Flowmeter manufactured by the Thermal Instrument Company, Cheltenham, Pennsylvania. It operates in the following manner: Two flow sensing tubes are mounted in a precisely machined heat sink. A relatively small constant amount of heat is provided by the measuring elements R1, R2, R3, and R4, shown in Fig. 2.13.

Tube Number One carries the flowing stream. Heat is removed from this tube in direct proportion to the mass flow through it. Elements R1 and R3 are cooled by the flow stream and change value in direct proportion to the flow stream. Tube Number Two contains the static fluid (no flow) and compensates for pressure changes in the flow system. Elements R2 and R4 are subjected to the same conductivity changes, due to pressure changes, as are elements R3 and R1. Therefore, the bridge remains balanced for all changes other than flow. Elements RT1 and RT2 are used for ambient temperature compensation of both heat sink and fluid stream. The response time for a liquid such as water is a matter of seconds for 63.2 percent of a step change.

The flow tubes are thin wall tubing. However, they are very small bore and may be used in high pressure applications. The tubes can be made of any metal required to meet a particular flow application (stainless steel, nickel, gold, platinum, etc.).

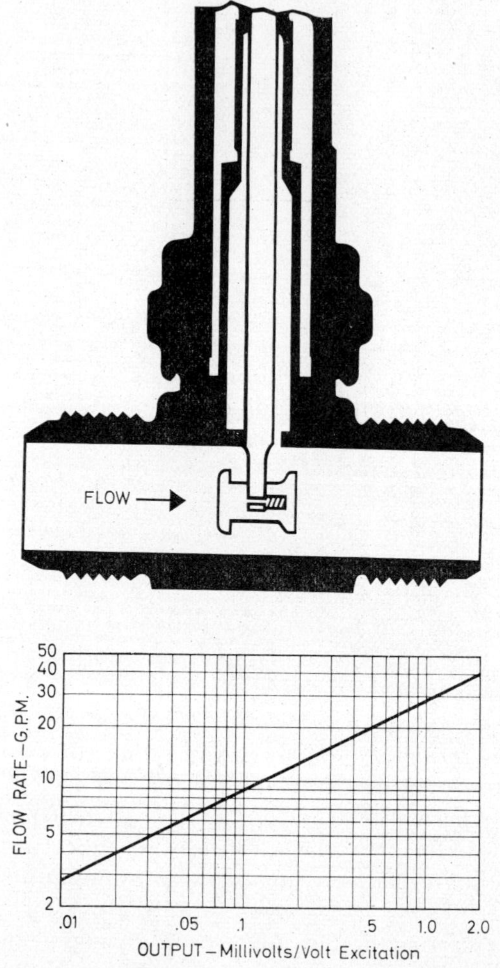

Fig. 2.12. Outline diagram and operating characteristics of the Ramapo Mark V Flow Transducer

VIBRATION MEASUREMENTS. The type of transducer required to record vibration is determined by the frequency range involved. Low frequencies, such as those encountered in seismology, require a magnetic-induction transducer. Magnetic-induction pickups, which are active transducers, are also designed to fit applications

68

having a frequency range to 1000 cps. Fig. 2.14 shows a cross-section schematic of a typical pickup. The construction is simple, and temperature stability is good.

Another type of vibration pickup is the proximity type. An example manufactured by Southern Instruments is illustrated in Fig. 2.15. It is particularly useful in the measurement of movements or vibrations of objects on surfaces which for practical purposes cannot be touched if the conditions of vibration are not to be altered.

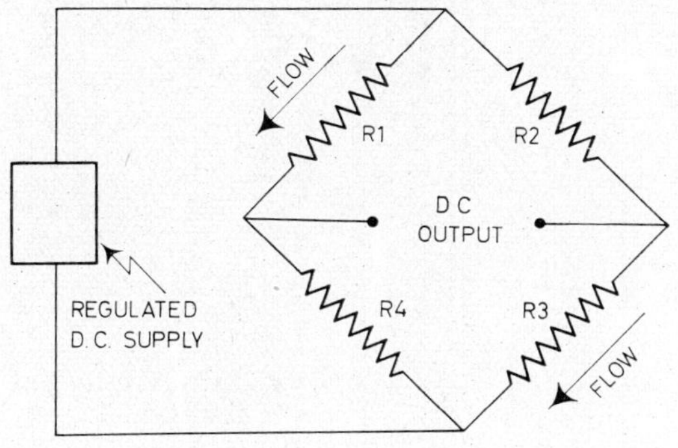

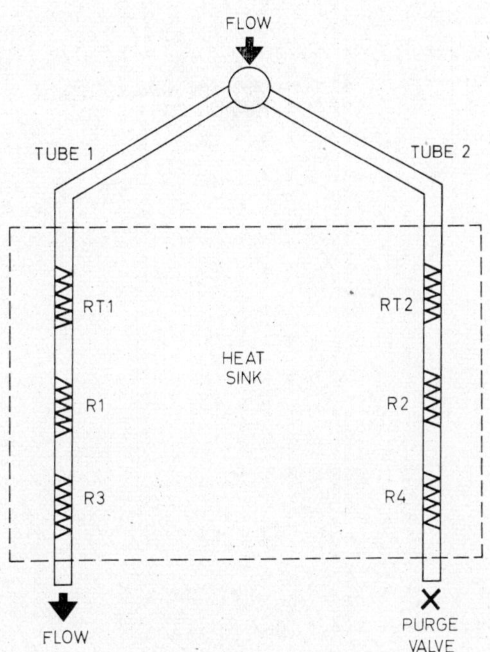

Fig. 2.13. Schematic flow measuring system used in the Thermal Instrument Company Model 59 electric flowmeter

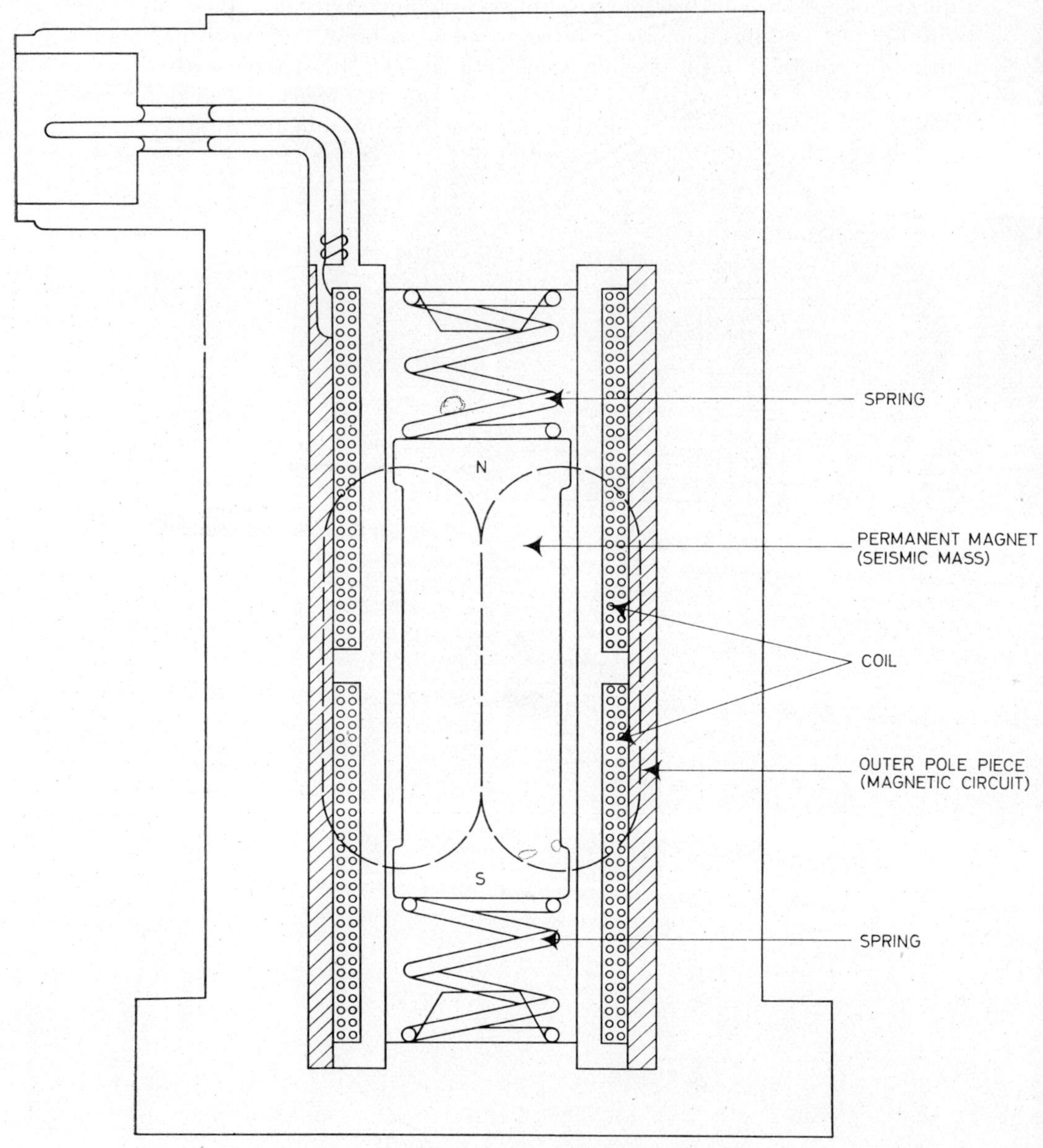

Fig. 2.14. Magnetic-induction pickup

The Southern Instruments Type G211 Proximity Pickup consists of a small high-frequency inductance forming part of the oscillator circuit of an FM system, which is brought close to the surface or object whose movement is to be measured. The object must be metallic and preferably of high conductivity material such as brass or copper. If the object is non-conducting, copper or aluminum foil must be attached to it. The effect of the metallic object is to reduce the inductance and hence modulate the oscillator frequency, the demodulated signal being amplified and recorded on the oscillograph.

Fig. 2.15. Proximity-type vibration pickup (Southern Instruments type G. 211)

The inductance is mounted on an extension through the spindle of an accurate dial type micrometer reading to 0.0001 inch (or millimeters if preferred) and a knurled screw is provided so as to advance the micrometer movement and inductance towards the object under measurement. The pick-up is therefore self-calibrating as the vibration amplitude, which is being indicated, can be directly compared with the dial micrometer readings. The comparison can either be made while the vibration is being recorded or the oscillograph deflection can be calibrated before the vibration effects take place.

The vibration measurement is independent of frequency as the FM carrier system gives sustained effects. However slow the movement of the object, the demodulated output is stricty proportional to that movement. The upper frequency is only limited by the band-width of the tuned circuits in the FM oscillator and amplifier.

The micrometer is in a cast aluminum box with a ball and socket fitting. This enables the inductance to be brought up to and aligned with the surface under measurement and then clamped rigidly in place. The base measures 4.5 by 3 inches (110×75 mm) and the height is 5 inches (127 mm). The weight is 2.25 lbs. (1.1 Kg). Two interchangeable inductances are provided. The larger can be arranged to give full output from the FM system for double amplitudes between 0.010 and 0.040 inches and the smaller covers double amplitudes between 0.020 and 0.100 inches. The ranges can be varied by adjusting the initial distance between the object and inductance. With normal conditions, movements of 0.0001 inch (0.0025 mm) can be detected, though this figure can be improved by increasing the amplification.

ACCELERATION MEASUREMENTS. Both active and passive types of accelerometer are in general use. See references[2.16] for further details.

Particular care should be exercised in handling the more sensitive accelerometers, as they can be damaged by excessive vibration or shock. The acceleration range for each pickup should not be exceeded, and special precautions should be taken to prevent the units being dropped. The units should always be stored in an upright position.

Active Accelerometers. These transducers utilize a piezoelectric crystal that produces an electrical output when strained in the correct direction. The principle

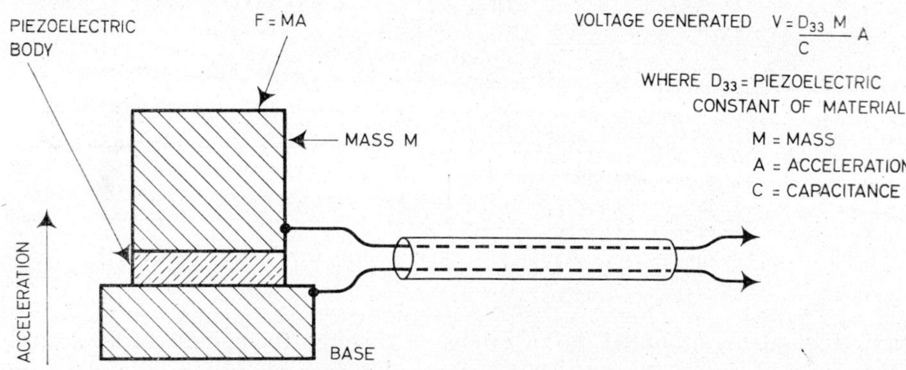

Fig. 2.16. Elements of a piezoelectric accelerometer

is similar to a phonograph cartridge being strained by its stylus. Fig. 2.16 shows schematically the elements of a piezoelectric accelerometer.

Fig. 2.17 shows a piezoelectric transducer manufactured by Endevco Corporation, Pasadena, California. The accelerometer features a single-ended compression design for complete physical separation of the active element from the cap and case wall. Two models in this series measure vibration from below 10 cycles per second to above 6 kilocycles per second without need for any special electronic equipment, and will produce oscillograph recordings by direct connection of the transducer to the oscillograph. The accelerometers, using Endevco Piezite Element VI, have response characteristics that are particularly independent of temperature variation.

Passive Accelerometers consist of two types: 1) those that operate on the resistance strain-gage principle; and 2) those that operate by variable capacitance.

In a strain-gage accelerometer, a seismic mass causes resistance to vary in the spring-type, unbonded strain-gage portion of the instrument. An example of a miniature, temperature-compensated, strain-gage accelerometer, manufactured by CEC, is shown in Fig. 2.18. The CEC Type 4–202 is a linear unbonded strain

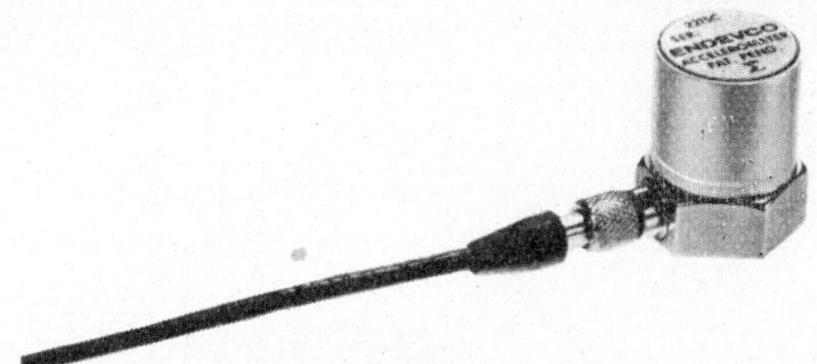

Fig. 2.17. Endevco Model 2215 Accelerometer

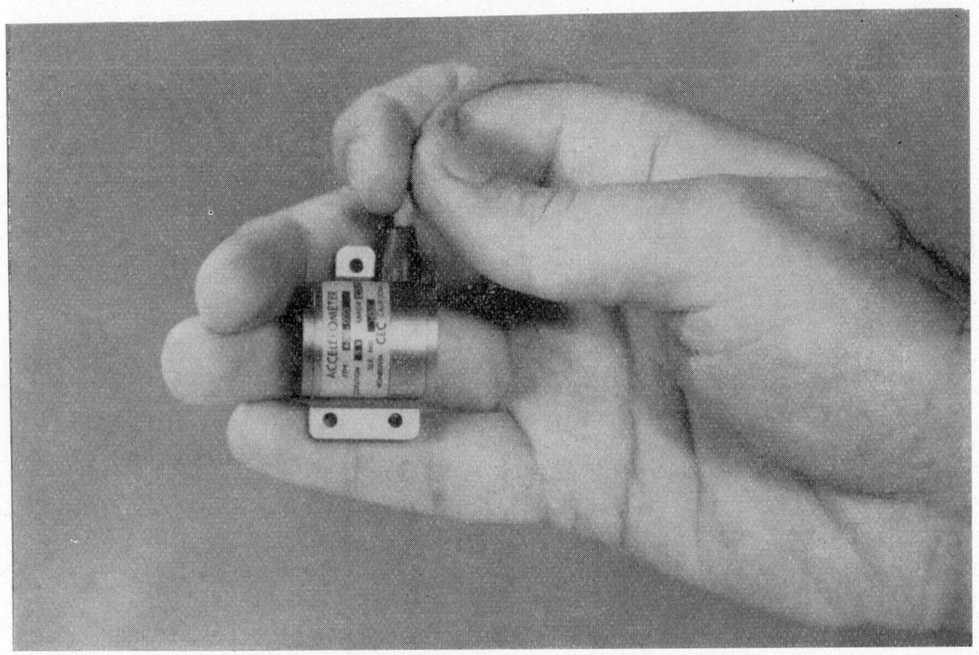

Fig. 2.18. Strain-gage accelerometer (CEC Type 4-202)

gage bi-directional instrument, designed for measuring accelerations perpendicular to the mounting surface. Weighing less than three ounces, the one-cubic-inch accelerometer's linearity and hysteresis qualities are conservatively rated at less than ±0.75 percent of full range output.

The non-pendulous type mass suspension used in the 4–202 permits exceptionally low cross-axis sensitivity of less than 0.01 g/g for an input of three times the rated range.

Damping is achieved by viscous oil shearing. A very low viscosity oil is used to keep viscosity changes to a minimum in both low- and high-temperature applications. Temperature compensation insures reliable static accuracy for the −65 °F to +250 °F temperature range.

Standard ranges are from ±5g to ±500 g. Over-acceleration up to 20 times the rated range is permitted by mechanical stops. Input voltage is five volts with a full range output of 40 mv (±20 mv).

An electrical compensation chamber allows external adjustments for bridge balance, temperature compensation, and sensitivity.

The CEC tri-axis strain gage accelerometer shown in Fig. 2.19, is a specialized sensor which measures acceleration along three mutually perpendicular axes.

Fig. 2.19. Tri-axis strain-gage accelerometer (CEC Type 4-204)

Dimensions of the Type 4–204 are 2.21 by 2.25 by 1.78 inches. Weight is approximately 7.0 ounces excluding mating connector.

Acceleration ranges are available from ±5 g's to ±500 g's. Cross axis response is less than 0.01 g per g for ranges through ±100 g. Operable temperature range is

−70 °F to +300 °F. Combined effects of linearity and hysteresis do not exceed ±0.75 percent of full range output for each axis.

The mounting structure of the 4–204 is fitted with three factory-interchangeable accelerometer modules. Each module consists of a four-active-arm, spring-type unbonded strain gage element and a seismic mass damped by the shear action of a viscous fluid. An integral connector affords maximum ease of electrical hookup.

Since all three accelerometer modules are self-contained, sealed units, the 4–204 can be supplied with any available acceleration range for any of its three axes.

The accelerometer is designed for in-flight and test-stand use with missiles and rocket engines. Other applications include wheeled vehicle dynamic-acceleration studies and the measurement of static acceleration in centrifuge and similar systems.

Fig. 2.20. Variable-capacitance accelerometer (Southern Instruments Type G.226)

An example of a variable capacitance accelerometer is shown in Fig. 2.20. It is manufactured by Southern Instruments. Limited, and is intended for use with their FM carrier system. It can measure directly the vertical or the horizontal acceleration of any object to which it is attached.

The type G226 accelerometer consists of a strong steel body that contains the condenser system, the mass, and, where fitted, the damping system. The condenser is similar in design to the type G.201 pressure gage and is located at the top of the body with a co-axial cable connector. The mass, which varies with the range required, is mounted on spiders which allow only axial movement and the center of the mass is held against the condenser diaphragm by a spring which applies a predetermined force. When the whole instrument is subjected to acceleration, the force on the diaphragm varies and alters the electrical capacitance. The FM system converts this into a voltage proportional to acceleration.

The incremental capacitance corresponding to full scale acceleration is 20 pF. The spring holding the mass against the diaphragm is set during manufacture so that the capacitance is initially half-way up the range enabling both positive and negative accelerations to be measured. The range required must be specified on order, the maximum being $\pm 1,000$ g and the minimum ± 5 g.

The resonance of the mass and diaphragm compliance is damped by oil for ranges of ± 40 g and lower. In this case, the lower end of the mass ends in a disc. A screw plug in the base is located near this disc so as to leave a small gap which is filled with silicone oil of suitable viscosity.

The frequency response increases with the acceleration range. For a ± 20 g range the undamped resonance occurs at about 1.2 kc per second, and when correctly damped the effective range is from 0 to 1 kc. This illustrates the chief advantage of a condenser device which is so sensitive as a detector of displacement that the whole mechanical system can be made very stiff and the frequency range consequently much higher than with any inductance or strain gage device. Calibration may be carried out on a vibrating table that can be operated at known amplitudes and frequencies. For the very high ranges the force-capacitance relation of the condenser diaphragm is first measured and the mass is calculated so as to cover the required range.

The base dimensions are 2.5 inches square (62 mm), the overall height 4 inches (100 mm) and the weight 2 lbs. (0.8 kgs).

PRESSURE MEASUREMENTS. Pressure is measured and recorded with: 1) resistance pressure transducers; 2) condenser pressure transducers; 3) inductance pressure transducers; and 4) piezoelectric pressure transducers.

Southern Instruments makes the following statements in their technical data concerning resistance pressure transducers:

"These gages are particularly suitable for applications where hig pressures are encountered or where minimum dead volume is important. In principle they consist of a metallic member which is mechanically strained by the application of fluid pressure. Measurement of this strain is carried out by a resistance winding bonded to the strain-member but electrically insulated from it. Expansion of the strain-member stretches the wire and increases its resistance. Most gages of this type have calibrations which are strictly linear and the hysteresis effects are negligible.

"Several types of resistance pressure gage are available. In some, the strain-member consists of a thinwalled steel tube to the inside of which the pressure is applied. This tube may be open at both ends, to permit the flow of fluid through it, or may be closed at one end so that the pressure is applied at the open end only. For application where higher sensitivity is required and in cases where the sensing element must be protected from high fluid temperatures, the fluid is sealed from the inside of the gage by means of a diaphragm and the pressure acting on this diaphragm is measured by a separate strain-member to which resistance wire gages are bonded.

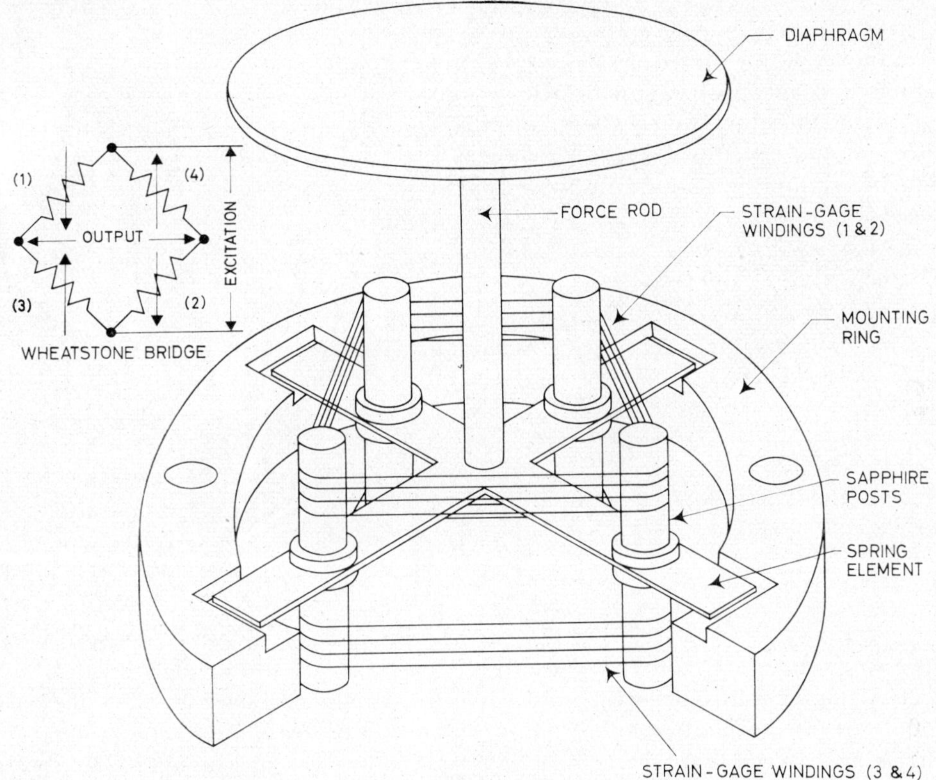

Fig. 2.21. Elements of a strain-gage pressure transducer

"The value of proportional resistance change for the nominal full-scale pressure is quoted for each type of gauge in the SI catalog data. In general, gages cannot be made for exact pressure ranges and the nearest is supplied that gives no less than 75 percent of the stated resistance change for the pressure range ordered.

"Mechanical resonances in resistance pressure gages usually occur at much higher frequencies than those appearing in the associated fluid passages. The rate of response of the gage then depends on the dimensions of the connecting passage and can be computed if these are known together with the relevant properties of the working fluid."

A strain-gage pressure transducer is illustrated in schematic form in Fig. 2.21.

This type of transducer utilizes the unbonded strain-gage principle with a four-active-arm spring type sensing element and a diaphragm force-summing area. With this sensing element attached to the force-summing diaphragm, pressure against the diaphragm produces a displacement which changes the resistance of the active arms. This in turn causes an electrical output precisely proportional to applied pressure. A special isolation design prevents erroneous outputs caused by distortion or vibration.

The CEC Type 4–326 strain-gage pressure transducer, shown in Fig. 2.22, is an example of the principle described above. Made in two models, one model features an overpressure stop which permits two times the rated pressure to be applied for three minutes without causing a zero set to exceed 0.5 percent of full range output. Ten times rated pressure, or 10, 000 psi, whichever is less, can be applied for three minutes without causing a zero set to exceed 1 percent of full range output.

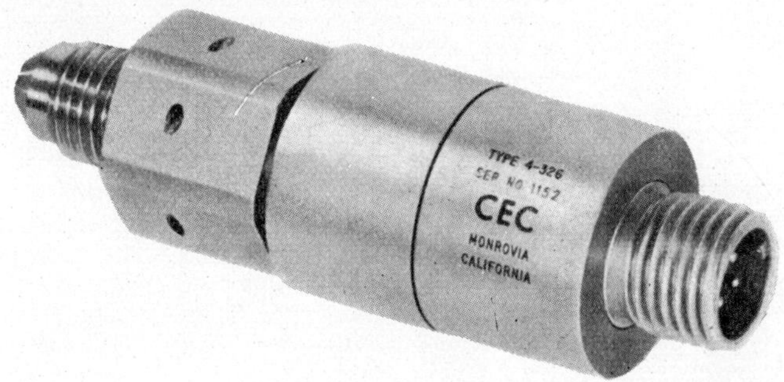

Fig. 2.22. Strain-gage pressure transducer (CEC Type 4-326)

Pressure ranges are from 0–100 psi to 0–5000 psi absolute and gage. Rated electrical excitation is 10 vdc or ac rms at a carrier frequency of 0–20 kc. Sensitivity is 40 mv at rated excitation and 77 °F. Combined effects of non-linearity and hysteresis do not exceed ±0.5 percent of full range output as measured from the best straight line through the calibration points.

The second model utilizes a special corrosion-resistant material, stainless steel type 17–4, for construction of the case and diaphragm. Outstanding corrosion resistance has been confirmed by actual service under exposure to a wide variety of corrosive media in the chemical, petroleum, and food process industries.

Pressure ranges from 0–100 psi to 0–10, 000 psi absolute and gage. Combined effects of linearity and hysteresis do not exceed ±0.35 percent of full range output as measured from the best straight line through the calibration points.

Signal Transmission and Conditioning

DIRECT-CONNECTION. Oscillograph manufacturers have specifically designed their galvanometers, transducers, amplifiers, and signal-conditioning equipment to be used with their recorders. Nearly all recording situations can be accommodated by selecting and matching these components to one another and to the variable to be measured. When using contemporary equipment, a direct hookup can generally be made from an active transducer to the recorder; however, a passive transducer will require a power source and in most instances this will be a low-voltage dc supply.

AMPLIFICATION. Satisfactory results are not always produced by such simple direct connections. Distance may be too great and adverse line losses may occur; characteristics of the transducer and the galvanometer may not match or may not be compatible with the physical effect applied to the transducer; or other factors may degrade the quality of the recordings. Where such conditions occur, a higher-voltage source or amplification may be necessary. Although relatively higher-voltages permit transmission without amplification for some types of transducer, lower-voltage, amplifying circuits permit compensation so that most pickups can be used. For example, compensating balance controls are needed with strain gages, since perfect bridge balance is impractical to achieve on a production basis.

CARRIER SYSTEMS. In many cases a carrier system is the preferred method of transmitting transducer signals. While straight amplification may provide sufficient power for recording purposes, either amplitude modulated (AM) or frequency modulated (FM) systems provide flexibility, control, matching capability, and response down to low frequencies.

In a typical AM or FM carrier system, an oscillator supplies a selected frequency. This is modulated by the variation of the transducer as it responds to the applied physical change. The signal is then demodulated, and is fed to the galvanometer. In the AM system, the *amplitude* of the electrical output of the transducer is proportional to the magnitude of the physical variation applied to the transducer. In the FM system, the *frequency* of the carrier changes proportionally to the change in the magnitude of the physical variation applied to the transducer. The potential accuracy of the FM system offers two advantages. These are: 1) changes in line voltage do not affect the measurement; and 2) an electronic counter can be used for monitoring purposes. Regardless of the amplifier system used, it should have a stable zero point, and should maintain a specific deflection for a sustained effect.

SERVO MONITOR PHASE SHIFTER. The Sanborn Phase Shifter is used with the 150–1200 Servo Monitor Preamplifier (with slight modification, it can be used with the 350–1200 or 850–1200 Phase Sensitive Demodulator Preamplifiers). The Model 150–1200-C7 may be used with 60 cycle or 400-cycle carrier systems.

It requires no external power. The error signal and reference are connected to the phase shifter and an output cable is provided to connect to the Preamplifier.

The Sanborn Servo Monitor, shown in Fig. 2.23, is a phase-sensitive demodulator. This valuable phase sensitivity permits the Servo Monitor to respond linearly to in-phase signals and simultaneously reject quadrature signals. It is this quadrature rejection characteristic which is important when using the Phase Shifter.

The Servo Monitor phase shift box, a transformer-type phase shifter requiring no external power, is inserted into the reference line and can shift the phase of the reference before it reaches the Servo Monitor. Since it is a transformer, it has the additional advantage of isolating the reference from all grounds in the preamplifier.

Theory of Operation. When the switch is in the lag position as shown, the combination of the resistor (R) and capacitor reactance (X) gives a variable phase shift from zero, when R is zero, to 90 degrees when R is equal to X_c. If R were increased to infinity, then phase shift would increase to 180 degrees. But since this is not practical, it was decided to stop at 100 degrees and use a coil and resistor to obtain the

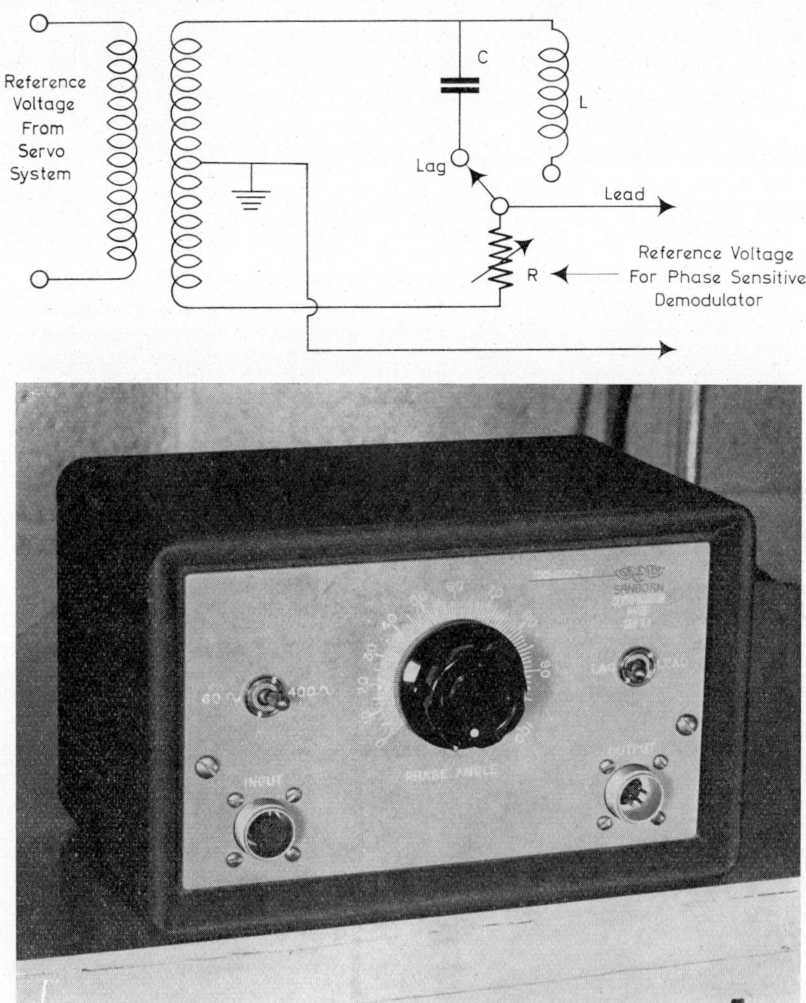

Fig. 2.23. Sanborn Servo Monitor phase shifter

quadrant corresponding to the area between 90 degrees and 180 degrees of the lag side of the switch. Both the lead and lag positions of the switch function in the same manner, i. e., an increase in phase shift from 0 to 100 degrees is produced with increase in R. The value of the capacitor and the coil with the correct Q are so chosen and held to close enough tolerance that the calibration of the dial is accurate within 2 degrees. It can be seen from Fig. 2.24, that a phase shift of 360 degrees can easily be achieved by using the lead lag switch for 180 degrees (Fig. 2.24a) and reversing the isolated reference leads for the other 180 degrees (Fig. 2.24b).

In many applications, the signal being sampled by the Servo Monitor is not in phase or in phase-opposition with the reference, but is at some unknown angle. In

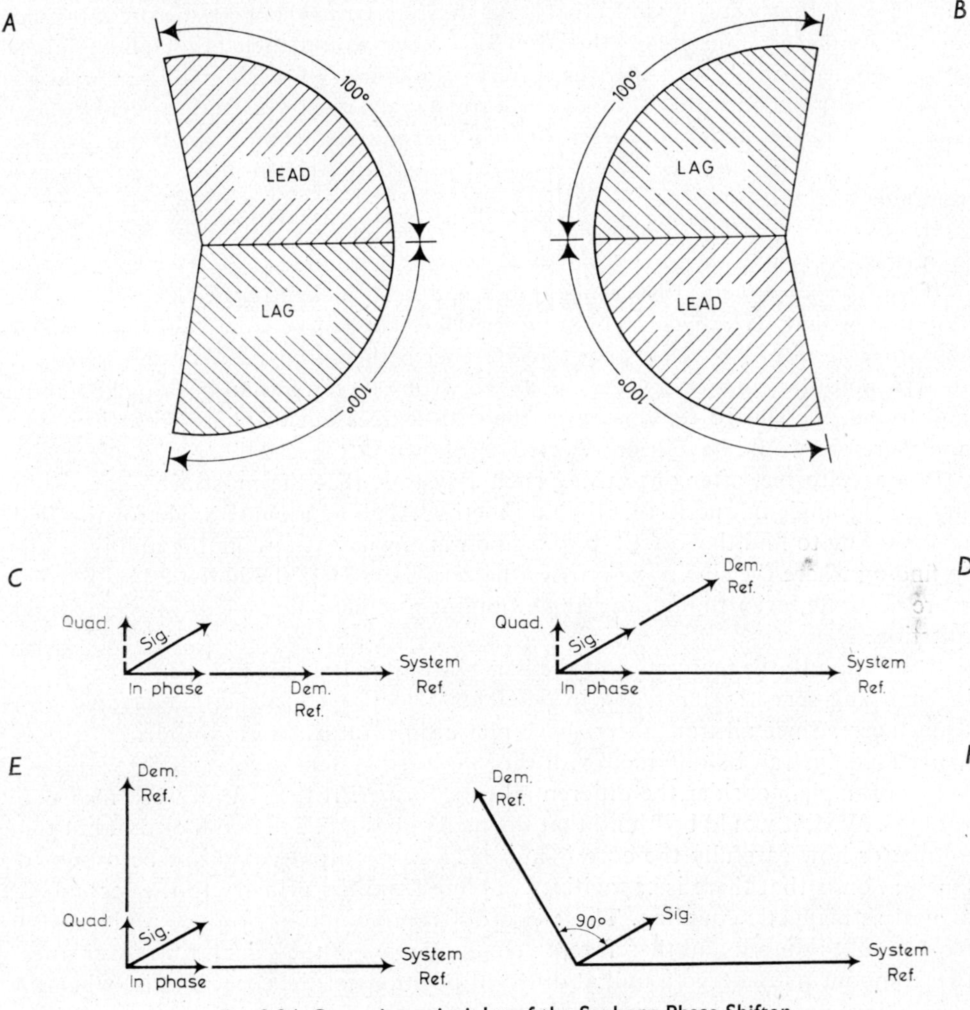

Fig. 2.24. Operating principles of the Sanborn Phase Shifter

such cases it is often necessary to know this angle. The Servo Monitor phase shift box employed with the Servo Monitor, enables the user not only to find this angle, but to find the magnitudes of the in-phase and quadrature signals which add up vectorially to produce this angle.

The in-phase and quadrature signal amplitudes can be found and recorded from these two values. The resultant can be calculated, or more easily, the phase shift dial can be tuned for maximum response giving a measure of the resultant.

The theory of the operation of this system is as follows: The demodulator of the standard Servo Monitor preamplifier responds only to signals which are in phase with its reference. If the system reference were not shifted at all, that component of the input signal would be measured which is in phase with the system reference, (Fig. 2.24c). If it were possible to rotate the reference 90 degrees before it reached the demodulator, Fig. 2.24d, this shifted voltage would then be in phase or 180 degrees out of phase with the quadrature component of the signal. As far as the demodulator is concerned, this quadrature signal appears as the in-phase signal, and the in-phase signal appears as a quadrature signal. Since quadrature signals are rejected, then, in this case, what was originally the in-phase signal is now rejected.

If, instead of rotating the reference by 90 degrees, it were shifted until a maximum or a null was found, the angle in question would be found.

If the correct direction for a maximum had been picked (either lead or lag) the resultant would have been found (Fig. 2.24e). This is because the whole signal looks like an in-phase signal since the reference is then in phase with the signal, and there is no other component present. If the wrong direction had been picked and a null found, the shifted reference was then 90 degrees out of phase with the signal and therefore would have been rejected, as shown in Fig. 2.24f.

It is easy to see, after once using such a system, that the most accurate way of finding the angle in question is the null method. Tuning a phase angle for the peak is like trying to find the exact top of a sine wave, whereas finding the null is similar to finding where the sine wave crosses the zero axis. Thus the null method is much more accurate, even though the angle obtained is the complement of the angle in question.

In addition to the many uses of the Servo Monitor Preamp and the Phase Shifter in analyzing servo systems, the instruments also provide a method of analyzing transducers. Since all transducers have phase shift, information not available from meter readings can be obtained with this Sanborn system.

For example, consider the differential transformer. If the output were measured with a VTVM, it would be found that the meter goes through a null or dip, but that no matter how carefully the core is moved, a zero output could not be observed. One reason is that there is capacitive coupling from the primary to the secondary, as well as magnetic coupling. These two different couplings produce a phase shift from input to output and therefore, two components of the signal. Moreover, these two components undergo a null at different positions of the core; that is, when the in-phase component is zero, the quadrature component is not zero. The Servo Monitor and its associated Phase Shifter, can aid in making better transducers by bringing these nulls closer together. If this is not possible, then very often the compensation necessary to make the phase angle zero can be determined.

LITERATURE REFERENCES

[2.1] TARBOX, J., and YOUNG, W. T. (deceased), *Multi-Channel Dynamic Recording*, presented at semi-annual meeting of the American Society of Mechanical Engineers, Los Angeles, California, July 2, 1953. Reprints available from Consolidated Electrodynamics Corp., Pasadena, California.

[2.2] KEINATH, GEORGE, *Recorder Survey: Recording Surfaces and Marking Methods*, U. S. Department of Commerce, National Bureau of Standards, U. S. Govermennt Printing Office, Washington 25, D. C., NBS Circular 601 (1959)

[2.3] NOLTE, C. B., "Marking Systems," *Instruments & Control Systems*, (three-part article) Vol. 38, May, June, July 1965

[2.4] OWENS, L., "New Ink-Writing Methods for Graphic Recording," *Instruments and Control Systems*, Vol. 38, No. 7, p 100, July 1965

[2.5] HALL, B., "Versatile X-Y Recorders," *Research/Development*, Vol. 14, No. 12, p 22, (1963)

[2.6] KAIZER, C., "Chart Recorders," *Systems Designer's Handbook*, Vol. 8, No. 1, p305, January 1964

[2.7] SIEGELMAN, A., "Direct-Writing Recorders — Which Writing Technique?" *Electronic Instrument Digest*, Vol. 1, No. 2, p8, July–Aug. 1965

[2.8] HOLZMAN, R., "Chart Recorders," *Electromechanical Design*, Vol. 6, No. 6, p35 June 1962

[2.9] FLEMING, L., "The Transducer...A Perspective View," *CEC Recordings*, Vol. 16, No. 2, p19 (1962)

[2.10] "Electromagnetic and Potentiometric Transducers," Part 1, *Electromechanical Design*, Vol. 3, No. 2, p41 (1959)

[2.11] Same, Part 3, *Electromechanica Design*, Vol. 3, No. 5, p35 (1959)

[2.12] TARBOX, J., *Strain Gages*, Consolidated Electrodynamics Corp., (360 Sierra Madre Villa, Pasadena, California)

[2.13] HATHAWAY, C., *The Measurement of Strain*, Hathaway Instruments Inc., (5800 East Jewell Avenue, Denver, Colorado) 1957

[2.14] BUCHBINDER, H. G.,"Electromagnetic and Potentiometric Transducers," *Systems Designers' Handbook*, Vol. 8, No. 1, p53, January 1964

[2.15] "Turbine Flowmeter Pickoff," *Electromechanical Design*, Vol. 3, No. 2 p14 (1959). Note: The publisher of references 2.6, 2-8, 2.10, 2.11, 2.14 and 2.15 is Benwill Publishing Corp., (167 Corey Road, Brookline, Mass.)

[2.16] A list of technical papers and literature on the measurement of vibration, acceleration, shock, force, pressure, turbulence and other conditions is available from Endevco Corporation, 801 South Arroyo Parkway, Pasadena, California. Individual papers selected from the list by title may be ordered by number from Endevco at no obligation.

General References

NEUBERT, H.K.P., *Instrument Transducers*, Oxford University Press, London (1963).

LION K.S., *Instrumentation in Scientific Research*, McGraw-Hill Book Company, New York (1959).

PERRY, C.C., and LISSNER, H.R., *The Strain Gage Primer*, McGraw-Hill Book Company, New York (1955).

III LIGHT-BEAM OSCILLOGRAPHY

Electro-optical Components

MIRROR GALVANOMETERS. Light-beam oscillographs are usually of the mirror-galvanometer type. Light from a source is reflected from the tiny mirror mounted on the galvanometer onto the strip of photosensitive material. The mirror is rotated in proportion to the electric current, and the reflected light records the varying current.[3.1]

In 1822, D'Arsonval introduced the moving-coil permanent magnet galvanometer. Basically, the galvanometer is an electromechanical transducer that accepts the electrical energy applied to it, and transforms it into mechanical rotational energy.

The principles that cause a galvanometer to function as an instrument for measuring current and voltage variation are as follows: (See Fig. 3.1) A coil is suspended in a magnetic field (supplied by a permanent magnet), in such a way that it is free to rotate. As electrical current flows through the coil, the coil develops a magnetic field strength proportional to the amount of current. The "poles" of the coil are arranged at right angles to the poles of the permanent magnet. Attraction of the

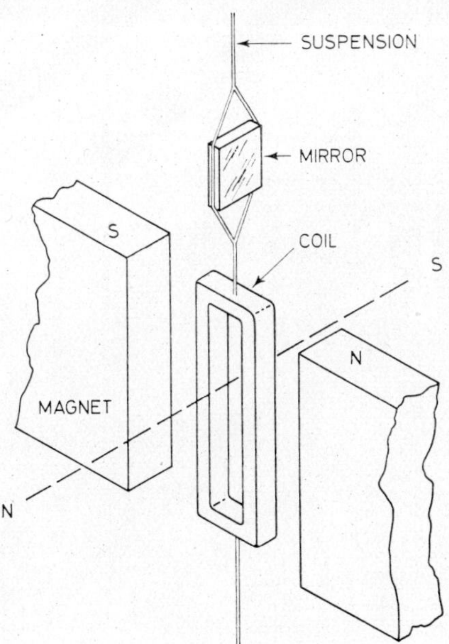

Fig. 3.1. Galvanometer principles

poles of the two magnetic fields causes a rotational force to be applied to the coil. As the coil rotates, the coil suspension twists, and the resulting spring action retards the movement of the coil. When the rotational force caused by the coil current equals the resisting force caused by the spring of the coil suspension, rotation stops. This position is a measure of the exact amount of current flowing through the coil.

Coil-Type Recording Galvanometers. More than 100 years elapsed between D' Arsonval's discovery and the application of his principle to dynamic light-beam recording. The present-day miniature pencil galvanometer, with a mirror attached

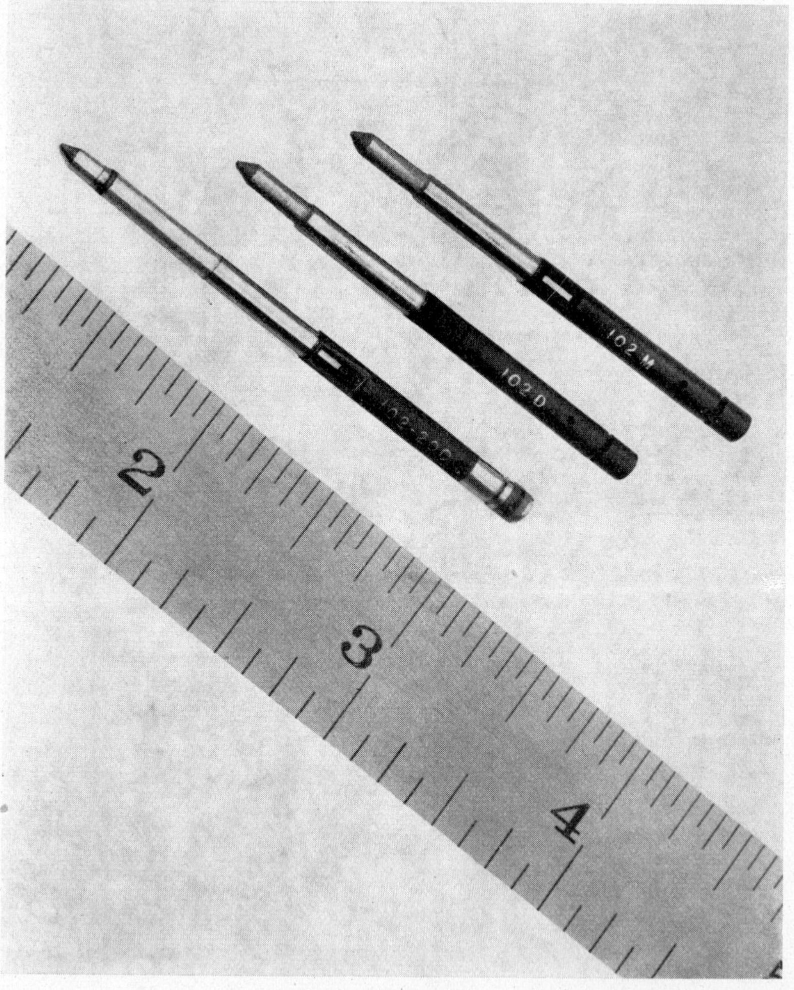

Fig. 3.2. Miniature coil-type mirror galvanometers. *Left:*

to its coil-suspension system, is a sophisticated descendent of D'Arsonval's prototype. Fig. 3.2 shows several galvanometers used in modern oscillographs.

Generally speaking, when a transducer is directly connected to a galvanometer, the frequency response is less than 1000 cps, and a frequency response of about 5000 cps is usually regarded as maximum for oscillograph recorders. However, useful frequency response from 0 to 13,000 cycls per second is claimed for at least one galvanometer design. To obtain a natural frequency as high as possible, the coil is made long and very narrow. A galvanometer that has a sensitivity sufficiently high to be driven directly from transducers achieves that result at a sacrifice of frequency response.

Bifilar Galvanometers. This type of oscillograph galvanometer is a simple mirror device capable of very high-frequency response. A typical bifilar-type galvanometer is illustrated in Fig. 3.3. The moving element consists of a single loop of metallic ribbon or wire stretched tightly between the poles of a magnet. A small mirror is cemented to the center of the loop.

Because of the low moment of inertia of its moving element, the bifilar galvanometer is particularly suitable for high-frequency recording. The element is characterized by large over-current capacity, extreme resistance to shock and vibration, low resistance, and high frequency possibilities. In addition, the simple element is easy to replace if it is destroyed by excessive current. Fluid damping is employed.

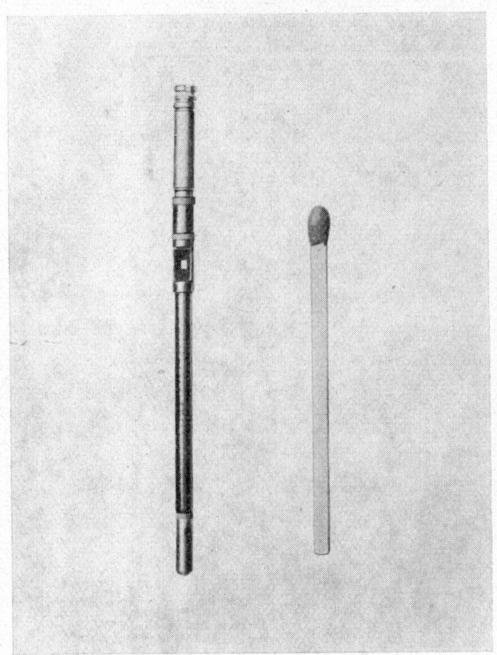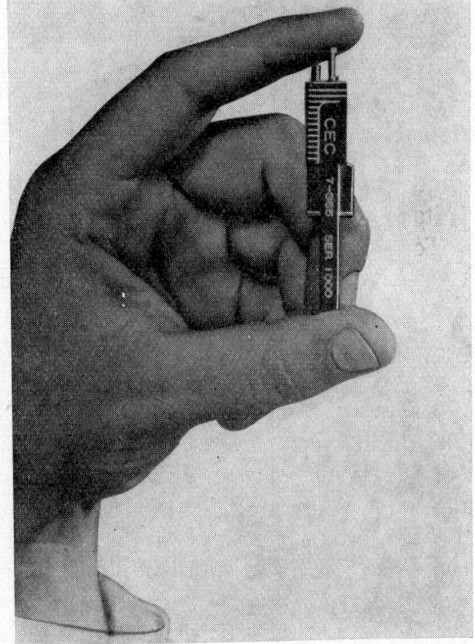

Midwestern; *Middle:* SEL; *Right:* CEC

Fig. 3.3. Bifilar galvanometer (Hathaway Instruments)

SELECTION OF THE PROPER GALVANOMETER. As previously described, the galvanometer converts electrical energy into mechanical motion. This motion is governed by the following characteristics: 1) frequency response; 2) sensitivity (current, voltage and wattage); 3) phase angle; 4) damping; and 5) power dissipation.

To obtain the desired performance for a given set of conditions, it is necessary to understand the various parameters of the instrument. The relation of each parameter to every other parameter and to the application must also be considered, for the galvanometer that is chosen is a compromise of all these characteristics.[3.2]

DAMPING. When current is applied to a galvanometer, the coil rotates, but its motion has a transient distortion. Because of the abrupt change in input level and angular momentum, the transient response of the galvanometer is to "overshoot" the exact proportional point. This action may be likened to the needle on a meter overshooting its final reading. Each galvanometer type has an inherent "natural frequency". For it to respond uniformly over a frequency range between zero frequency (d-c) and 60 percent of the natural frequency, it must be intentionally and properly damped. Either fluid (air or liquid) or electromagnetic damping may be employed. Damping plays an important role with respect to transient inputs.

88

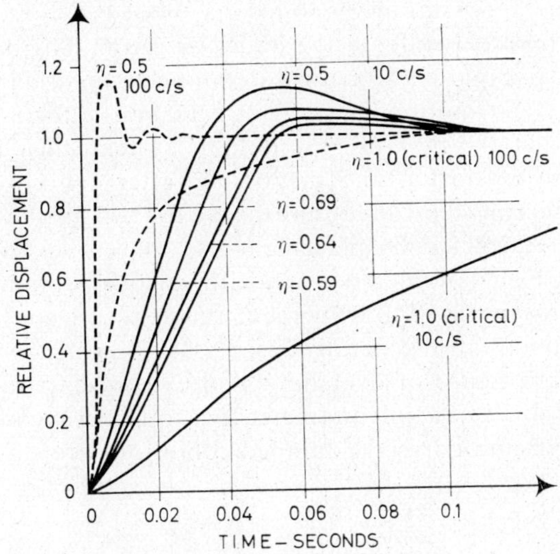

Fig. 3.4. Step-function response

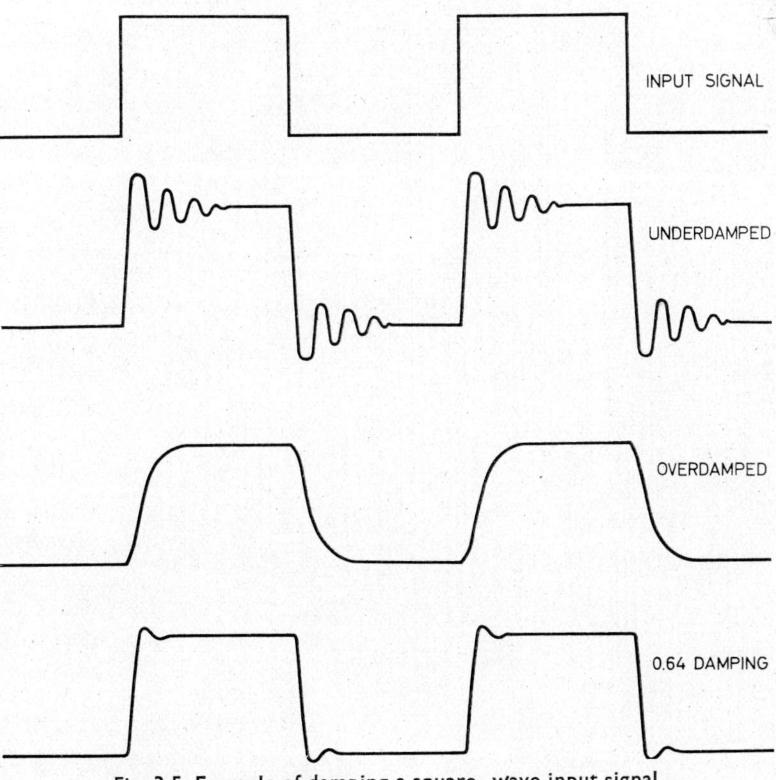

Fig. 3.5. Example of damping a square- wave input signal

The manner in which a galvanometer follows a sudden current rise is called "step-function response", and depends on the damping ratio. (The damping ratio is the ratio of the actual damping to the critical damping; critical damping, also defined as aperiodic or deadbeat motion, may be said to be the fastest response without overshoot condition of the moving system. The system must reach but not pass through this point in as short a time as possible after deflection begins.) Fig. 3.4 illustrates the difference in step-function response for two galvanometers of different undamped natural frequencies. Displacement-rise time curves have been plotted for 10-cycle and 100-cycle galvanometers, each with various damping-ratio values.

With damping ratios in the 59 to 69 percent range, the response curves are almost identical. The damping ratio generally used is 0.64 which is the best practical compromise between absence of overshoot and extended frequency response. Fig. 3.5 illustrates the response of a galvanometer to a square-wave input signal. Under-damped, overdamped, and correctly damped conditions are shown.

Table 3.1

DAMPING VALUES AS A FUNCTION OF PERCENT OVERSHOOT

Percent overshoot	η	Percent overshoot	η	Percent overshoot	η
0.1	.911	5.0	.690	10.8	.577
0.2	.893	5.2	.685	11.0	.575
0.3	.888	5.4	.681	11.5	.567
0.4	.879	5.6	.676	12.0	.560
0.5	.861	5.8	.670	12.5	.551
0.6	.853	6.0	.666	13.0	.544
0.7	.845	6.2	.662	13.5	.537
0.8	.838	6.4	.658	14.0	.530
0.9	.831	6.6	.654	14.5	.524
1.0	.825	6.8	.650	15.0	.518
1.2	.815	7.0	.646	15.5	.511
1.4	.805	7.2	.642	16.0	.504
1.6	.796	7.4	.638	16.5	.497
1.8	.788	7.6	.634	17.0	.491
2.0	.780	7.8	.630	17.5	.485
2.2	.772	8.0	.626	18.0	.479
2.4	.765	8.2	.622	18.5	.473
2.6	.759	8.4	.619	19.0	.467
2.8	.752	8.6	.615	19.5	.461
3.0	.745	8.8	.612	20.0	.455
3.2	.739	9.0	.607	20.5	.499
3.4	.733	9.2	.605	21.0	.444
3.6	.727	9.4	.601	21.5	.439
3.8	.722	9.6	.597	22.0	.434
4.0	.716	9.8	.593	22.5	.429
4.2	.710	10.0	.591	23.0	.429
4.4	.705	10.2	.587	24.0	.414
4.6	.700	10.4	.584	25.0	.404
4.8	.694	10.6	.581		

The required damping of a galvanometer can be determined from the percent overshoot on the leading edge of a square wave. Table 3–1 shows the various damping values as a function of the percent overshoot. The value "η" in the table in the ratio of actual damping to critical damping.

Damping is applied to the galvanometer by mechanically or electromagnetically retarding the motion. The higher-frequency galvanometers are fluid-damped and utilize vanes or disks that work against air or liquid in a confined space. Low natural frequency types are more conveniently electrically damped, using the generator action of the moving coil.

Electromagnetic damping is more easily understood if the galvanometer is considered to be a generator. The generator action takes place as the moving coil of the galvanometer cuts the lines of force of the magnetic field. When a resistance is connected across the terminals of the coil, the generated voltage causes a current to flow in the circuit, and power is dissipated by the circuit resistance. This current flow, or power dissipation, causes the movement of the galvanometer coil to be retarded. This retarding force supplies the damping.

The resulting damping is dependent upon the magnitude of the voltage developed by the "generator" and the circuit resistance. If the sensitivity of the galvanometer is high (i.e., the generator voltage is high) and the circuit resistance is low, the damping force is considerable. As the circuit resistance is increased, the damping is reduced. For electromagnetic damping the optimum condition of 0.64 of critical damping exists when the resistance of the driving source for the galvanometer is equal to the specified damping resistance. When these resistances are not equal, a series or parallel resistor must be used so that the equivalent resistance across the galvanometer is equal to the specified damping resistance.

Where the resistance of the source is low compared to the damping resistance required by the galvanometer, a resistance must be connected in series with the source and the galvanometer (Fig. 3.6a). The value of this resistance plus the value of the resistance of the driving source must equal the required damping resistance.

If the source resistance is high compared to the required damping resistance, a shunting resistor will be needed (Fig. 3.6b). The parallel value of the source resistance and shunt resistor must equal the required damping resistance.

Sometimes, more complex networks are required to provide the proper magnetic damping, or to provide damping while satisfying other requirements of the system. For example, a single damping network can be derived to provide an optimum load for the signal amplifier, some signal attenuation, and the proper damping resistance for the galvanometer (Fig. 3.6c).

If other than 0.64 damping is desired for magnetically damped galvanometers, the damping resistance requirement will be different from that shown in the standard specification. To increase damping, a lower value series resistance is required; to decrease damping the series resistance must be increased.

The damping ratio may be checked in the following manner. Connect a variable resistance (R_d) across the galvanometer. With a separate dc source deflect the galvanometer to some large value, for example, 4 inches. Observe limitations on safe deflection of fluid-damped galvanometers during this procedure. Note the deflection of the first overshoot when the current is removed instantaneously. Adjust R_d until deflection is exactly 25 percent of the original deflection. This value will correspond to 40.2 percent damping-ratio. Taking the value of the coil resistance

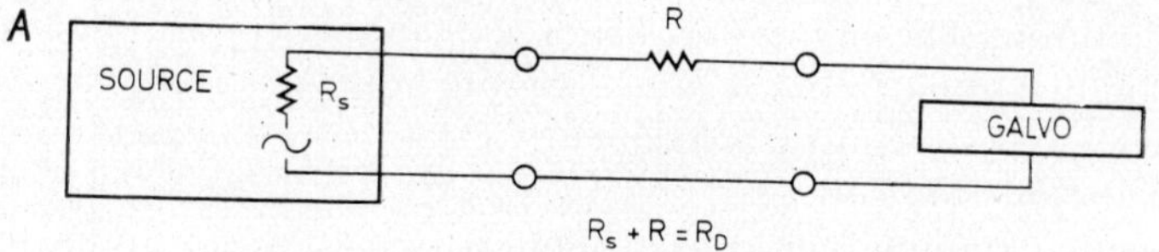

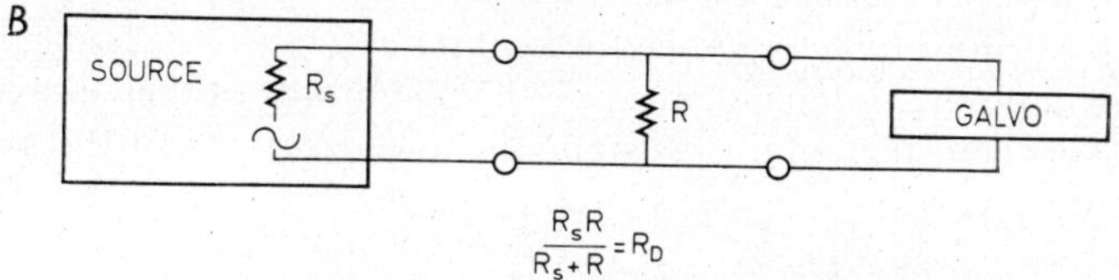

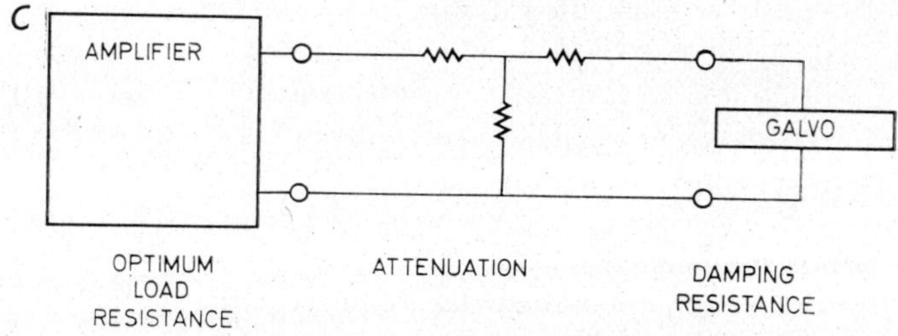

Fig. 3.6. Galvanometer damping

(R_g) from the specifications, compute the factor K/f_o. The damping-ratio and corresponding damping resistance may be determined mathematically by solving the equation

$$K/f_o = \eta(R_g + R_d)$$

in which K is a constant, f_o is the undamped natural frequency, and η is critical damping. The relationship is also shown Fig. 3.7.

FREQUENCY RESPONSE. Galvanometers are used primarily to record alternating current, and it is necessary that the spot deflection be proportional to the instantaneous current over the frequency range of interest. With galvanometers employing a damping value of 64 percent of critical, most galvanometers will have a frequency response which is flat to within ± 2 to ± 5 percent to 60 percent of the natural (undamped) frequency.

To select a galvanometer for a specific requirement, estimate the frequency response needed and multiply this by 1.6. The resulting figure will be the minimum natural frequency of the galvanometer that can be used and that will still provide

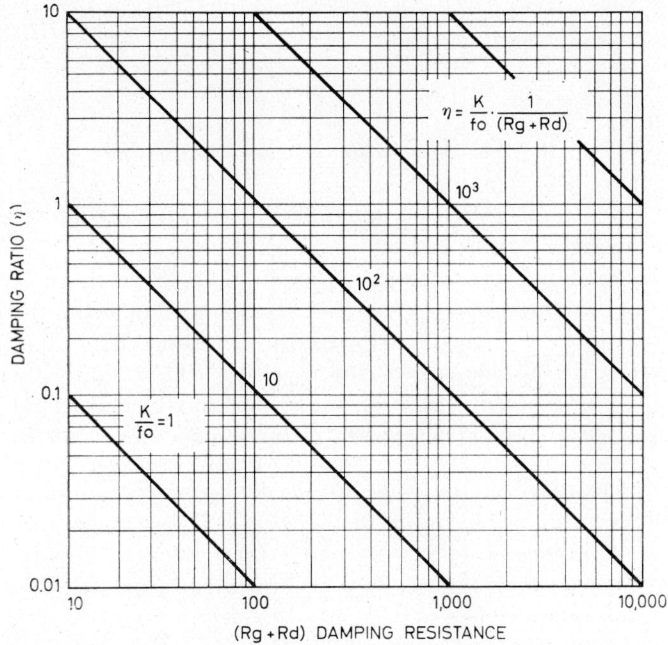

Fig. 3.7. Damping relationships

a frequency response flat to within ± 5 per cent. Therefore, if sensitivity is of prime importance, the undamped natural frequency of the galvanometer selected should be no greater than that calculated above. When a galvanometer is used that has a natural frequency higher than that determined by the preceeding rule, a lower sensitivity will be the result of the accompanying higher frequency. Response to frequencies beyond twice the natural frequency will be small and essentially filtered out. Hence, galvanometers with low natural frequencies make excellent low-pass filters to eliminate undesirable higher frequencies.

The curves in Fig. 3.8 represent the theoretical deflection of a galvanometer responding to a sinusoidal current, constant in magnitude but variable in frequency.

93

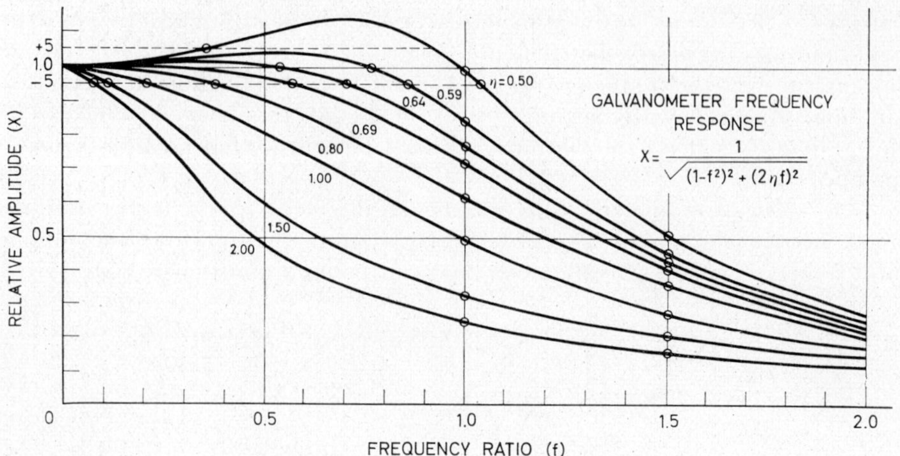

$$X = \frac{1}{\sqrt{(1-f^2)^2 + (2\eta f)^2}}$$

Fig. 3.8. Galvanometer frequency response

To make the curves applicable to a galvanometer of any undamped frequency and sensitivity, the frequency of the current is shown as the ratio of this frequency to the undamped natural frequency of the galvanometer. The deflection is also dependent upon the amount of damping in the system, electromagnetic or fluid, and the curves in Fig. 3.8 are plotted for several ratios of electromagnetic damping. For all practical purposes, these same curves can be used for fluid-damped galvanometers. The deflection itself is shown as a ratio and is the peak deflection at a given frequency ratio divided by the deflection at "zero-frequency" i.e., d-c.

SENSITIVITY. Sensitivity is the amount of deflection per unit current, and is expressed by the reciprocal in terms of μa/inch (or μa/cm or μv/cm in European equipment). Since the amount of deflection for a given current is directly proportional to the distance from the mirror to the screen (optical arm), this distance must be specified when quoting a sensitivity.

System voltage sensitivity is generally used to define the requirements of the galvanometers relative to the signal source. Ohm's law can be applied to calculate the amount of current which flows through the galvanometer with any given open-circuit source voltage. This is shown schematically in Fig. 3.9.

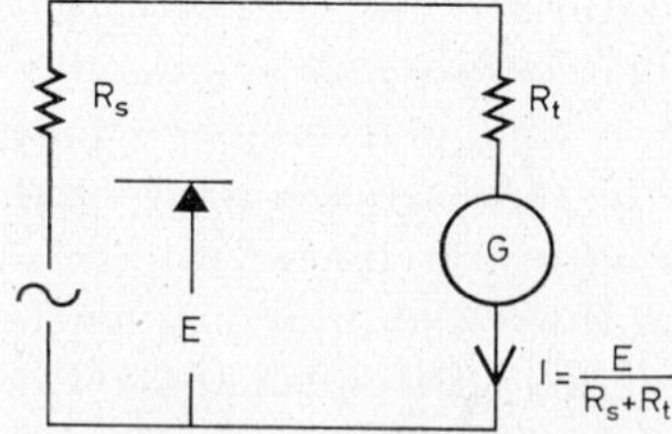

$$I = \frac{E}{R_s + R_t}$$

Fig. 3.9. Typical driving source/galvanometer circuit

94

LINEARITY. A galvanometer that has a deflection proportional to the current through the coil is said to be linear. However, coil design, field distribution, the optical system and other factors may affect linearity. The major percentage of non-linearity is the result of the spot deflection being recorded on a flat surface rather than on a curved surface, so that any point would be equidistant from the mirror. This non-linear effect is called "tangential error". An opposing influence, "cosine error", tends to correct tangential error, and occurs because the galvanometer suspension rotates in a parallel magnetic field. Fig. 3.10 illustrates linear and non-linear effects.

Another non-linear effect occurs when deflection is different on either side of the straight-ahead position of the galvanometer. This condition is described as "non-symmetry".

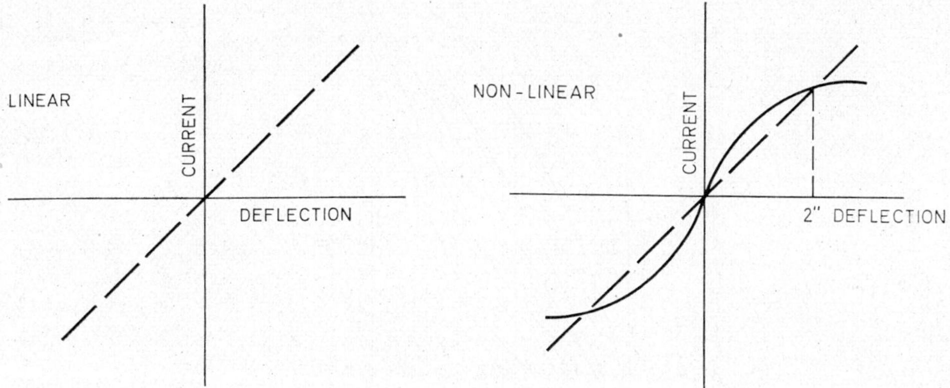

Fig. 3.10. Linear and non-linear effects

PHASE RELATIONSHIP. Phase shift will occur in a moving coil galvanometer between the applied frequency of the input current and the frequency of the coil deflection and appears as a time shift. Since phase shift is inherent in the galvanometer it is essential that it should be kept linear for easy analysis.

In a complex wave many harmonic frequencies exist and since each component frequency will be subject to a different time delay, subsequent analysis can be made easier by choosing an optimum damping ratio of 0.64. A nearly linear frequency phase relationship will then be obtained as shown in Fig. 3.11.

The preceding galvanometer description summarizes technical data and includes excerpts from information published by Consolidated Electrodynamics Corporation, Hathaway Instruments, Honeywell, and SE Laboratories. Further data may be obtained from references[3.3] and [3.4].

TYPES OF GALVANOMETER. Each instrument manufacturer has a wide variety of galvanometers from which the user may make a selection. After evaluating the requirements as outlined in the previous sections, the user can select the galvanometers that will produce optimum results for his particular set of conditions. Standard galvanometers are available in high and low frequencies, and special purpose galvanometers can be obtained to perform a specific function. For example, integrating, computing and event-marking galvanometers are available. The integrating device records the time integral of the signal voltage, and is used primarily to integrate velocity-sensitive transducer outputs to provide trace displacements proportional to vibrating amplitude. The computing galvanometer performs addition,

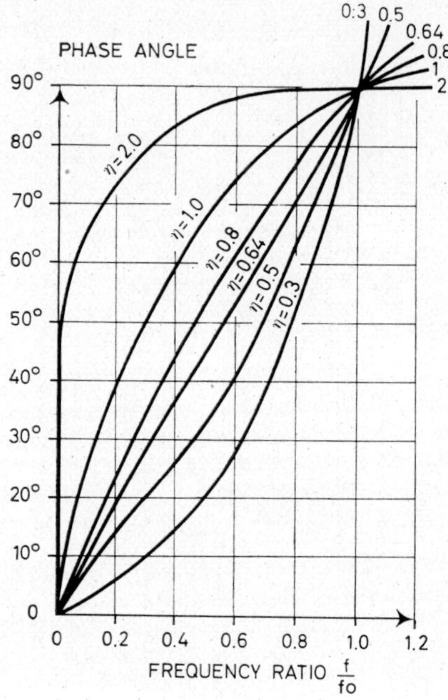

Fig. 3.11. Phase angle

subtraction, or multiplication of two input signals, recording a single trace corresponding to the mathematical result. Event-marking galvanometers allow on-off information to be recorded as a discontinuous trace.

Other types of galvanometers may be classified as follows:

1) Dynamic reference—to record the precise time of occurrence of significant events during the recording.

2) Static reference—used to record a measurement reference for an active trace.

3) Dummy galvanometers—may be inserted in vacant slots when less than a full complement of active galvanometers is used in a magnet block.

In addition, trace-blanking kits are available to permit a galvanometer to be retained in the magnet block during recording without making a trace.

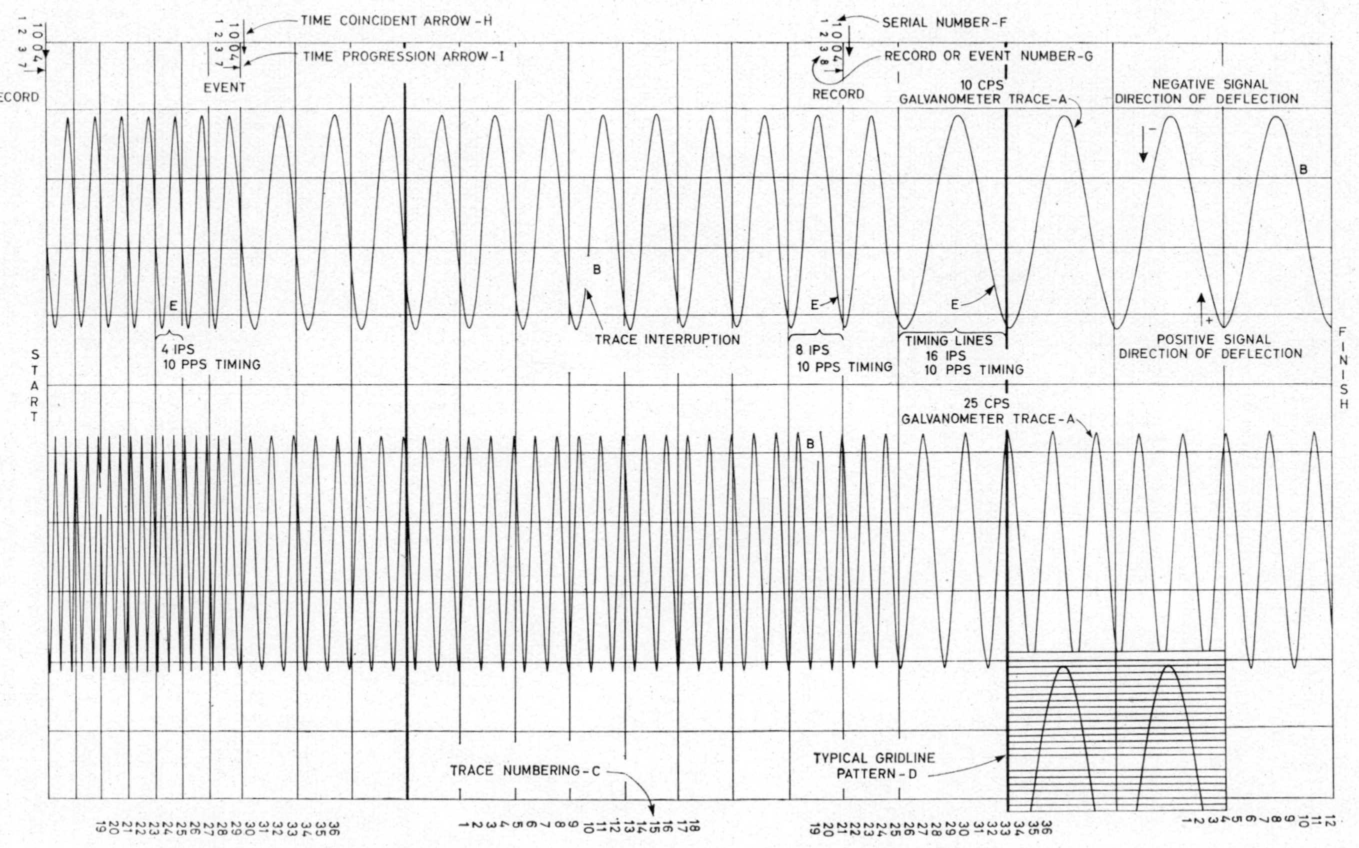

Fig. 3.12. Sample record produced on a CEC 5–133 Datagraph Oscillograph

Other Optical Systems. A study of the sample oscillogram illustrated in Fig. 3.12 will reveal that there are optical devices in the modern oscillograph recorder other than galvanometers. In addition to the recorded galvanometer traces, the serial number, record or event number, trace numbering, grid lines, and timing lines have all been recorded. Other features include trace interruption and time-coincidence and time-progression arrows. Each recorder manufacturer has developed

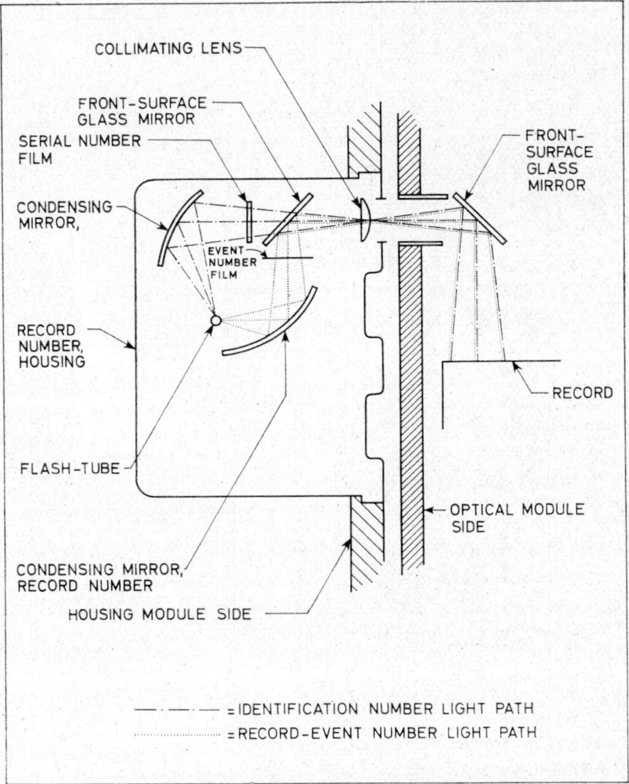

Fig. 3.13. Schematic of the record or event-numbering optical system used in the CEC 5-133 Datagraph Oscillograph

individual designs for each optical system.[3.5, 3.6] A review of the equipment descriptions in this chapter will acquaint the reader with the many features offered. The following discussion will serve as an introduction to the general types of systems in common use.

Record or Event Numbering System. A record or event numbering system (as used in the CEC 5-133 Datagraph Oscillograph) contains a lamp, numbering films, mirrors and a lens, and is shown in the self-explanatory optical schematic, Fig. 3.13. The system provides the following data at the beginning of each record and automatically thereafter at intervals of 40 inches of paper travel, regardless of the paper speed. The system may also be operated manually by closing the PRINT switch.

1) The record number identifies the record for future reference. When the knob is set at RECORD the number advances one digit and prints each time recording is started. When the knob is momentarily rotated to PRINT the number prints without advancing. The number is referred to as an Event Number when used to mark a special event, in which case the same number repeats at each closure of a remote switch. (See "G" Fig. 3.12.)

2) The serial number of the oscillograph identifies the record for future reference after it has been removed from the instrument. (See "F" Fig. 3.12)

3) A time progression arrow provides immediate orientation as to the direction of time progression when the record is used for study after it has been removed from the instrument. (See "I", Fig. 3.12.)

4) A time coincident arrow identifies the exact occurrence of an event so that calculation of the number of seconds the instrument was in operation, and the approximate time a special event occurred will allow the operator to find quickly the desired special event. (See "H" Fig. 3.12.)

Trace-Identification System. A combination of trace interruptions and trace numbers on the margin of the record corresponding to the channel numbers on the galvanometer block, provides rapid identification of any trace even though it has crossed or intermixed with other traces.

1) Trace interruption is caused when the trace identifier momentarily blocks out the recording light beam of each channel in turn, creating a sharp void in the printed trace. A void for each trace appears at certain intervals of paper travel regardless of paper speed, and appears on consecutive traces at a slope of approximately 45 degree. (See "B", Fig. 3.12.)

2) Trace numbers are printed along the bottom margin of the record (or at the operator's right hand side) in sequences, so that a line drawn perpendicular to a trace interruption will intersect a number which corresponds to the position of that particular galvanometer in the galvanometer block. (See "C", Fig. 3.12.) Light from the right side of the recording lamp is projected through a rotating film strip containing the appropriate numbers of the galvanometer positions.

Traces can also be identified by color oscillographic methods which use photographic color materials and colored filters. The technique and the materials are discussed in Chapter Five.

Timing-Line Generator. Timing lines are usually produced across the full width of the record. The light may originate from a continuous-source incandescent lamp or from a flash lamp, depending on the timing system. With incandescent systems, the recording material is exposed to a periodically-interrupted light. Light interruption is achieved by a rotating shutter (such as a multiple-sided mirror) driven by an accurate synchronous motor. Light reflected from one of the mirror surfaces passes through a collimating lens onto the sensitized material, exposing a line across its width.

Flash-tube timing systems use intermittent high-intensity flashes of light to produce timing lines. Flashes from the lamp are directed by a mirror and a collimating lens to the recording material. A flash-tube timing system used by one manufacturer is illustrated in Fig. 3.14.

Some equipment is capable of external flash synchronization or slaved operation. An oscillator or external time-base generator provides the timed triggering pulse to flash the tube. This is an advantage in multiple-recorder installations because the common signal will provide time coincidence on all recordings.

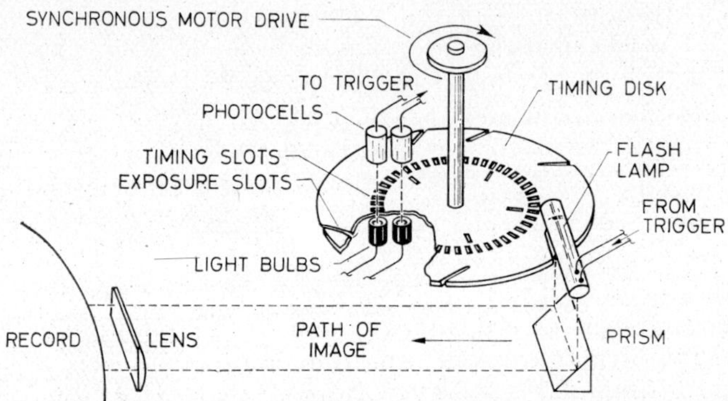

Fig. 3.14. Operation of the Model 11–2104–00 Time Line Generator used in the Brush Model 2300 Oscillograph

Grid-Line System. The grid-line system photographically prints a predetermined pattern of grid lines which subdivide the amplitude of the traces. The lines serve as points of reference for accurate measurement of trace amplitude without the use of a scale or reading glass.

Recording Equipment

OSCILLOGRAPH RECORDERS. Although the mirror-galvanometer is perhaps the most critical component of the lightbeam oscillograph, recording cannot occur unless other components have been provided. Besides the light-sensitive material and a transport to drive it through the recorder, the light source, optical systems and controls must be installed in a light-tight enclosure. The resulting unique "camera" is the photographic oscillograph.

Photographic oscillographs may be classified in two categories: develop-out and print-out recorders. In addition to the general-purpose instruments available in these categories, there are special-purpose recorders for seismology, medicine, and electric-utility functions. Both general-and special-purpose equipments are described in the following sections.

Develop-out recorders normally use an incandescent lamp as a light source. After recording, the exposed sensitized material is processed under safelight conditions in a photographic darkroom. Alternately, the exposed material may be transferred in a darkroom to a processor magazine. The magazine is then attached to a continuous processing machine capable of white-light operation, and the photographic material is chemically processed and dried.

Print-out recorders eliminate the chemical processing by using an intense light source, and paper that does not require chemicals to produce a visible image.

NEED FOR QUICK-ACCESS RECORDS. In the course of research projects, it is desirable to monitor, evaluate, and determine the next step of an experiment from data obtained in the previous test. When photo-oscillography is used for data-instrumentation purposes, the need for a quick look at its output is self-evident. Time and money can be saved, and schedules can be improved by a quick-access system—providing it does not reduce the performance characteristics of the oscillograph or sacrifice record quality. Development of oscillographs was directed toward a system that would fulfill these requirements. Many attempts were made to shorten the time lag between the recording and the visual-presentation stages. These studies included stabilization and high-temperature processing, mono-bath processing, the use of xerography, and the use of print-out materials. Of the various systems studied, just three have resulted in widespread commercial distribution of equipment and products. They are integral exposure-chemical development, print-out, and a specialized print-out method[3.7] using a heated platen in conjunction with latensification lamps.

INTEGRAL EXPOSURE — CHEMICAL DEVELOPMENT. In this system, the development process is conducted in an accessory magazine-processor unit, integral with the recorder. Because the system uses highly sensitive silver-halide emulsions amplified by chemical reduction, the inherent high performance of develop-out photo-oscillography has been retained. Thus, a high writing speed can still be attained, and the undegraded performance of the recorder is available with instantaneous readout.

An integral exposure-chemical development accessory is exemplified by the Consolidated Electrodynamics Datarite magazine processor. The CEC equipment and process are described in a later section of this chapter. Also see references[3.8, 3.9].

PHOTOGRAPHIC PRINT-OUT RECORDERS. The second type of quick-access oscillograph is the print-out recorder. This instrument has been available since the mid-1950's. In the United States, print-out recorders are produced by several firms, among whom are Honeywell Denver Division, Century Electronics and Instruments, Consolidated Electrodynamics (CEC), Sanborn, Midwestern and Brush. European manufacturers include SE Laboratories, and Savage and Parsons in England; Siemens & Halske, and Hartmann & Braun in Germany.

Specialized print-out recorders are also built; an X-Y recorder is manufactured by Sanborn, and seismograph equipment is manufactured by Geotechnical Corporation. Both types of instruments use print-out recording paper as a chart. All of these instruments will be described following a discussion of the basic print-out technique.

Principles of Print-out Recorders. There are three characteristics of print-out oscillographs that distinguish them from develop-out instruments: 1) the use of a high pressure mercury vapor or xenon recording lamp; 2) the use of print-out photographic recording paper; and 3) the need for a second low-intensity exposure, usually supplied by a fluorescent-type exposure lamp, mounted over the record exit section of the oscillograph. On some instruments there is a fourth distinguishing characteristic, a heated platen.

MERCURY AND XENON LAMPS. Because print-out paper is a relatively slow material, it cannot receive sufficient exposure with the ordinary tungsten lamp used in the conventional develop-out type oscillograph. This is due to the comparatively low intensity of the filament lamp, and to its spectral output which does not match the spectral sensitivity of print-out paper. (NOTE: Papers with extended spectral sensitivity have enabled print-out recording to be done with high-intensity tungsten

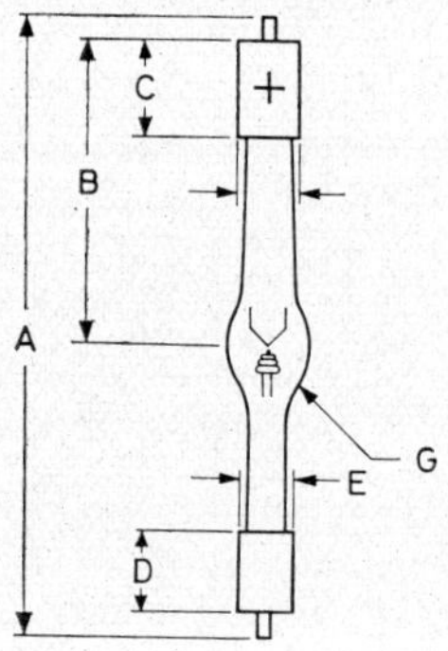

TYPE	PEK X-75
A	3.400
B	1.700
C	.500
D	.450
E	.300
F	.350
G	.500

LAMP VOLTAGE	14±2 VDC
LAMP CURRENT	4.6 – 6.2 ADC
LAMP POWER	75 WATTS
ARC SIZE	.015 x .015 INCH
LUMINOUS FLUX	1400 LUMEN
AVERAGE BRIGHTNESS	100,000 $^{Cd}/_{CM^2}$
AVERAGE LIFE	300 HRS
OPERATING POSITION	ANODE HORIZONTAL OR UP VERTICAL
BASE	SLEEVE

Fig. 3.15. PEK X-75 Xenon-arc lamp operating characteristics

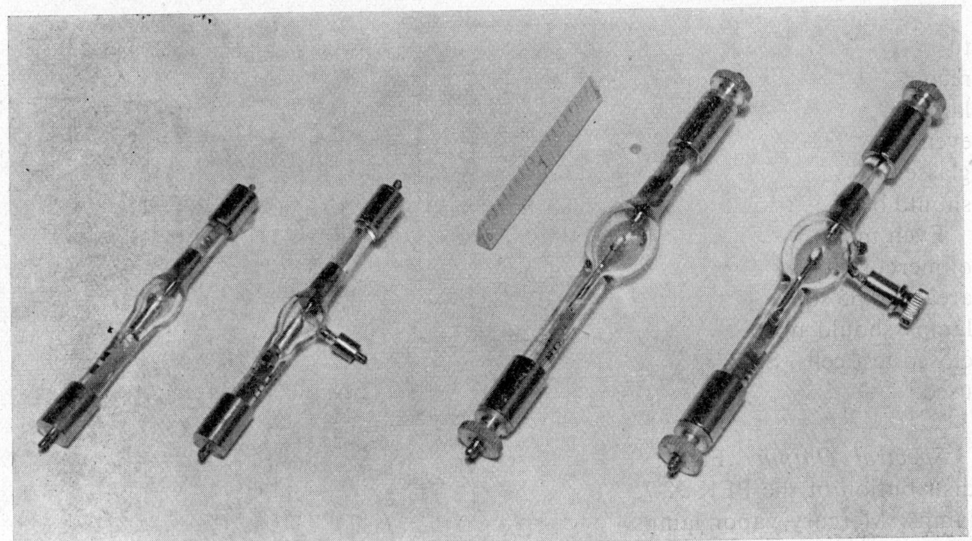

Fig. 3.16. PEK High-pressure vapor arc lamps. *Left to right*: lamp numbers 107, 109, 202 and 203

sources.) The paper is sensitive only to the short-wavelength end of the spectrum, blue and ultraviolet and a light source having the greater percentage of its output in this region is required. The low sensitivity of the paper makes a very intense light source necessary.

The two specifications of high intensity and correct spectral output are met by the miniature, high-pressure mercury vapor arc or xenon arc lamp.[3.10, 3.11, 3.12] A typical xenon arc lamp, together with its principal operating characteristics, is shown in Figure 3–15. Because these two types of lamp operate under pressure as high as 80 atmospheres, and because their operating temperatures are very high, they must be completely enclosed in the recorder, and must be fan-cooled for optimum performance and long life. Fig. 3.16 shows several lamp of this type manufactured by PEK Labs, Inc. Sunnyvale, California.

Lamp Envelope and Optics. Quartz, rather than glass, is used in the manufacture of the high-intensity mercury and xenon lamps. The clear, fused quartz transmits the maximum amount of ultraviolet radiation, and does not discolor with age from solarization; therefore, effectiveness is not diminished over a period of time. In addition, the physical characteristics of quartz—its strength and low coefficient of expansion—provide a margin of safety for the high pressures and extreme temperatures encountered in this application.

Collector lenses and other optical components of the print-out recorders are sometimes made of quartz for the above reasons.

The PEK company recommends that because of the ultraviolet radiation and high operating pressure of the X-75 lamp, it must be mounted in a housing completely enclosing the lamp. The housing should vented to insure the anode cap temperature does not exceed 200 °C. If a small housing is required, forced-air cooling is recommended. The manufacturer issues this word of caution for handling this lamp: "The X-75 lamp contains xenon gas under several atmospheres pressure even when not in operation. The lamp should be handled only while wearing safety glasses. This lamp should never be operated except in a safe enclosure. Spent lamps should be disposed of by wrapping in heavy canvas and smashed."

Even more precautions are indicated for mercury lamps because of the toxicitiy of mercury if the lamp should explode. A lamp enclosure having a minimum surface area of 0.65 square inches per watt is recommended. Recorders having mercury-arc lamps should not be operated in confined areas such as submarines, mines, and closed test cells. Safety venting should be installed where mercury vapor lamps are used.

Spectral Output. Fig. 3.17 and 3.18 show respectively the spectral energy distribution of the PEK X-75 xenon arc and the Osram HBO 200 W mercury arc lamps. Mercury vapor lamps have a higher actinic light output, but newer print-out papers having orthochromatic sensitivity (such as Du Pont Linowrit 7), utilize more of the xenon spectrum than papers sensitive to ultraviolet only.

Similar lamps are manufactured by Engelhard Hanovia, Inc., in the United States, by Osram in Germany, and by other firms. An Osram lamp is shown in Fig. 3.19.

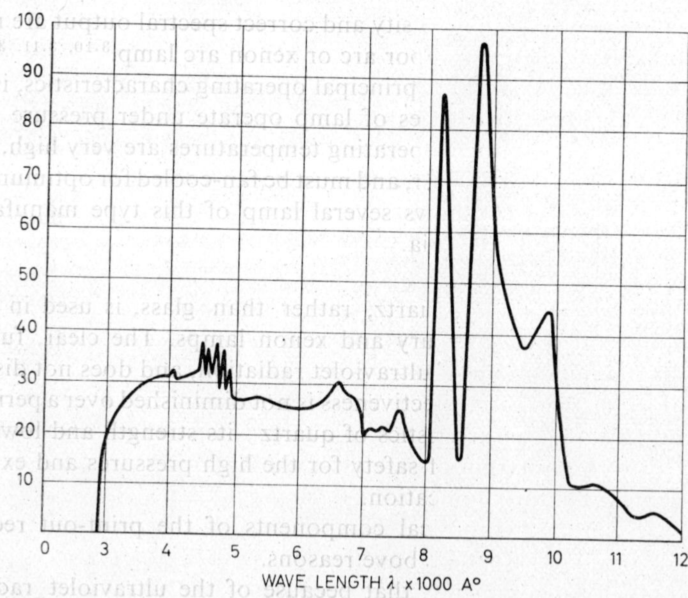

Fig. 3.17. Relative spectral energy distribution of the PEK X-75 xenon-arc lamp

104

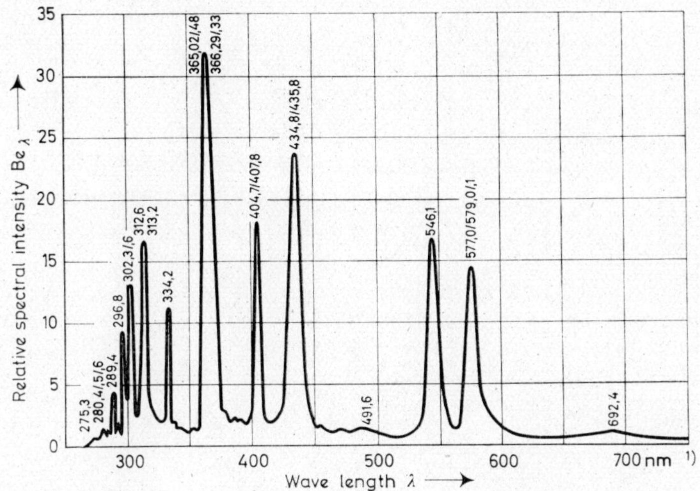

Fig. 3.18. Relative spectral energy distribution of the Osram HBO 200 W mercury vapor arc lamp

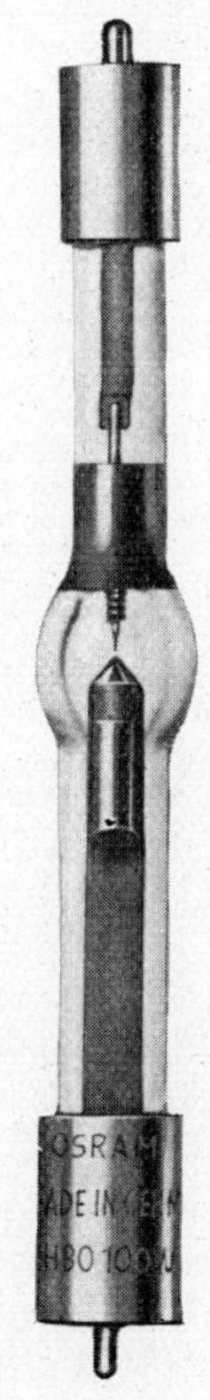

Fig. 3.19. Osram HBO 100 W mercury arc lamp

Power Supply. A typical power supply for a PEK X-75 xenon arc lamp is shown in Fig. 3.20. The power supply for the X-75 should provide 30vdc minimum open circuit voltage, 9 amp maximum starting current and a high voltage starting pulse.

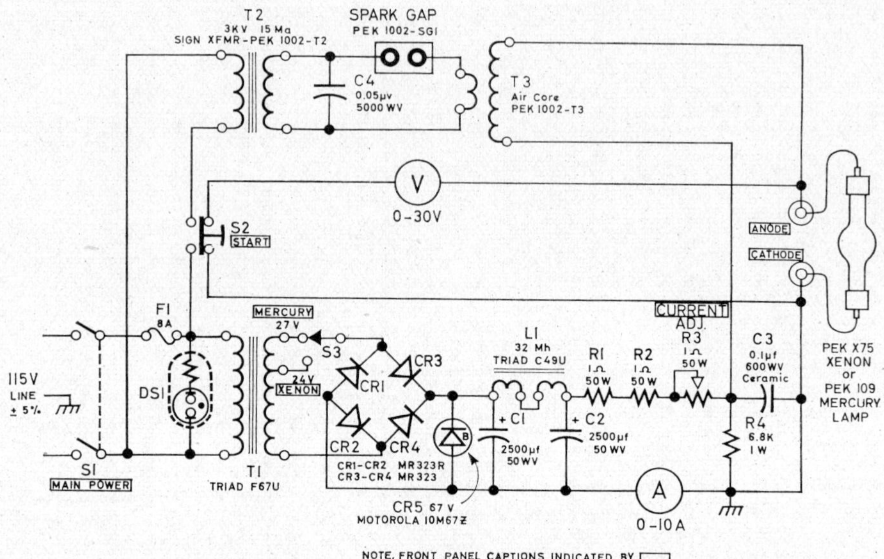

Fig. 3.20. Typical power supply for a PEK mercury or xenon arc lamp

HEATED-PLATEN LATENSIFICATION. The application of heat to print-out paper while it is being latensified is capable of allowing faster access by a factor of up to 60 times. This principle, explained in Chapter 5, was initially developed and marketed as the CEC Dataflash system. Dataflash units consist of a heated platen and latensification lamps and are available on several CEC recorder models, described in later sections of this chapter. (Equipment that is similar is now available from competitive manufacturers.)

Visible traces are provided within one second after exposure, and complete latensification can be obtained at speeds up to 16 inches per second. The process eliminates ambient light as a factor in data access time and retains the advantages of the print-out process, such as high writing speed and no chemical processing.

To illustrate the advantage of using Dataflash in a typical test, assume that data is being recorded in the conventional print-out manner, that is, without Dataflash, at a recording speed of 16 inches per second. With good fluorescent room lighting, latensification of the data traces will take place in approximately 1 minute. At the recording speed indicated, 80 feet of record will pass before the first data is fully visible. Conducting the same test with Dataflash, only 1 second will elapse before full latensification and only 16 *inches* of record will be exposed.

106

THE BRUSH SERIES 2300 OSCILLOGRAPH. The Brush Series 2300 eight or sixteen channel oscillograph, manufactured by Brush Instruments, 37th & Perkins, Cleveland, Ohio, and shown in Fig. 3.21, produces print-out recordings by an incandescent light source. Writing speed is said to be greater than 30,000 inches per second. Records are visible directly on exposure to ambient room light. They appear in fine traces at all writing speeds, uncluttered by UV "jitter" or RF inteiference. Eight record speeds, controlled by pushbuttons, may be changed during operation

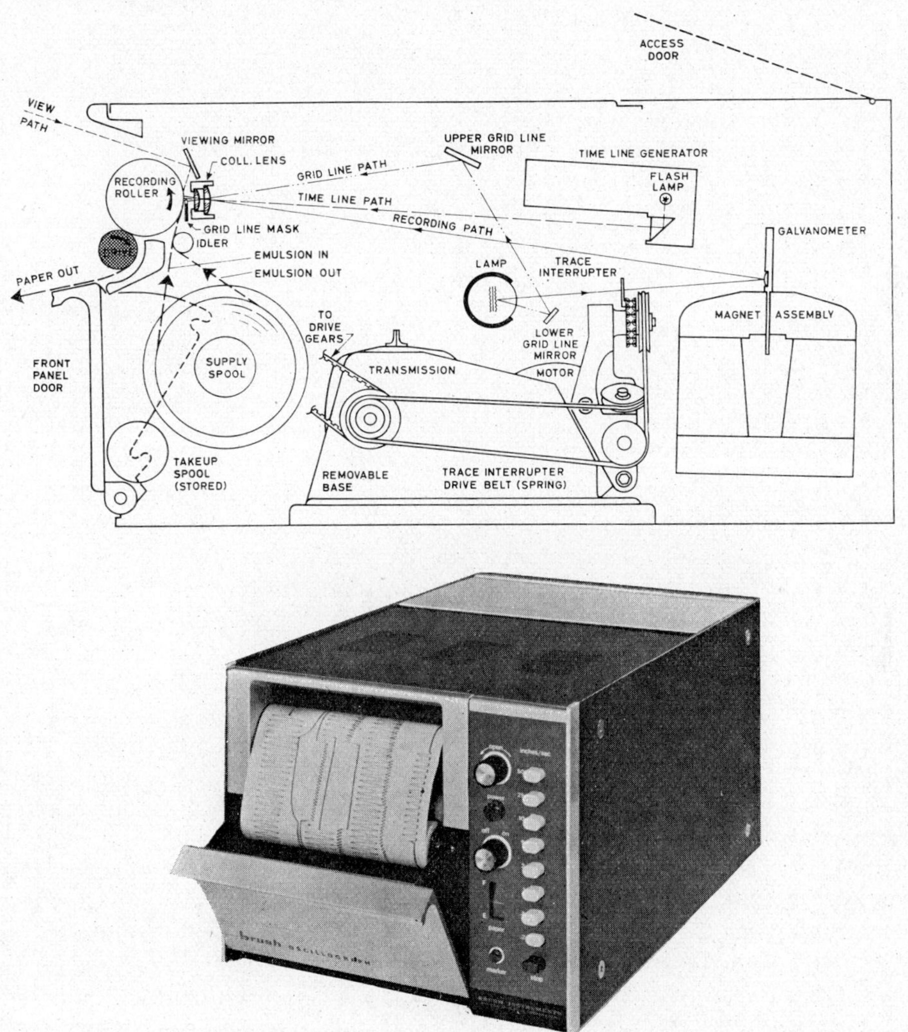

Fig. 3.21. The Brush Instruments Series 2300 oscillograph

with no loss of data. An internal takeup accepts a full roll of paper and the record may be fed outside for on-the-spot inspection. Paper loads from the front, it is simply dropped in and threaded.

Galvanometers are simple plug-in units with no setscrews, requiring no adjustment other than spot-zeroing. The basic galvanometer model 11–2111–10, performs the great majority of recording tasks. Matched solid state galvanometer amplifiers, optional, further augment application range and provide simplified signal control. Galvanometers also are available for a number of specialized applications.

Accessories include flash timers, amplitude grids, a trace-identification interrupter, a 10-channel event recorder kit using no galvanometer positions, provision for remote control and for rack mounting. An exposure hood permits storage of paper or films for wet processing when desired.

The portable unit weighs 45 pounds. Provisions are made for rack mounting. The unit, complete with six amplifier channels occupies only 10–1/2 inches of panel space. The Series 2300 recorder operates on 140 watts maximum, plus 7 watts (max.) for each amplifier. A six channel system draws approximately 180 watts maximum power. Special low speed units drawing less than 60 watts at 12 and 24 vdc are available.

CONSOLIDATED ELECTRODYNAMICS EQUIPMENT. The following sections describe equipment manufactured by Consolidated Electrodynamics Corporation (CEC), 360 Sierra Madre Villa, Pasadena, California.

THE CEC TYPE 5–114 OSCILLOGRAPH. Since its introduction in 1947 to industrial and scientific fields, the 5–114 Recording Oscillograph, shown in Fig. 3.22, has had numerous design improvements.

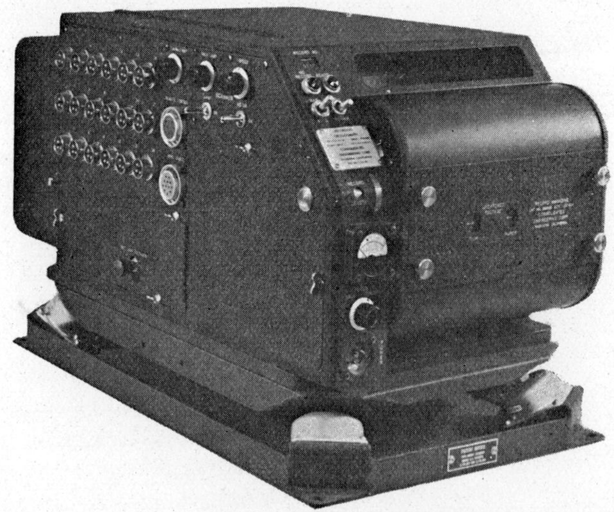

Fig. 3.22. CEC Type 5–114 oscillograph

Fig. 3.23. The CEC Type 5–119 oscillograph

This oscillograph uses an incandescent light source and is available in either 18- or 26-trace models. Such features as remote control, choice of ac or dc models, multiplex operation, viewing screen with scanning system, automatic record-length control, automatic record numbering, and a variety of magazine types combine to make the 5–114 a versatile instrument. Timing lines as well as a five-digit number are photographed on the record automatically. The 5–114 uses 7-inch-wide develop-out paper or film, and is available with magazine capacity of 225 feet. Record speeds from 1/2 inch to 115 inches per second are obtained by simple gear changes. The 5–114 oscillograph will accept a Datarite magazine which is available as an accessory.

The CEC Type 5–119 Oscillograph. The Type 5–119 Oscillograph, illustrated in Fig. 3.23, is available in 36- or 50-trace models. With ± 2 inch trace deflections, data value can be measured and identified directly from the record. The record transport employs a positive geardrive system with a range of recording speeds from 0.16 to 160 inches per second. A front panel jump-speed switch allows instantaneous speed changes in a 10 to 1 ratio to be made at any time during a test. The Type 5–119 Oscillograph is available in several configurations, and with many matched accessories and optional features.

The 5–119 oscillograph features warning and test circuits combined with reserve lamps to provide the utmost dependability in recording dynamic phenomena in the 0 to 5000 cycles per second frequency range. Records of 12 inches or narrower, by 400 feet in length, can be made at record speeds as low as 0.10 and as high as 160 inches per second on either film or paper. A jump-speed switch allows an instant 10-time increase in recording speed while data are being recorded. Timing lines are photographed on the record at 0.1 or 0.01 second intervals. Static reference traces, record-length pre-selection, trace identification, visual monitoring system, sequential event numbering, and shock mounting are standard equipment. All controls and indicators are located on a single front panel. Remote-control unit and vertical rackmounting are also available. Type 5–119 Oscillographs are available for operation from 115-volt, 60 cycles per second, single-phase; 200-volt, 400 cycles per second, 3-phase; and 28 volt power sources.

The Type 5–119 Oscillograph was designed to accept three types of record magazines. Through the use of these magazines, various models of the instrument can utilize any of the photographic recording media known: conventional papers, films, and print-out papers.

The 5–006 Standard Record Magazine produces oscillograms in the conventional manner; records are chemically processed after the record run. Two other modes of operation offer rapid-access records; the special model and the accessories required are discussed further on in this chapter.

With the standard Type 5–119 Oscillograph, 36-trace instrument, four static and one dynamic reference traces are provided. Two static reference traces are available on the 50-trace type. Each trace of the 5–119 record is positively identified for easy data reduction by a 1/32-inch gap which appears on the trace at 12-inch intervals. Coincident with each interruption, the trace number is printed on the edge of the record.

There are two separate recording-galvanometer banks, and each individual magnet block has its own galvanometer light source. This permits separation of high writing speed and low writing speed traces. Lamps can be individually adjusted for brillance—to yield uniform trace width and density on all channels. Magnet blocks are automatically maintained at 100 °F by a thermostat.

If a galvanometer lamp fails during a recording run, a reserve lamp is automatically switched on, insuring uninterrupted data recording. To assure uniformly good trace contrast and density, regardless of the speed selected, a corresponding change in galvanometer lamp brilliance is automatically made when the 10:1 jump-speed switch is operated. Speed range of the Type 5–119P3 model oscillograph is 0.1 to 100 inches per second.

Consolidated's Type 5–119 Recording Oscillograph uses the Slotted-Disc Timing System. The recording medium is exposed to a periodically interrupted light from an incandescent lamp. Timing lines are photographed across the record at 0.01 or 0.1 second intervals, with every 10th line emphasized.

For applications where external synchronization or slaved operation of several oscillographs is required, CEC's Electronic Flash Timing System is available as an optional accessory. In this system, a strobe-flash tube provides the timing light source.

THE CEC TYPE 5–119P4 OSCILLOGRAPH. The Type 5–119P4–36V and –50V Recording Oscillographs employ an internal high-actinic light source—rather than the standard tungsten lamps used in the standard 5–119 model previously described.

The new light source and the Type 5–051 Slot-Exit Magazine make the 5–119P4 Type Oscillographs efficient, rapid access instruments. Records emerge from the magazine slot continuously, and latensification of the trace images takes place in ordinary room light. Oscillograms of high quality and trace contrast are produced on standard print-out papers without need for processing of any kind. Traces are clearly resolved at writing speeds in excess of 50,000 inches per second. Frequencies from d-c to 5000 cps are accurately reproduced, and 12-inch wide records up to 475 feet in length can be made continuously at speeds variable from 0.16 to 160 inches per second (0.1 to 100 inches per second are provided on special request). The "V" type oscillographs operate from 115 volt a-c, 60 cps power.

Following are standard features of the 5–119P4 models:

> 36- or 50-Trace Magnet Blocks
> Twin Galvanometer Banks
> Electronic Flash Timing
> Automatic Record-Length Control
> Visual Monitoring System
> Automatic Warning and Indicator Systems
> Vibration Isolator
> Provisions for Remote Control
> Automatic Record and Event Numbering

The 5–036 Datarite Magazine loaded with develop-out paper, can also be used with the "V" Models if light filters are first installed over the high-actinic light source to prevent fogging of the develop-out silver halide emulsions. Fig. 3.24

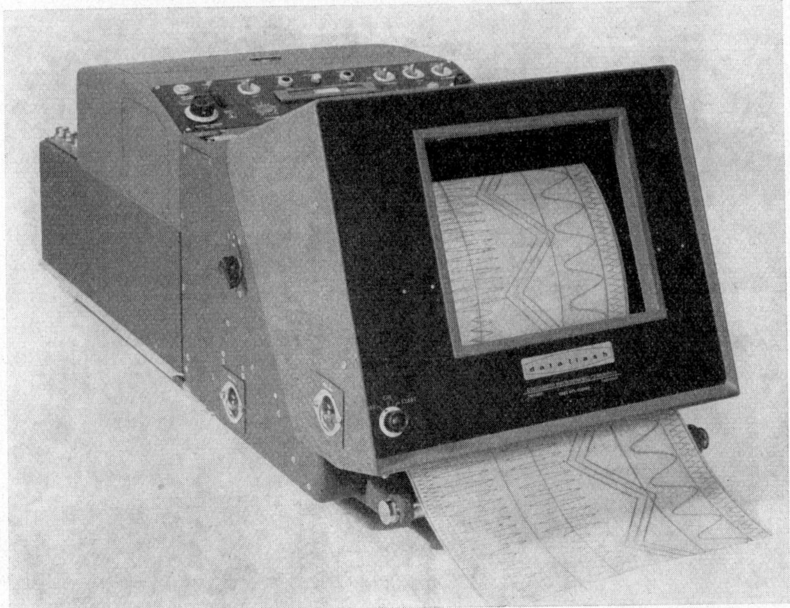

Fig. 3.24. The CEC Type 5–119P4 oscillograph with Datarite and Dataflash attachments

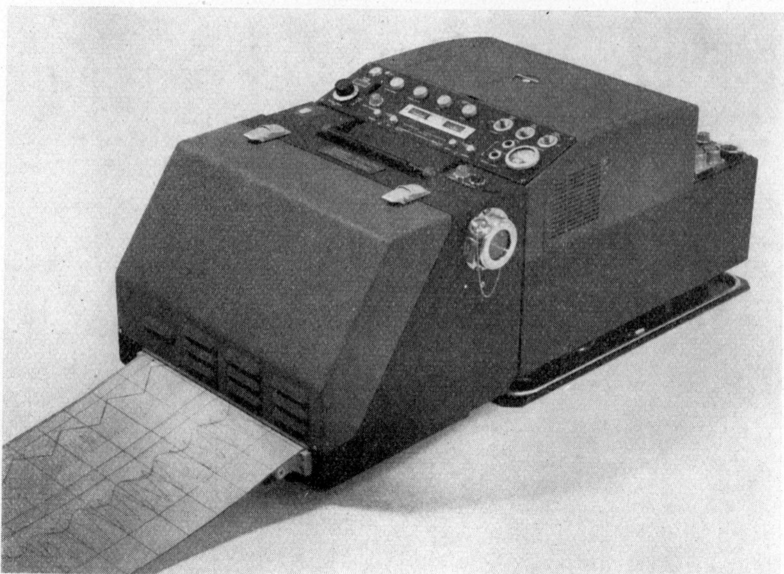

Fig. 3.25. Datarite magazine. Developer solution is added to the magazine through an easily accessible filler neck

shows the 5–119P4 oscillograph with Datarite Magazine and Dataflash Cover accessories.

The conventional 5–006 Record Magazine also can be used with the "V" models. This combination provides high writing speeds with the use of low-sensitivity emulsions.

THE CONSOLIDATED DATARITE SYSTEM. The integral recorder-magazine-processor developed by CEC and shown in Fig. 3.25, is called the Datarite system. It features the shortest continuous record access time of any known oscillograph process. Oscillograms are automatically developed and dried within the magazine as quickly as data are recorded. The completed, permanent record emerges from the Datarite almost simultaneously with the occurrence of events under study. Recording speeds are variable from 0.1 or 0.16 to 25 inches per second, depending on the speed range of the CEC Type 5–119 Oscillograph used. At the highest speed, the oscillogram emerges within 0.8 seconds after exposure. The Datarite holds 400 feet of commercially available, thin-base photographic papers.

The Datarite Magazine attaches directly to the CEC Type 5–119 Recording Oscillograph and is interchangeable with the standard 5–006 Magazine.

The record produced by Datarite combines many features sought by the instrumentation engineer. The small amount of developer required yields only one ounce of water vapor for every 40 feet of record dried and the processing method is not hazardous to operating personnel, because it uses no toxic chemicals, sprays, or vapors.

Datarite Operation. The Datarite Magazine performs the following operations, which occur almost simultaneously:

1) Recording paper is exposed to the input data signals.
2) The record is flash-developed.
3) Heat to dry the record is applied.

The flash processing technique used in Datarite can develop the record in less than 50 milliseconds as it passes through the magazine. Developer solution is applied to the record by a gravity feed system: a thin film of developer is applied to the emulsion side only of the record, leaving the paper base dry. By wetting the gelatin emulsion and not the paper base, the amount of moisture to be evaporated in the drying stage is minimized. Also minimized is distortion of the record, which would result from wetting and drying the paper base.

Fig. 3.26 shows the path the exposed paper follows in its travel through the magazine-processor. Paper from the supply roll 1, travels around the metering roller 2, and is exposed to the galvanometers beams through the collimating lens 3. (Actually, the lens is a part of the oscillograph). The exposed paper passes a guide roller and goes through the developer applicator 4. The applicator slit spreads a layer of developer a few tenths of a mil thick on the emulsion. The paper then passes over a curved platen 5, maintained at a constant temperature by thermostatically controlled electric heaters. The hot platen, at a temperature somewhat above the boiling point, completely dries the record simultaneously as it accelerates developer action. The dry record emerges through the exit slot, 6. The heating platen is shown in Fig. 3.27. The total paper path from the exposure point to the exit slot is approximately 20 inches.

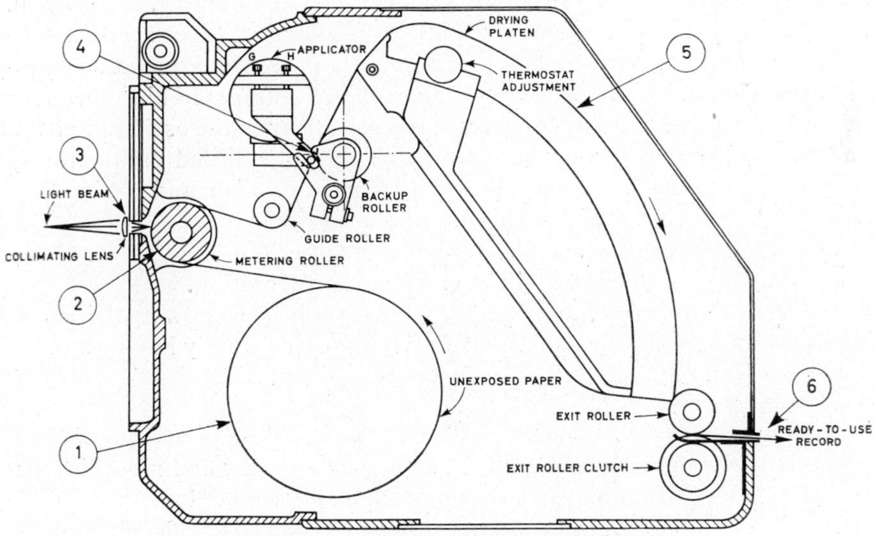

Fig. 3.26. Datarite magazine. Paper-threading schematic

Fig. 3.27. Datarite with cover removed showing heating platen over which the record is drawn

USE OF THE DATARITE AS A RECORD PROCESSOR. Oscillograms, recorded in the conventional CEC 5–006, 5–006A, or 5–032 Magazines at paper speeds up to 160 inches per second, may then be developed in a Datarite Magazine at speeds up to 25 inches per second. This technique combines the advantages of high speed recording and rapid access developing to increase the versatility of the oscillograph system.

To adapt the Datarite Magazine for processing pre-recorded oscillograms, it is necessary to mask off the Datarite Magazine record number and collimating lens aperture slots to prevent double exposure of the record.

Developing speeds are the same as for regular Datarite oscillograms. Generally, the slower developing speeds will yield greater trace definition.

When used as a processor, the Datarite Magazine may be powered either by the oscillograph or by a specially constructed drive-motor assembly having a minimum of 1/5 horsepower.

DATARITE CHEMISTRY. Chemicals used in Datarite's flash processing were developed especially for the magazine by CEC. These chemicals, and recording papers, are available from Consolidated through its many field offices throughout the country. The capacity of Datarite's developer supply tank is sufficient to process a 400 foot record.

114

Basically, the developer formula uses Phenidone as a developing agent instead of metol and/or hydroquinone. The Phenidone was found to be capable of producing records free of the fog encountered with other developing agents when used at high temperatures.

The accelerator may be one of several alkalies, such as tetra-alkyl ammonium hydroxide, but must decompose from heat. Therefore, by the time development is complete, the activity or pH is sufficiently low to produce a record having good contrast and stability.

Another manufacturer, Du Pont, makes two Lino-Writ Rapid-Access Photo-recording Developers for processing in magazine processors. The solutions, supplied in 8 ounce plastic containers, are known as White Cap Developer and Blue Cap Developer. The White Cap formula produces maximum speed and trace density, while the Blue Cap developer was formulated to yield records of high image stability under high relative humidity.

STABILIZING DATARITE RECORDS. The Datarite processed chart is reasonably stable as it leaves the magazine-processor. The stability increases as the residual developer is de-activated by exposure to air and light, and a record which has been exposed to ordinary laboratory conditions for a day or two becomes relatively insensitive to moisture. Normal precautions should be taken against re-wetting the record immediately after it leaves the Datarite. If wetting occurs, it may reactivate the dry residual developer and cause blemishes.

The record may be made permanent by an additional stabilization treatment. This can be done in a CEC 23–109 Processor using a Datarite stabilization chemical kit. The kit, CEC No. 217123, consists of a sodium thiosulphate bath for all four tanks. No water rinse is used.

ACCESSORIES. Fig. 3.28 shows a CEC 5–119 Oscillograph equipped with the Datarite and several accessories. Faster visual record access time at paper speeds below 4 inches per second can be obtained by using a special accessory magazine cover for the Datarite. The cover incorporates a ruby colored window to permit earlier viewing of the recorded traces as the oscillogram passes over the drying platen. Another accessory, Takeup Reel, Catalog No. 5–046, attaches to the front of Datarite and is used to hold the processed record as it leaves the magazine. The reel holds a full 400 foot roll of thin-base paper and provides a firm, smooth surface beneath the oscillogram for making notations on the record. The reel is self powered and operated from 115 volt, 60 cycles per second power.

THE CEC TYPE 5–123 OSCILLOGRAPH. Using 12-inch-wide print-out recording papers, the 5–123 Oscillograph, shown in Fig. 3.29, provides up to 50 individual channels of data on visible records without chemical processing of any kind. For ultra-fast access to data traces, the optional Dataflash module provides up to 60 times faster record-access time than standard print-out processes. The 5–123 uses standard CEC Type 7–300 Galvanometers and provides a range of dynamic measurement from d-c to 5,000 cycles per second.

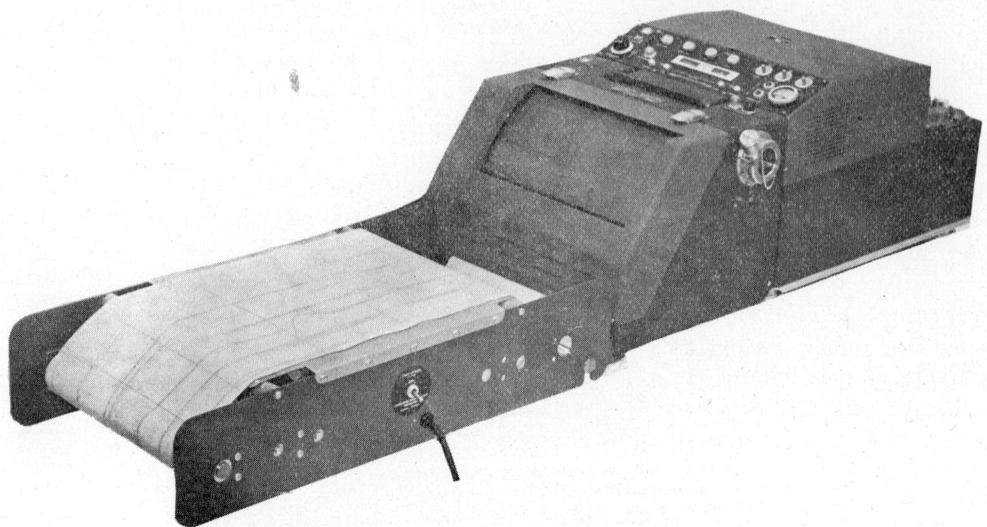

Fig. 3.28. The 5-119 oscillograph with Datarite magazine and takeup reel attached. Special viewing window in Datarite magazine cover permits earlier examination of signal traces

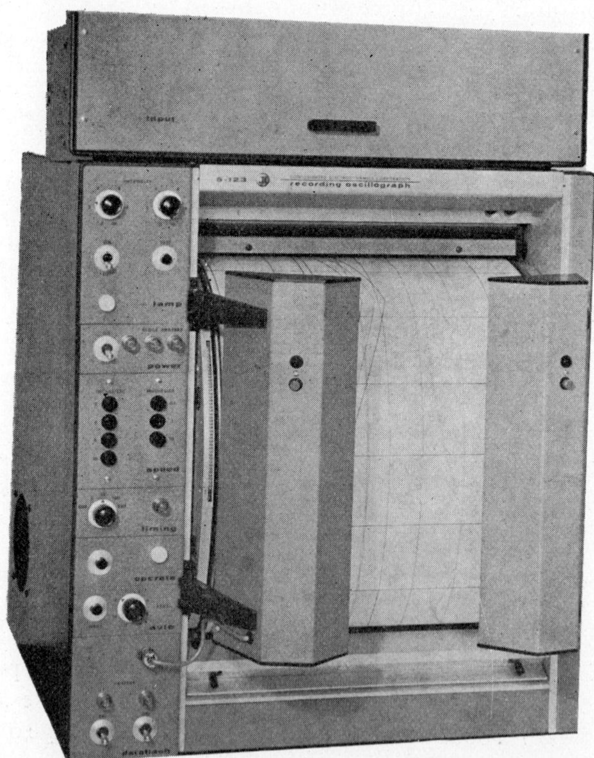

Fig. 3.29. The CEC Type 5-123 oscillograph with Dataflash

The 5–123 can be used in any vertical ground-environment installation. It is designed, however, to mount directly into a standard RETMA, 19-inch relay rack or cabinet without accessories or modifications of any kind. Operation, adjustment, and routine maintenance can be performed from the front of the oscillograph. After initial installation, rear access is never required.

The transmission system consists of gears, a self-energizing spring clutch, and actuating solenoids. Pushbuttons change the recording speeds, even while a test is in progress.

The need for conventional external record magazines has been eliminated from the design of the 5–123. The oscillograph's transport module supplies, transports, and exposes the record and provides internal takeup. In addition, it forms a large viewing surface for examining records as they latensify.

Lowering the record-transport module provides full front accessibility to the magnet block, easy insertion and adjustment of galvanometers. The oscillograph slides out from its rack-mounting enclosure for inspection or maintenance.

Modular design also permits easy removal of the entire optical module, which contains the twin magnet blocks, high-energy light source, and all optical components.

THE CEC TYPE 5–124 OSCILLOGRAPH. The 5–124 recorder, shown in Fig. 3.30, is a multichannel direct-writing instrument that produces 7-inch-wide records. Eighteen channels, each with an individual input, can be used to record static or dynamic data from d-c to 5000 cps, using CEC 7–300 series galvanometers. The complete recorder weighs only 40 pounds and measures 7 5/8 inches high by 13 inches wide by 15 1/4 inches long. Despite its compactness, the 5–124 incorporates all of the standard and optional features needed for the most critical data recording work. These include trace identification, timing lines, grid lines. To permit high accuracy in the resolution of data traces, trace widths are maintained at 0.010 inches or

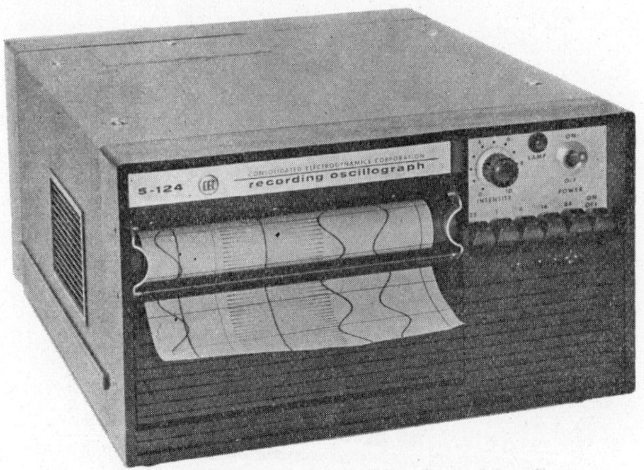

Fig. 3.30. The CEC 5-124 oscillograph

less. Using presently available print-out papers, writing speeds to 50,000 inches per second are easily achieved. To provide even greater record accuracy, galvanometer block heaters are available to maintain block temperature at 100 °F.

Recording speeds are selected through a pushbutton-operated transmission. Five record speeds can be selected instantly before or during the recording process: 1/4, 1, 4, 16, and 64 inches per second. A mechanical interlock prevents operation of the pushbuttons until the drive motor has been started.

An optional drawer-slide accessory is available for mounting the oscillograph in a standard 19-inch RETMA rack. Two versions permit the oscillograph to be mounted either in the center of the panel or to the right side, so that the accessory flash timing generator can be installed. Timing intervals can be selected without withdrawing oscillograph from the rack. Both panel versions are 10 1/2 inches high, 19 inches wide, and require a rack depth of 22 1/2 inches.

THE CEC 5–133 DATAGRAPH OSCILLOGRAPH. The 5–133 Datagraph Oscillograph is a rack-mounted or bench-mounted direct-writing instrument producing 36 or 52

Fig. 3.31. CEC 5-133 datagraph oscillograph

channels of data on 12-inch-wide light-sensitive-paper. Twelve recording speeds, ranging from 0.1 to 160 inches per second, may be selected by pushbutton. An optional remote control unit provides complete control of all electrical functions at distances up to 1000 feet. An ambient light shield permits recording with no latensification if it is desired to chemically process records for archival quality.

Design of the 5–133 Datagraph, shown in Fig. 3.31, includes the following features:

1) Magnetic Regulated Lamp Power Supply—provides proper power to lamp regardless of input voltage variations, allows start/restart times of less than one second.

2) Adjustable Grid Line Intesity—continuously variable vernier control.

3) Timing Line Generator—electronically flashes timing lines at intervals of 10, 1, 1/10, 1/100, and 1/1000 second.

4) Record/Event Numbering—selected by front panel switch.

5) Automatic Record Length Control—continuously variable from 0 to 15 feet; multiplier extends range to 0 to 150 feet.

6) Twelve Recording Speeds—pushbutton selectable speeds of 0.1, 0.4, 0.8, 1, 1.6, 4, 8, 10, 16, 40, 80, and 160 inches per second.

7) Trace Identification—trace interruption and trace numbering.

8) Vibration Isolation—four isolator mounts on recorder and four on the drive motor/transmission assembly.

9) Galvo Light Intensity Controls—manual and automatic controls provide optimum trace quality for each block.

10) Filtered Air Cooling—cools and pressurizes the optical module for maximum cleanliness.

11) Module Construction—all modules removable as single assemblies.

12) Dataflash Facility—1 second access time

Century Equipment. Century Electronics and Instruments, Inc., 6540 E. Apache Street, Tulsa 15, Oklahoma, makes the recording equipment described in the sections that follow.

Century Model 404 Oscillograph. This special-purpose recorder is described in a later section of this chapter under the heading *Seismographic Equipment*.

Century Model 408 Oscillograph. The Model 408 Recorder accommodates up to 30 Century Model 210 galvanometers, and records on 8-inch paper. Magazine capacity is 150 feet of regular weight paper, or 100 feet of film. Paper travel, from 0.5 to 125 inches per second in two ranges, is selected by a single knob. Record-length control, full-width timing lines, voltmeter, and switches for power, timing, trace viewing, scanning, and automatic or manual recording are provided. Remote control unit, ac power supply, multiplex galvanometer and multiplex cables are optional.

The recorder measures 8–7/8 by 12–1/2 by 20 inches, exclusive of cable projections, shock-mount and ac power supply and weighs approximately 65 pounds. The ac power supply increases the length to 25 inches and the weight to 82 pounds.

The Model 408E1 operates at 22–28 vdc. With the Model 408D900 power supply attached, 105–125v, 60–400 cycle power can be used. The unit requires 300 watts.

CENTURY SERIES 409 OSCILLOGRAPH. This compact, rugged 14–channel recorder, shown in Fig. 3.32, has operated under severe environmental conditions, including over 20 g's of acceleration, and altitudes above 100,000 feet in balloon tests.

Fig. 3.32. Century Series 409 oscillograph

Direct-writing models are now available, and standard models can be modified to provide direct-write capability. Several models of the 409 offer various recording speeds, timing frequencies, magazines, galvanometers, and connection locations. A shock-mount base is available.

Information is recorded on 3–5/8 inch paper, contained in four types of magazines equipped with both feed and takeup spools. All 409 models with 50 foot magazines are 5 by 5 by 11–5/16 inches; with 100 foot magazines, 5 by 6–15/16 by 12 inches. A magazine having a capacity of 600 feet of thin-base paper is available.

The 409 will record as many as 20 traces in two ranges of recording speeds: 0.5 to 6 inches per second and 2 to 24 inches per second. Standard features include trace identification, trace viewing, paper supply indicator, and internal timing. All models require 3 amperes of current at 22 to 28 vdc.

Century Model 414 Oscillograph. The 414 instrument shown in Fig. 3.33, is a compact, lightweight, 14-channel recorder, rugged enough to withstand the severe environmental conditions of rocket, missile, aircraft, and torpedo testing. The unit will withstand 20 g's constant acceleration. Forty feet of film or 50 feet of Kodak 809 (or equivalent) paper 3–1/2 inches wide can be accommodated in the magazine. Paper-travel speed ranges from 1/8 through 1–5/16 inches per second.

The size of the recorder is 5 by 5–1/2 by 7–1/2 inches and the weight is 8 pounds. The power requirement is 28 volt dc at 3 amps maximum.

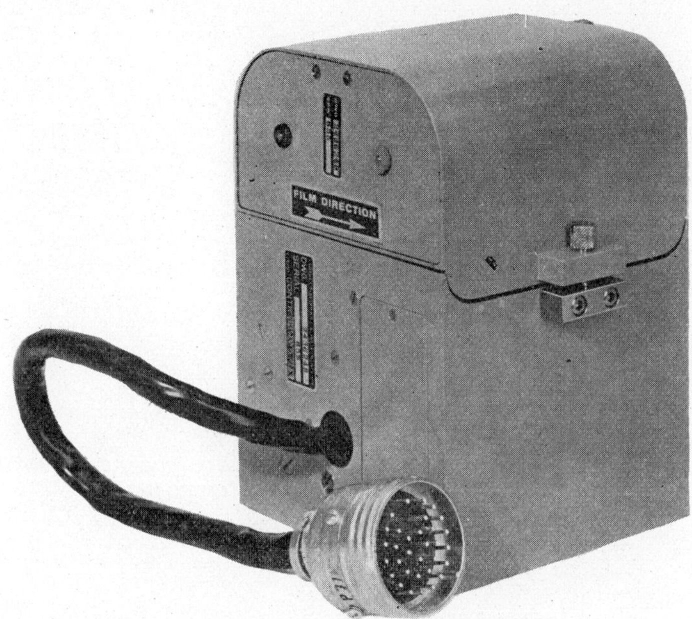

Fig. 3.33. Century Model 414 oscillograph

Century Model 444 Wide-band Oscillograph. The Model 444P Ultragraph, shown in Fig. 3.34, is a portable, precision bench tool that will record 6 channels of information at frequencies to 2000 cps ± 5 perc ent. Up to 150 feet of 3–5/8 inch wide thin-base direct-print paper can be daylight-loaded; the cover is simply raised and the roll dropped in. The instrument is specifically designed to use incandescent-sensitive direct-print materials, such as Du Pont Lino-writ 7. The tungsten lamp used in the Ultragraph is a low-cost bulb that requires only 75 watts. The standard model operates from a 115v, 60 cycle ac supply. Twelve and 24 volt dc models are optionally available.

Recording speeds of 1, 5, 10 and 50 inches per second can be selected by a rotary switch. Other models are available in inches per minute and inches per hour.

Optional features include grid lines, full-width timing lines, and selection of galvanometers.

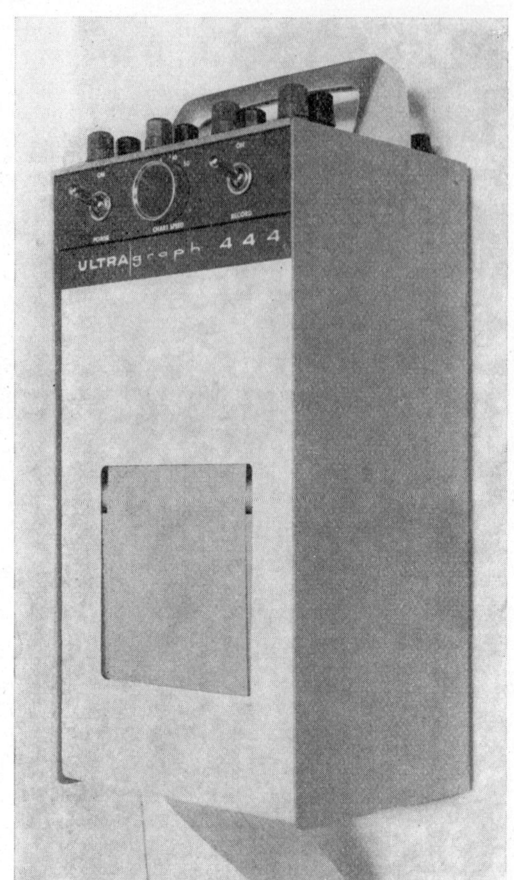

Fig. 3.34. Century Model 444 wide-band oscillograph

Fig. 3.35. Century Model 447 Ultragraph

The recorder measures 6–1/4 by 5–5/8 by 12–1/2 inches, and weighs 15 pounds.

CENTURY MODEL 447 ULTRAGRAPH. The Model 447 direct-writing oscillograph, shown in Fig. 3.35, will provide up to 36 active traces on 8 inch paper. Paper capacity is 475 feet of thin-base stock. The standard Model 447 is equipped with a xenon source that provides a writing speed of 25,000 inches per second. The 477 is also available with a tungsten light source, with a writing speed of 10,000 inches per second. Standard paper travel speeds are 1, 5, 10, and 50 inches per second; other speeds are also available.

The 10 by 13 by 15 inch unit can be used on a bench, or, with an adaptor, it can be rack-mounted. The basic recorder weighs 35 pounds, and operates on 115 volt, 60 cycle power. The unit features front-panel paper loading, with no threading. All controls are located on the front. Sensitivities range from 5 microamps per inch, and 4800 cycles per second flat frequency response is obtained.

Grid lines and full-width timing lines are furnished as optional selections.

HARTMANN AND BRAUN OSCILLOGRAPHS. The following data were supplied by Messrs. Hartmann & Braun AG., Frankfurt am Main, Germany, who produce four models of Lumiscript print-out oscillograph instruments. All models of Lumiscript Oscillographs have identical optical systems. Maximum pressure mercury vapor lamps are used as light sources. Thermostatically temperature-stabilized magnet blocks accommodate the subminiature galvanometers.

Some other identical features of all Lumiscript models include the following items:

1) Fifteen adjustable chart speeds operated by push-buttons on the control panel (1.25; 2.5; 5; 10; 20 to 2000 mm/sec) (approximately 1/20 to 79 inches per second).

2) Maximum writing speed of 2000 m/sec, (approximately 78, 740 inches per second).

3) Continuous paper supply indicators.

4) Full width electronic timing device for 0.01; 0.1; 1; and 10 seconds flash sequence. Model 150–8 includes the first three intervals only.

5) Controlled chart speed, pre-selectable recording length of 0 to 3 m (approximately 0 to 9.84 feet), except for model 150–8.

6) Mirror relay, except for model 150–8.

7) Gridline system: 2 mm raster with separate intensity control.

8) Switching device for remote operation and impulse control, except for model 150–8.

9) Built-in power supply for 220 volts, 50 cycles per second.

10) Various devices for rack mounting.

The four models differ mainly by the width of their charts, the number of galvanometers, individual equipment, and accessories.

LUMISCRIPT MODEL 150 AND 300 OSCILLOGRAPHS. Chart width on the Model 150 is 150 mm maximum (although 60, 90 and 120 mm paper can be accommodated).

The model 300 records on 300 mm print-out paper, and can also use 120, 150 and 200 mm wide material. Model 150 is equipped with one thermostatically temperature-stabilized magnet block that will accommodate either up to 25 subminiature galvanometers or up to 10 only, if a design for a rated voltage of 650 volts is preferred to permit direct across-the-line oscillography.

Model 300 can be equipped with either one or two similar magnet blocks.

Both models include a built-in power supply for 220 volts, 50 cycles per second. Other voltages or battery operation are available on request.

Sizes and weights are as follows:

Lumiscript Model 150: 300 by 340 by 435 mm; weight approximately 30 kg.
Lumiscript Model 300: 445 by 380 by 450 mm; weight approximately 42 kg.

Lumiscript Model 150–8 and 200–25 Oscillographs. Model 150–8 was intentionally constructed with certain limitations on the number of channels and accessories, to permit the instrument to be sold at a very reasonable price. Its magnet block accommodates up to 8 subminiature galvanometers and its chart width is limited to a maximum of 150 mm. It measures 300 by 340 by 435 mm and weighs approximately 29 kg.

Model 200–25 was especially constructed for rack installation and can be considered a standard model for high technical requirements. Many new accessories are

Fig. 3.36. Hartmann & Braun Lumiscript Model 150 print-out oscillograph

Fig. 3.37. Hartmann & Braun Lumiscript Model 200–25 print-out oscillograph

now available for this model, among which are current and voltage dividers, adapters for resistance thermometers, and transducers for mechanical variables, such as pressure, stress, force, distance, acceleration, etc.

Dimensions of Model 200–25: 483 by 210 by 432 mm; weight approximately 36 kg.

HONEYWELL EQUIPMENT. The print-out oscillograph recorders described in the following sections are manufactured by Honeywell, Denver Division, 4800 East Dry Creek Road, Denver 10, Colorado.

HONEYWELL MODEL 906 VISICORDER. First introduced at the Instrument Society of America exposition at New York City in September 1956, the Honeywell Model 906 Visicorder combined the high frequency response of oscillographs using develop-out paper with the readout features of directwriting equipment, and overcame the chief disadvantages of both types.

The original 906 Visicorder oscillograph provided a maximum of 8 recording channels with frequency response to 2000 cycles per second. Later 906 models increased the number of channels, extended the frequency range to 5000 cycles per second and added self-starting recording lamps, grid lines, and other improvements. The latest version of the Model 906C Visicorder has all of these features plus a built-in flash-tube timing system, and 8- or 14- channel capacity on 6-inch wide paper. One hundred foot standard or 150-foot extra thin paper rolls may be used.

The 906C-1 Visicorder, shown in Fig. 3.38, utilizes the high sensitivity subminiature plug-in galvanometers and magnet assembly, which is interchangeable among Visicorder Models 906A-1, 906B-1, 906S-1, 906C-1, 906T-1, 1108, 1012, 708C and 712C.

Fig. 3.38. Honeywell Model 906C-1 Visicorder print-out oscillograph

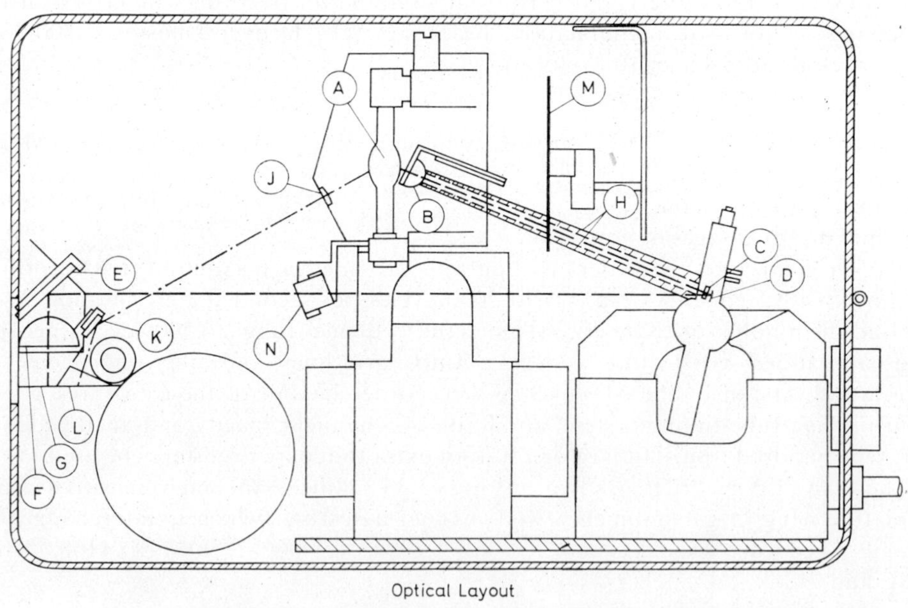

Fig. 3.39 illustrates the optical system. The galvanometer light spots are focused on the recording point and are visible on an amber calibration screen. This allows the operator to check the light spot while the galvanometer is being adjusted, thus permitting precise calibration. The data can also be monitored at the recording point, while the record is being made. Calibration is accomplished by adjusting the galvanometers which are located under the top rear of the instrument.

The transport speed of the recording paper may be changed during recording by adjusting the knobs on the control panel. Several paper drive systems, which can be interchanged in a few minutes without tools, are optionally available in the following speed ranges:

> 0.2, 1, 5, 25 inches per hour
> 0.2, 1, 5, 25 inches per minute
> 0,2, 1, 5, 25 inches per second
> 0.4, 2, 10, 50 inches per second
> 5, 25, 50, 100 millimeters per second

The single point-source recording lamp is a 2 electrode high-pressure mercury-vapor lamp. It ignites automatically through a starter circuit when the LAMP switch is closed.

A connector for remote operation is located on the back panel and includes POWER, DRIVE, LAMP, and TIMING controls.

Optional features which include a timing system, grid line system, trace identification, and recording intensity control. A record takeup, latensifier, and relay rack adapters are available as accessories.

HONEYWELL MODEL 1108 VISICORDER. The Model 1108 Visicorder, shown in Fig. 3.40, was designed to fulfill those requirements in direct recording that lie between the capabilities of the Model 906 and the Model 1012. The 1108 records up to 24 channels of data on 8-inch direct-print paper, and incorporates most of the convenience features of the larger Model 1012. These features include front-surface paper loading, operation, and control; visible recording point; easy galvanometer adjustment; provisions to change speeds during operation; record forward and reverse; and automatic record-length.

The record numbering system and the control timing line systems are illustrated in Fig. 3.41, and the gridline trace optical system is shown in Fig. 3.42.

Fig. 3.39. Optical system of the Honeywell 906C Visicorder

The galvanometer shown is the solid frame type used in the 906C-2. The sub-miniature type used in the 906C-1 is in the same position.

(A) Light source; (B) Collector lens; (C) Galvanometer lens; (D) Galvanometer mirror; (E) Recording mirror; (F) Recording lens; (G) Recording plane; (H) Band of intense light; (J) Gridline collector lens; (K) Gridline mirror; (L) Gridline paper bar; (M) Trace identifier; (N) Recording intensity control

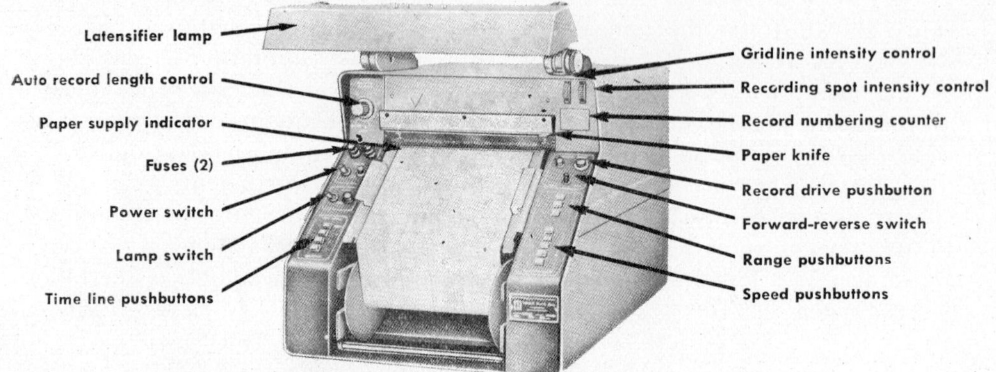

Fig. 3.40. Honeywell Model 1108 Visicorder print-out oscillograph

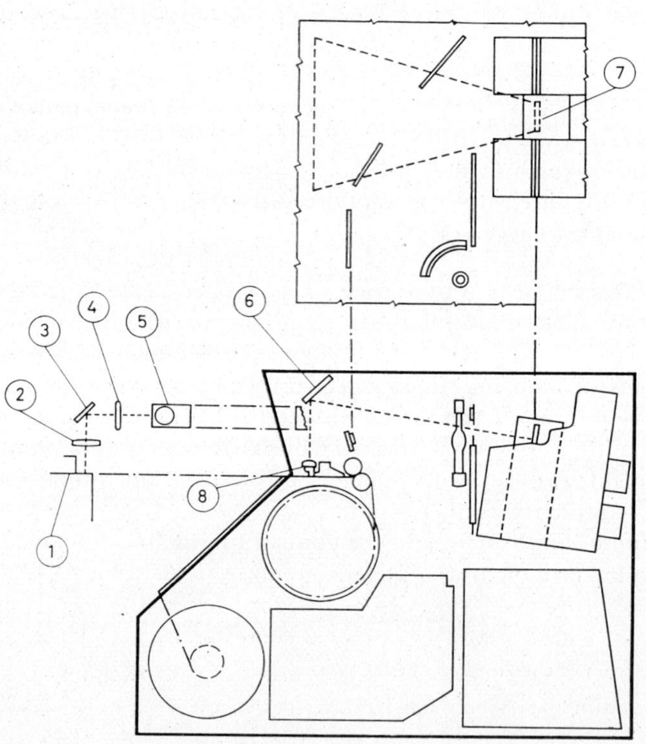

Fig. 3.41. Record numbering and timing line systems of the Honeywell Model 1108 Visicorder

1. Recording plane; 2. Record number lens; 3. Record number mirror; 4. Record number flash tube; 5. Record number counter; 6. Recording mirror; 7. Time line flash tube; 8. Recording lens

128

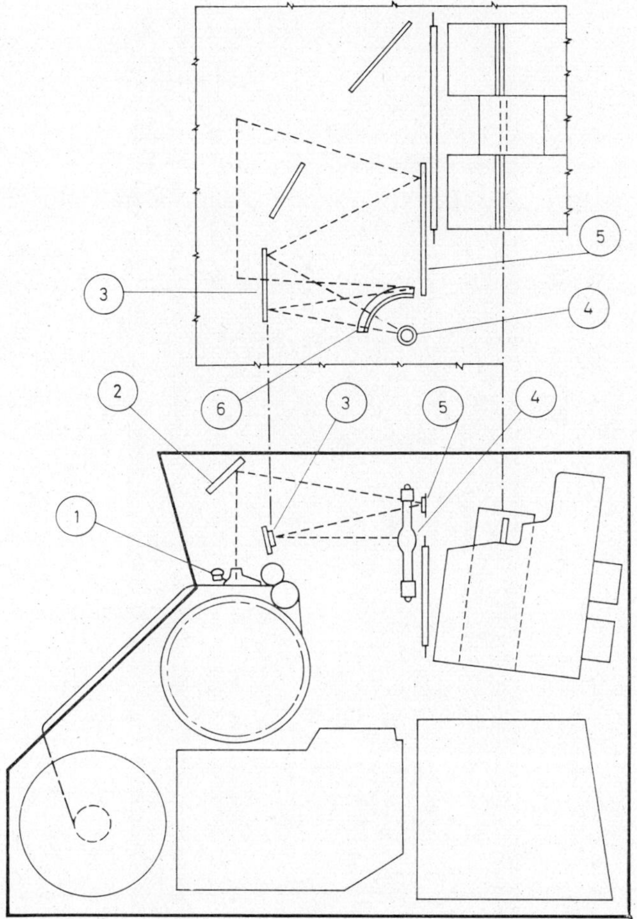

Fig. 3.42. Honeywell Model 1108 Visicorder gridline trace optical system

1. Gridline aperture bar; 2. Recording mirror; 3. First gridline mirror; 4. Recording lamp; 5. Second gridline mirror; 6. Collector lens

Optional equipment for the Model 1108 includes a latensifier unit, and various accessories.

HONEYWELL MODEL 1508 VISICORDER. The rack-mounted Model 1508 Visicorder, illustrated in Fig. 3.43, is a 24-channel oscillograph that uses 8-inch wide direct-print paper. The chart width permits an 8-inch deflection peak to peak, with traces recorded at writing speeds in excess of 50,000 inches per second.

Accessibility to the interior of the recorder for galvanometer adjustment, servicing, etc., is accomplished by pulling the instrument forward on its slide mount. The

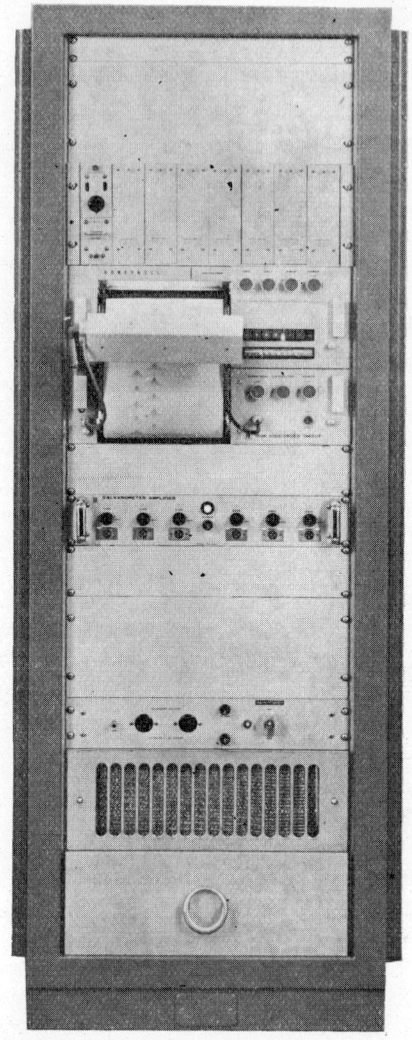

Fig. 3.43. Honeywell Model 1508 Visicorder print-out oscillograph

cover is designed to remain in the rack, and the recorder will operate while in its extended position.

Any one of twelve paper-transport speeds may be selected by pushbutton during operation. All operating controls and loading are accomplished from the front of the rack.

HONEYWELL MODEL 1612 VISICORDER. The 1612 direct-recording Visicorder, shown in Fig. 3.44 is a rack-mounted model that makes use of the Visicorder direct-recording principle. It has a built-in heated platen, and two plug-in latensifier lamps, to provide immediate development of records at speeds up to 16 inches per second.

The Model 1612, designed specifically for systems use, takes only 22 3/4 inches of rack height, yet it records up to 36 channels of static or dynamic data simultaneously on 12-inch chart paper at frequencies from dc to 5000 cycles per second. Writing speed exceeds 50,000 inches per second. Galvanometers may be deflected as much as 8 inches peak to peak and maintain specified linearity.

There are 15 forward recording speeds from 0.1 to 160 inches per second, all selected by push-button controls. There are 10 reverse speeds for easy review and annotation after tests have been run. All controls are on the front surface, and paper may be loaded in a matter of seconds.

All sub-assemblies are either "plugged in" automatically as they are put into their proper places or are connected by means of plug-in cables. There are no screw-type terminal blocks on the 1612. The power supply can be removed completely by turning two fasteners. A convenient side door provides easy access to fuses and internal adjustments.

Available as an optional built-in accessory, the heated platen permits the operator

Fig. 3.44. Honeywell Model 1612 Visicorder print-out oscillograph

to apply heat to the paper to develop the traces faster. Platen heat improves the quality of the record and provides better contrast. When turned off, the platen stays in place and serves as a regular platen.

All controls are located on the front surfaces of the instrument. Paper loading is accomplished by pulling the paper cradle forward, dropping a roll of paper in place, and engaging the paper to the takeup spool, as shown in Fig. 3.45. The 15 forward and 10 reverse paper speeds are all push-button selected.

Fig. 3.45. Loading the Model 1612 Visicorder

The Model 1612 Visicorder is used primarily in military and aerospace applications. Because of this, several options have developed so that the instrument will meet various specifications. One option includes special attachments which qualify the Model 1612 for radio-frequency interference (RFI) supression. This capability is especially needed in telemetry control stations where stray radio signals from instrumentation must be kept at a minimum. Another option includes special filters so the Model 1612 may be converted in only a few seconds from a direct-recording oscillograph to one which utilizes the various developing-out papers.

MIDWESTERN EQUIPMENT. The following section describes oscillograph recorders manufactured by Midwestern Instruments, Inc., 41st and Sheridan Road, Tulsa, Oklahoma.

MIDWESTERN SERIES 560 OSCILLOGRAPHS. The Model 560, shown in Fig. 3.46, is a compact, light-weight, ruggedized unit capable of operation under severe environmental conditions. The instrument records 14 channels on 3–5/8 inch wide paper, at speeds ranging from 0.390 to 8.0 inches per second. Manufactured in various configurations, the approximate overall dimensions of the 9 pound recorder are 5–3/32 by 6–5/8 by 7–3/16 inches. Power requirement is 2 amperes maximum, 24 volt d-c. The capacity of the magazine is 50 feet of paper; 40 feet of daylight-load film.

Fig. 3.46. Midwestern Series 560 oscillograph

The Standard 560 will withstand constant accelerations up to 15 g. The rugged construction of the 560B has been subjected to 2-phase shocks in excess of 3000 gravities for 2 milliseconds and 300 gravities for 25 milliseconds.

MIDWESTERN 561 OSCILLOGRAPH. Slightly larger and heavier than the 560 Series previously described, is the Model 561 Recorder, shown in Fig. 3.47. It measures 5–13/16 by 5–3/16 by 9–3/16 inches and weighs 15–1/2 pounds. Designed for "on-board" data recording, the instrument has been subjected to constant acceleration above 25 gravities, and to shock accelerations above 1500 gravites for 20 milliseconds. Applications include use on torpedos, missiles, rockets, aircraft, test sleds and other test vehicles.

Fourteen channels of data can be recorded on 3–5/8 inch paper or 3–1/2 inch film. The magazine capacity is 95 feet of paper, and the recording speed may be varied from 0.5 to 80 inches per second by gear changes. The recorder operates on

Fig. 3.47. Midwestern Series 561 oscillograph

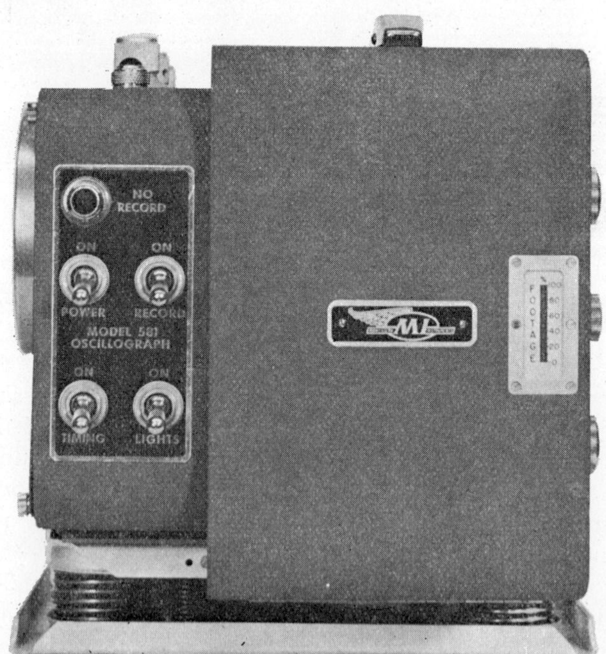

Fig. 3.48. Midwestern Model 581
oscillograph

134

28 vdc at 12 amps maximum. Environmental conditions are −65° to 165 °F for temperature, and altitudes up to 100,000 feet above sea level.

MIDWESTERN 581 OSCILLOGRAPH. The Model 581 is a 14–channel recorder designed for the rugged environmental conditions associated with flight-test applications; however, it is equally useful for other field work as well as laboratory use. The instrument, of rigid, cast aluminum-alloy construction, measures 8 inches wide, 7–1/2 inches high, and 15–3/4 inches long overall, and weighs 22 pounds without a shock-mount base. It operates at 24 to 28 vdc at 12 amps maximum. The instrument is illustrated in Fig. 3.48.

Gear changes permit recording speeds to be varied from 0.446 to 44.750 ips. Two magazines are available; the standard magazine which accepts 3–5/8-inch Spec. 1 paper in lengths to 125 feet for light-weight base; and the high magazine which accepts 400 feet maximum of lightweight-base paper.

The recorder features a beam-interrupter type trace identifier; no-record indicator; standby heaters for low-temperature operation; ventilation system for high-temperature operation; easily accessible galvanometer-damping resistor board; unexposed footage indicator; automatic record numbering; automatic record-length control; and burn-out indicators for recording and timing lamps.

MIDWESTERN 591 OSCILLOGRAPH. The Model 591 instrument, designed for either laboratory or extreme environmental operation, will record static or high-frequency transient data with equal accuracy. The unit, shown in Fig. 3.49, measures 16–13/16 by 11–1/6 by 27–7/16 inches and weighs 130 pounds. Plug-in power supply units enable the Model 591 to operate on 28 vdc or on 115 vac. A maximum of 50 active traces can be recorded on 12 inch-wide paper, at speeds ranging from 0.0812 to 129.9 inches per second, for a total of 20 different speeds. Five basic

Fig. 3.49. Midwestern Model 591 oscillograph

speeds are obtained by coded change gears, and electric clutches multiply the basic speeds by factors of 1, 4, 16, or 64.

A thermostatically-controlled heating and ventilating system permits opertion under temperatures ranging from $-65\,°F$ to $160\,°F$. Other environmental conditions include 15 gravities along any major axis, zero to 100 percent relative humidity, and altitudes to 100,000 feet above sea level.

Standard magazine capacity is 250 feet of normal weight paper or 400 feet of lightweight paper.

Full width electronic flash timing is provided for external synchronization of two or more oscillographs. Timing intervals are 0.01 and/or 0.1.

Records are identified in three ways: 1) by successive numbering of each run; 2) interval numbering; and 3) by information written on the record. Traces are identified by a light-interrupting device.

The 591 also features automatic record length control, unexposed footage indicator, viewing scale, trace adjusting screen, standby filament indicator, no-record indicator, and shockmount base. Optional equipment includes remote control and rotary scanning units.

MIDWESTERN DIRECT-WRITING RECORDERS. Midwestern Instruments also manufactures several direct print-out recorders described in the following paragraphs.

MIDWESTERN 603 AND 607 OSCILLOGRAPHS. These two models, illustrated by Fig. 3.50, record up to 50 data traces on a 12-inch record. Thin-base paper capacity is 350 feet maximum which may be daylight loaded. The instrument is equipped with forward and reverse record transport. Three speed ranges of 15 speeds each provide drive speeds from 0.050 to 170 inches per second.

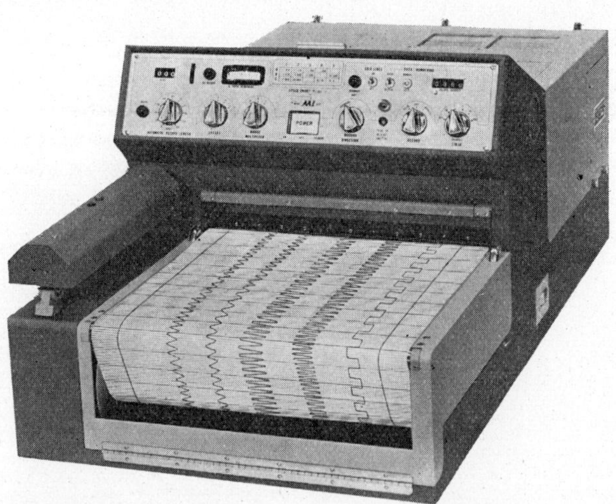

Fig. 3.50. Midwestern Models 603 and 607 oscillographs (same appearance)

136

The major difference between the two recorders is that the 603 produces timing lines mechanically and that the 607 features electronic timing. Up to five 607 units may be operated synchronously from pulses generated by an external source. Full width timing lines and longitudinal grid lines, which may be eliminated at the descretion of the operator, are produced optically and exposed simultaneously with the galvanometer traces.

As the record emerges from the recorder proper, it is presented on a 12-inch viewing table, and taken up on a reel at the front below the table surface. An adjustable latensifier lamp swings out of the way when not in use.

The 603 and 607 Direct Recorders also feature beam-interrupter type trace identification, automatic control of record length, and modular construction of plug-in components such as the data numbering unit and the power supply.

MIDWESTERN 616 DIRECT RECORDER. The Midwestern Model 616 Direct-Recording Oscillograph, shown in Fig. 3.51, is a smaller version of the Midwestern Model 603. The principle of recording is identical to that used in all Midwestern direct recorders. The recording paper is a 7-inch wide print-out material that provides a visible record without chemical processing. The emulsion is given a short duration, high intensity exposure as the paper travels through the instrument, thus forming a latent image. This image is formed by the collimated galvanometer spot and is made visible by exposing the record to a lighting condition of lesser intensity, such as normal fluorescent room lighting.

The magazine is capable of ejecting the record directly, where developing accomplished by ambient room lighting, or storing the record on the internal takeup spool. A viewing window is provided so that the record may be stopped at any time

Fig. 3.51. Midwestern Model 616 oscillograph

during a recording period and examined when a quick look is required. A mirror system enables the operator to view the galvanometer and timing information as they are being recorded.

The Model 616 instrument measures 9–1/2 by 10–3/4 by 20–5/16 inches excluding the shock-mount base, and weighs approximately 75 pounds. The power supply is a self-contained plug-in unit designed for operation at 115 vac, 50–400 cycles per second.

The 36 channel galvanometer magnet structure is thermostatically controlled, and a ventilation system provides cooling air for the power supply and galvanometer lamp.

MIDWESTERN 621 DIRECT RECORDER. The Model 621 oscillograph, shown in Fig. 3.52, records up to 14-traces on 6-inch wide direct-print photographic chart paper. Paper capacity is 200 feet of thin-base material that can be transported at speeds ranging from 0.2 to 60 inches per second, provided by gear changes selected electrically from the control panel.

Three models in the series are available: the 621-HT, a horizontal table model; the 621-VT, identical to the HT model except that the exposed record is displayed on a vertical table; and the 621S, containing no table nor takeup spool. The HT and VT models have forward and reverse record drives for review, analysis, and marking of the record while it is on the machine.

Optional features include full-width timing lines, longitudinal grid lines, and beam interrupter type trace identification.

Fig. 3.52. Midwestern Model 621 oscillograph

Fig. 3.53. Midwestern 800 Series oscillograph

MIDWESTERN 800 SERIES DIRECT-RECORDING OSCILLOGRAPH. The Midwestern 800 Series Direct-Recording Oscillograph (shown in Fig. 3.53) is the first Midwestern print-out oscillograph to be designed as a general-purpose, industrial-laboratory type recorder. It is available in bench and rackmounted versions, easily converted in the field at a later time. The Model 800 design uses the module concept for basic component assemblies as well as for optional accessories.

Standard Specifications. The Model 800 provides eight positive-drive recording paper speeds covering a range of 1000 to 1 from the highest to the lowest speed. Eight push buttons, located on the front panel, allow direct speed selection without separate range and multiplier controls. The primary transmission unit has speeds of 0.1, 0.4, 1, 2, 5, 20, 50, and 100 ips. Optional speed ranges may be ordered.

Either 25 or 36 active data recording channels are available for factory installation. Standard Midwestern Model 158 Magnetic Structures are installed complete with thermostatically controlled heater units.

The standard trace identifier is a light beam interrupter system which provides a simplified method of interpreting the visual record regardless of trace amplitude or overlapping. An optional trace numbering system is available.

All standard Midwestern Optical Galvanometers may be used in the Model 800 Oscillograph covering a range of frequencies from d-c to 6000 cycles flat response. Galvanometer recording spots make direct contact with the recording paper in full view of the operator. A convenient spot-setting scale, graduated in 0.1 inch increments, permits accurate adjustment of the spot position. Galvanometer trace blanking is accomplished by rotating the galvanometer barrel 180 degrees in the magnetic structure so that the mirror faces the rear of the instrument.

Damping Resistors. Provision is included for electrical damping of low-frequency galvanometers. The damping resistors may be mounted on a terminal board located at the top rear of the oscillograph under the main cover.

Galvanometer illumination is provided by a single high-intensity, three-electrode arc lamp for maximum spot stability and reliable lamp starting. Either a 100-watt mercury vapor or 75-watt xenon arc lamp may be used interchangeably by means of a simple field wiring change.

Lamp starting is automatic for either mercury vapor or xenon arc lamp. A manual lamp intensity control is located on the front panel to adjust trace intensity. Trace width may be varied from 0.02 inch to 0.002 inch depending on spot velocity desired. The maximum recording spot velocity, or writing speed, is in excess of 75,000 inches per second.

Longitudinal reference grid lines are optically produced on the record to provide convenient amplitude measurement of the recording traces. The intensity of these lines may be manually increased, decreased or turned completely off by means of a control located under the galvanometer access door. The grid line spacing is provided in either 0.1-inch or 2-mm increments with each fifth line accentuated.

All commercially available print-out recording papers may be used in the Model 800 Oscillograph. Eight-inch-wide recording paper, spooled to manufacturer's specification 1, is standard, with up to 250 feet of standard base or 475 feet of thin base paper accommodated. Narrow paper widths may be used with an adapter spool available as an optional accessory.

Recording paper loading is accomplished from the front of the instrument by dropping the roll in place behind a swing-out cover door.

Environmental Capability: The oscillograph is designed to function normally over an ambient temperature range from plus 32 °F to over plus 120 °F. It will survive non-operational vibration magnitudes of 0.15 inch double amplitude at 20 cycles per second, decreasing linearly to 3 g's at 110 cycles, per second and non-operational shock up to 10 g's.

The Model 800 may be operated from a 50/60 cycle, 105/125 volt ac source. The power supply is a self-contained unit, of modular construction, located in the lower rear portion of the oscillograph. Maximum operational power is 400 watts. A 230-volt ac input power adapter is available as an optional feature.

MIDWESTERN MODEL 1210 OSCILLOGRAPH. The Model 1210 recorder, shown in Fig. 3.54, is of modular design, and may be flush mounted in standard equipment racks or used as a table-top recorder. Fifty active data channels are provided by dual, 25 trace, magnetic structures that are thermostatically stabilized. All standard Midwestern instrumentation galvanometers may be used.

A separate modular paper transport system provides positive forward and reverse record drive with internal takeup. Record ejection is possible with an optional ejection roller.

The recording speed is selectable by pushbutton from 0.01 to 160 inches per second. The speed tolerance is within ± 2 percent over the entire speed range. Speeds may be changed during operation either by manual pushbutton control or through remote-control inputs.

Any 12-inch wide by 475-foot roll of thin base or 250-foot roll of standard base, commercially available print-out paper, spooled to specification 28, may be loaded from the front of the model 1210. Narrower paper rolls may be used with an adapter. Latensification methods range from normal room lighting to forced latensification utilizing the heated platen and dual high-intensity lamps. The record

Fig. 3.54. Midwestern Model 1210 oscillograph

may be run in either forward or reverse direction, to monitor previously recorded traces. Provisions allow for optional selection of calibrated automatic record length. An indicator on the upper control panel shows the percentage of full roll remaining on the supply spool.

An electronic flashtube timing system, with external timing synchronization provisions, produces full width timing lines in five intervals of 0.001 to 10 seconds.

Longitudinal reference grid lines are optically produced, with each fifth line accentuated. Grid line intensity is controlled automatically. A separate ON–OFF control for grid lines is also provided.

Trace identification is provided by a sequential beam interrupter driven by the record transport system. Sequential numbers are photographically imprinted on the edge of the record to coincide with the sequential trace breaks of the recorded beam.

The record identification system utilizes a four-digit counter to record a record or event number on the oscillograph paper. A second counter displays the same number on the control console. For record numbering, the counters advance one number automatically. When used as an event numbering system, the counters may be advanced independently in conjunction with an external event. Counter reset is provided. The record/event identification may be flashed automatically or independently at a maximum rate of three pulses per second.

Power requirements for the Model 1210 Oscillograph are 115 ±10 vac, 60 ±1 cycles per second, 1660 watts maximum power with heated platen, 700 watts without heated platen.

Dimensions of the rack-mounted unit are 25–1/2 by 19 by 21 inches. Weight is 190 pounds. The bench-mounted unit is slightly smaller and weighs 175 pounds.

Numerous optional accessories are available.

SANBORN EQUIPMENT. The Sanborn Company, 175 Wyman Street, Waltham 54, Massachusetts, manufactures recorders for many purposes that use various marking methods. The following recorders use photographic-marking.

SANBORN X-Y RECORDER. The Sanborn Model 670B X-Y Recorder (see Fig. 3.55) will record a 100th-of-a-second X-Y variable directly. It has a writing speed of 2500 inches per second, produces 8 by 8 inch charts, is flat to 1300 cycles per second within 3db, and has 1 percent linearity. Its capabilities make it particularly valuable in such applications as rapid recording of diode and transistor charac-

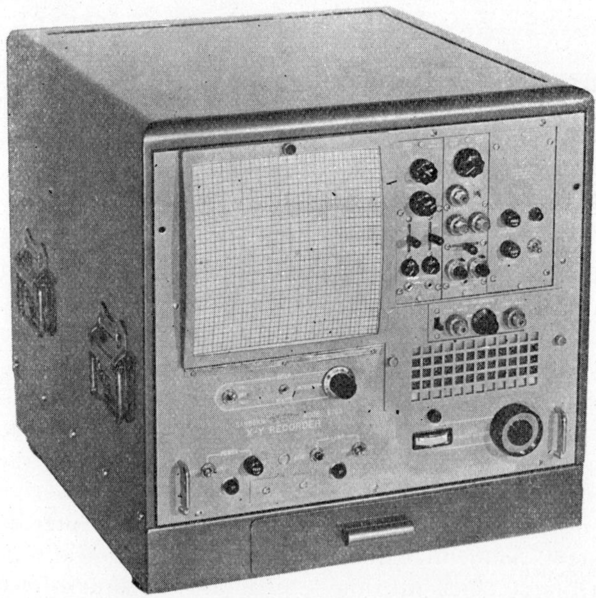

Fig. 3.55. Sanborn Model 670B X-Y photographic print-out paper recorder

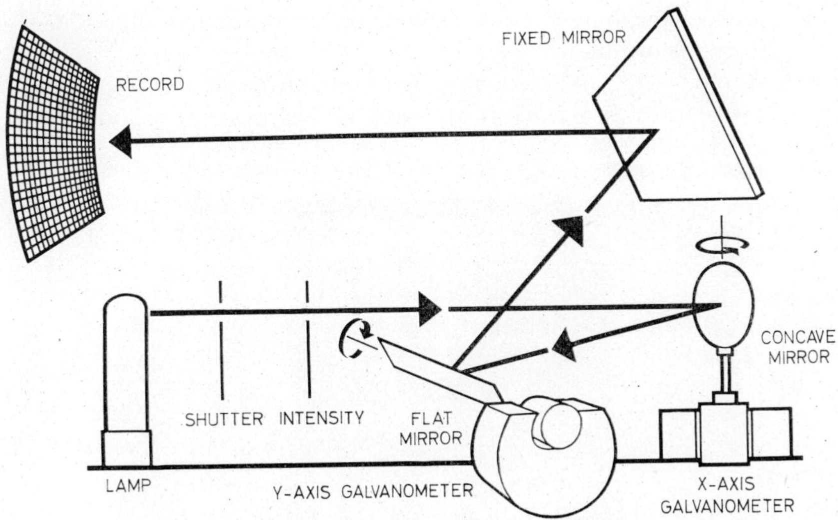

Fig. 3.56. Schematic of light-beam path in Sanborn X-Y Recorder

teristics, hysteresis curves for control systems, gyros, servo valves, magnetic coils; velocity or deceleration vs. vibration of mechanical elements.

Recordings are made on 8 by 8 inch daylight-loading, ultraviolet-sensitive charts by an optical (light beam) recording system, shown in Fig. 3.56. Traces are visible almost immediately after recording and require no chemical developing. Exposure in normal room light will fully develop the recordings. With normal care, the recordings will last for long periods; they may also be fixed using standard photographic techniques. A lamp current control and monitor enables the user to achieve optimum life from the ultraviolet light source.

SANBORN 650 OPTICAL OSCILLOGRAPH. The Sanborn 650 Optical Oscillograph, shown in Fig. 3.57, is an optical recording system composed of amplifier, galvanometers (8 channels standard, 1 to 24 available), and recorder which has been totally designed as an integrated, compatible system. The amplifiers are designed to drive 2KC optical elements. Inherent in this design are frequency boost and compensation circuits which extend the response of the amplifier and galvanometer combination to 5KC (–3db). In addition, each amplifier channel has a position control which allows positioning of the galvanometer trace anywhere on scale without resorting to mechanical adjustments. Thus, the Sanborn system concept means that one amplifier and optical galvanometer combination can be used for all recording applications. The 2KC element is a relatively inexpensive and rugged galvanometer, and with this combination of amplifier and galvanometer, greater deflection is possible with a 5KC than is possible at 5 KC galvanometer by itself. It is no longer necessary to stock an assortment of galvanometers with a variety of frequency responses; nor is it necessary to install and align a new galvanometer for every recording job.

The transistorized "650" Amplifier Assemblies consist of six or eight channels of amplification with a common power supply. There are four types of "650"

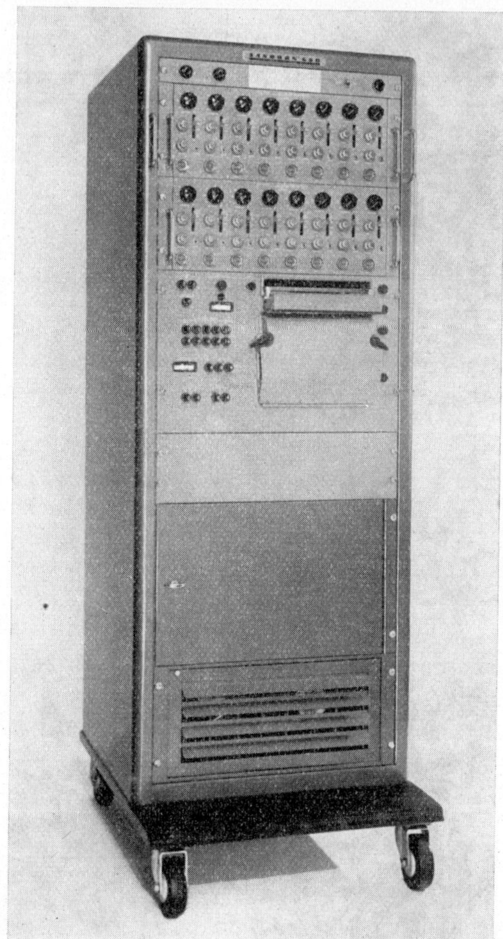

Fig. 3.57. The Sanborn 650 optical oscillograph

Amplifiers available: A high gain amplifier (2.5 mv per inch); a medium gain amplifier (50 mv per inch); a low gain amplifier (0.5 volt per inch); and a carrier amplifier (0.5 mv per inch). Amplifier components for each channel are mounted on a single plug-in circuit board. Controls for each channel are located on a common front panel. A quickly accessible transfer chassis at the rear houses the electrical components, on plug-in boards, which are required for proper matching

144

of the amplifier to optical galvanometer elements. A current feed-back provision on this card obviates the requirement for temperature-controlled galvanometer magnet blocks. Response of each galvanometer is precisely stabilized. These plug-in compensation boards may be readily changed to provide compensation for almost any optical galvanometer. Therefore, existing installations can readily be brought up to date by taking advantage of the integrated-system concept offered by Sanborn.

Recordings on the Sanborn "650" Recorder are made on eight-inch ultraviolet sensitive, self-developing photographic paper which is easily loaded in normal room light through a hinged front panel door. Pushbuttons permit convenient selection of nine paper speeds from 0.25 to 100 inches per second. A fluorescent lamp speeds development so that the trace is fully developed within milliseconds after exposure to normal room light. Traces may overlap at amplitudes up to eight inches. Additional features include: full width time lines, amplitude lines (removable over part or all of paper), sequential light beam interruption for trace identification, event marker, and a lamp power control (monitoring total lamp power extends the life of the lamp). All recorder controls may be operated remotely.

SAVAGE AND PARSONS EQUIPMENT. Savage and Parsons Limited, Watford, Hertfordshire, manufactures a print-out oscillograph described below.

SAVAGE AND PARSONS DIRECT-WRITING RECORDER. The Type RG 32–12/15 recorder, illustrated in Fig. 3.58, employs an ultra-violet light source, and uses pencil type galvanometers focused on to 6-inch wide ultra-violet sensitive recording paper.

It has been specifically designed as a 12 channel unit although 15 galvanometers are normally fitted in the magnet block. The three extra galvanometers are intended

Fig. 3.58. The RG 32—12/15 direct writing oscillograph recorder

for use as time, or event markers, or datum reference lines, but may be used as additional signal channels if desired. A two-position switch on the front panel permits two of the three extra galvanometers to be fed internally with a timing mark derived from the mains input frequency, or to be fed with an external signal.

The recording paper used is six inches wide ultra-violet sensitive (e.g., Kodak R. P. 12), exposed by a high-pressure mercury-vapor lamp.

The recorder can take a roll of up to 200 feet of recording paper, loaded through a door in the front panel. Once the paper is placed against the moving drive roller, it will feed through automatically.

Permanently meshed gears in the transmission provide eight speeds, giving a range of 1800: 1. The following speeds may be push-button selected from the front panel:

150 inches/second and/minute.
50 inches/second and/minute.
15 inches/second and/minute.
5 inches/second and/minute.

The push-buttons control solenoid-operated clutches. The selected speed is shown by indicator lights adjacent to the appropriate push-buttons. If more than one speed selector button is depressed at once the slowest speed is automatically engaged.

The paper-drive roller can be started and stopped by operating push-button switches on the front panel, or by remote push-buttons which may be selected by a switch on the recorder front panel.

A footage selector has been incorporated which will automatically stop the paper drive after a pre-determined length of paper (2, 5, or 10 feet) has been exposed. There is also provision for running continuously.

The optical system is pre-focused and requires no adjustment. The only adjustment provided is for positioning of the arc in the lamp laterally so that maximum intensity is obtained on the galvanometers. This should rarely be needed except when changing lamps. It is also necessary to check the lamp current periodically. This may be achieved without interrupting the lamp by using the monitor socket provided. Access for both of these adjustment is provided at the rear of the case. Fig. 3.59 shows the recorder with the cover removed.

The galvanometers are easily accessible through the top of the case and special clamping arrangements allow easy and rapid changing.

The spots formed by the galvanometers on the recording paper are visible through a window in the front panel. This makes positioning of the galvanometers simple and allows a check to be made on signal amplitude.

Since there is no mechanical limitation to the lateral movement of the galvanometer traces, unwanted channels can be directed off the edge of the recording paper, thus leaving it clear for those channels with signals.

The maximum voltage allowed between any two galvanometers and between any galvanometer and earth is 500 volts.

146

Since the galvanometer traces may sweep the whole width of the paper it is desirable to have some means of identifying each galvanometer trace. This is achieved by the inclusion of a trace interrupter which is directly geared to the paper drive roller. This device interrupts the light from each galvanometer sequentially, the gap in the trace being approximately 1/16 inch and the interruptions occurring at approximately 2 feet intervals independent of paper speed.

Fig. 3.59. Savage and Parsons recorder with cover removed

Savage and Parsons manufactures the following related equipment for use with the RF 32–12/15 recorder:

1) Pencil type galvanometers for use with Direct Writing Oscillograph; 2) Driver amplifier DA. 35/1 for driving pencil galvanometers;

3) Power unit PU. 35–1/3 for stabilized power source; and 4) A gage selector for use with strain gages.

S. E. LABORATORIES OSCILLOGRAPH RECORDERS. The four models of recorders manufactured by SE Laboratories (Engineering) Ltd., 606 North Feltham Trading Estate, Feltham, Middlesex, are described in the following paragraphs. The model SE 2800 is illustrated in Fig. 3.60.

General Description. The SEL product line of UV galvanometer recorders uses a high intensity light beam generated from a high-pressure mercury vapor lamp to record on print-out paper. The recording galvanometers are of the "pencil plug-in" type, and for zero alignment the whole body is rotated rather than the suspension. (The galvanometer is illustrated in the galvanometer section at the beginning of this chapter.) As the body diameter of the galvanometer is only 1/8 inch, as many as 50 channels of information can be recorded on some models.

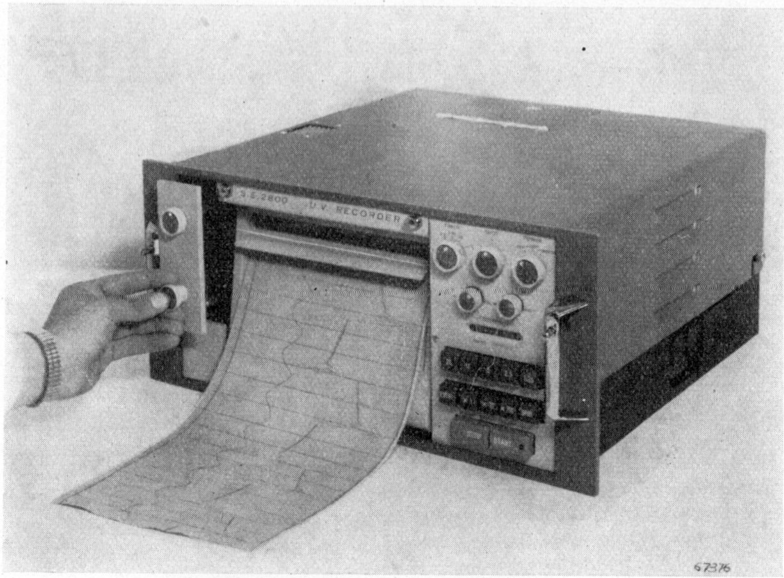

Fig. 3.60. SEL Model 2800 oscillograph

All SEL recorders will accommodate print-out paper having a minimum width of 60 mm; maximum width varies from 6 inches to 12 inches on various models. All SEL oscillographs feature front-panel paper loading, accomplished by a quick-release catch that allows the hinged, flush-mounted cassette to be lowered.

A choice of 15 paper speeds can be selected and changed during recording by means of front-panel controls. Selectable speeds range from 3 inches per minute to 80 inches per second. A paper speed "jump" switch for X1, X10, and X100 speeds is mounted on the control panel, and remote-control provision is available on some models. Record length, independent of time may be selected from a "shot length" front panel-mounted control.

A trace intensity control permits trace velocities in excess of 50,000 inches per sec to be recorded. Records may be made with or without reference grid lines. The separate grid-line intensity control can reduce intensity or eliminate the grid lines completely. The UV lamp is mounted in a plug-in assembly so constructed as to allow lamp focusing to be carried out from the front panel.

An electronically-controlled flash-tube timing system prints timing lines across the full width of the paper. There is a choice of four time intervals selected by a front panel-mounted control with facilities to accentuate the tenth pulse, and provide paper speed identification. In addition, the flash tube may be operated at a repetition rate of up to 250 pps from any externally generated frequency. When information is "played back" from magnetic tape, the tape timing signals may be used to drive the SEL timing system, thus achieving true time coincidence. Three types of 100 cycles per second timing generators are available on all models except the SE 2005.

For ease of identification of individual and overlapping traces, a light beam interruptor is provided.

A paper-length indicator gives a direct percentage reading of paper remaining on the feed spool. A "paper low" warning lamp signals the operator when the indicator reaches 0.1. A microswitch automatically stops the main drive when the paper on the spool reaches zero.

Recorder ancillaries include heated magnet blocks, timing units, oscillators, takeup unit, event numbering, and 24 vdc recorders.

Siemens and Halske Equipment. The recorders described in the following paragraphs are manufactured by Siemens & Halske AG, Karlsruhe, W. Germany.

The firm also makes the Oscillomink, CRT recorders, and processing units described in other chapters.

Oscillogrand Recorder. This instrument, shown in Fig. 3.61, is a six-channel, or alternatively a twelve-channel oscillograph recorder, that uses 120 mm paper. The Oscillogrand allows the option of two different recording techniques made possible through the use of separate magazines—one is a continuous drive magazine and the other is a drum-type magazine.

Galvanometers. The Oscillogrand accommodates up to six or twelve galvanometers. The galvanometers are small diameter plug-in units. Up to six zero mirrors can be inserted in the galvanometer assembly for recording the zero lines.

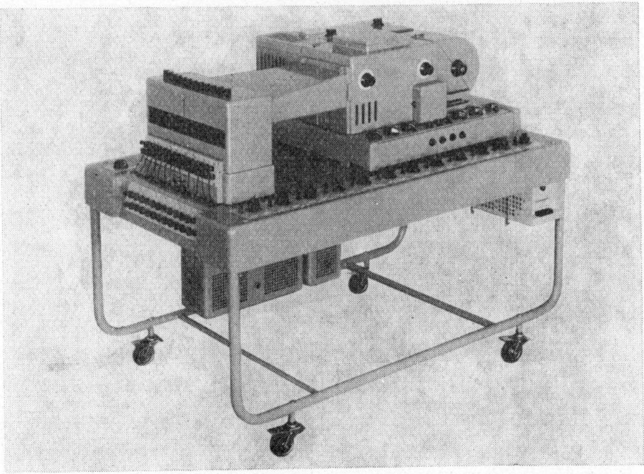

Fig. 3—61. Siemens & Halske Oscillogrand oscillograph

All galvanometer types use the universal adjusters to permit the voltage measuring range to be extended up to 500 v and the current measuring range up to 10 amps. The control on each adjuster allows the measuring potential to be removed from both terminals of the galvanometer and also permits polarity of the latter to be reversed.

The user has the option of using a high intensity or a low intensity light source. To attain sufficient exposure at the highest recording rates possible with this equipment (above 1,000 meters per sec, approximately 40,000 inches per second) a mercury-vapor extra high-peessure lamp with an automatic adapter for current regulation may be chosen as a light source. For lower recording rates it is possible to use an automobile-type headlamp bulb. The two lamps can be interchanged with only a few hand motions. The optical system, shown in Fig. 3.62, consisting of condenser lenses, adjustable slits, and adjustable mirrors, is designed for six or twelve light beams.

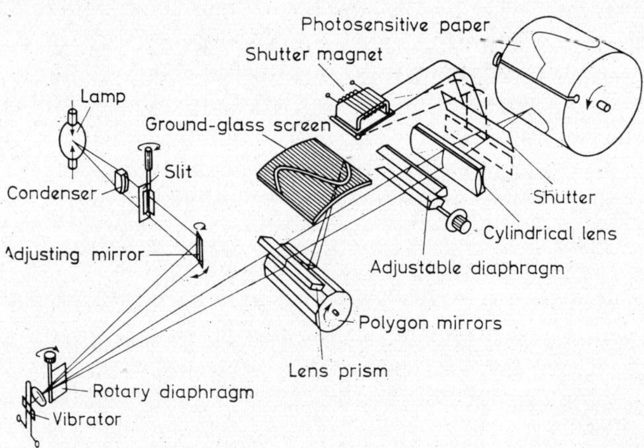

Fig. 3.62. Optical systems of the Oscillogrand oscillograph

The optical viewing system comprises a lens prism, a twenty-sided polygon mirror for scanning with respect to time, and a ground-glass screen. Centimeter calibrations in the center of the screen serve for measurement of deflections, while twelve marks on its right edge are for adjustment of the zero lines.

The optical recording system comprises an adjustable diaphragm, an achromatic cylindrical lens, and a high-speed shutter with diaphragm controlled by an extremely fast-acting electromagnet.

The recording light spot is produced by the following imaging process: the condenser forms an image of the arc or coiled lamp filament on the mirror of the galvanometer to obtain maximum brightness. An image of the slit is projected from the lens window of the galvanometer on to the photo-sensitive paper and "concentratted" by the cylindrical lens in that it images the height of the galvanometer mirror on the paper. The same applies with regard to the ground-glass screen.

150

The Oscillogrand is equipped with a dual drive consisting of a d-c shunt-wound electric motor (1,200 to 6,000 rpm) and a three phase synchronous induction motor (3,000 rpm). The gearing is designed for four speeds. The speeds can be changed by way of electromagnetic clutches with the motor running. Shifting is possible at any time without the motor first having to be shut down. A tachometer gives a direct reading of the prevailing paper feed rate.

Magazines. Two different types of magazines can be used: 1) a continuous drive magazine for long recordings up to 30 m, (approximately 100 ft.), in length at speeds selectable between 8 mm/s and 2.5 m/s, (approximately 0.3 to 100 inches per second); and 2) a drum-type magazine which permits speeds up to 10 m/s, (approximately 400 inches per second) but with the length of recording limited to 600 mm (approximately 2 ft.).

With the continuous drive magazine, recording begins some 5 ms after the start of the feed roller. A time relay enables the recording to be terminated automatically at the end of a set period adjustable between 40 ms and 2 sec (recording-time limitation). As an alternative to a time limit it is also possible to limit the length of the oscillogram to 0.15–0.30–0.45 to 1.5 m meters (recording-length limitation).

Prior to each recording a serial number is automatically projected on to the photo-sensitive paper and can at the same time be read externally on a counter. The end of the oscillogram is also mechanically marked automatically with a pressure pin operated by an electromagnet.

The drum-type magazine secures an entirely even time scale on the oscillogram because the photo-sensitive paper reaches the rated speed before recording begins. In addition, the drum-type magazine has the advantage that it allows the same section of the chart to be exposed more than once (multiple-exposure recording). Quite apart from saving paper, this method offers excellent opportunities for comparisons.

The recording drum has a circumference of 600 mm (approximately 24 inches). The drum will also accept strips of paper only 200 and 400 mm in length. Feeding rates between 3 cm/s and 10 m/s, (approximately 1.0 to 400 inches per second) i.e., four times as fast as those with the continuous drive magazine, are possible. Paper width is 120 mm for both magazines.

Recording Technique. When the drum-type magazine is used, the recording will not start until the shutter opens (approx. 5 ms response). With the continuous drive magazine, the shutter will be opened and the high-speed clutch actuated simultaneously. With instantaneous takes of periodic phenomena and also with controlled takes using the drum magazine, the shutter opens the instant that the point where the paper is clamped to the drum passes the entrance slit for the light beams, and closes again—irrespective of the feeding rate—after one revolution of the drum.

Controls. The Oscillogrand can be remote-controlled for all recording functions. The fact that the shutter can also be operated when motor and magazines are not running introduces the possibility of saving paper by taking stills. For instance, for calibration purposes it is possible to record the position of stationary light spots or the crests of curves. It is further possible to expose the oscillogram section by section at low speed and to choose at option the time sequence of dc throws or periodic phenomena to permit convenient comparisons.

Time Marking. A pulse sender, consisting of two squarewave generators working in synchronism with frequencies in a ratio of 1:10, are superimposed and fed to one of the galvanometers. On the resulting time scale, each tenth bar division is accentuated.

SIEMENS & HALSKE OSCILLOMAT. The Oscillomat recording oscillograph, shown in Fig. 3.63, is available as either an 8-channel or 12-channel instrument. Recordings are made on either 12-cm or 20-cm (approximately 4–3/4 or 7–7/8 inches)

Fig. 3.63. Siemen & Halske Oscillomat recorder

wide paper contained in a 100 foot delivery (feed) magazine. During exposure the paper is fed at speeds which can be continuously adjusted from 0.0033 to 33 feet per second (in two ranges with an intermediate gear). The unit has "head-start feeding" which permits rapid acceleration to the maximum preset speed. The exposed paper is collected in the takeup magazine. Up to 33 feet of paper can be accepted in the takeup bellows located at the bottom of the recording section (100 feet capacity is available on special order).

The exposures can be electrically triggered and remote controlled—by the phenomenon under examination itself if required. The unit features automatic preparation for the next exposure, and has the capability for making at least two exposures per minute.

152

In the galvanometer section, the standard light source is a 15-volt, 60-watt incandescent lamp, which is automatically adjusted for exposure. For paper speeds exceeding 500 feet per second, provision is made to replace this lamp with a mercury-discharge tube. The optical path is from the lamp through condenser lenses, through

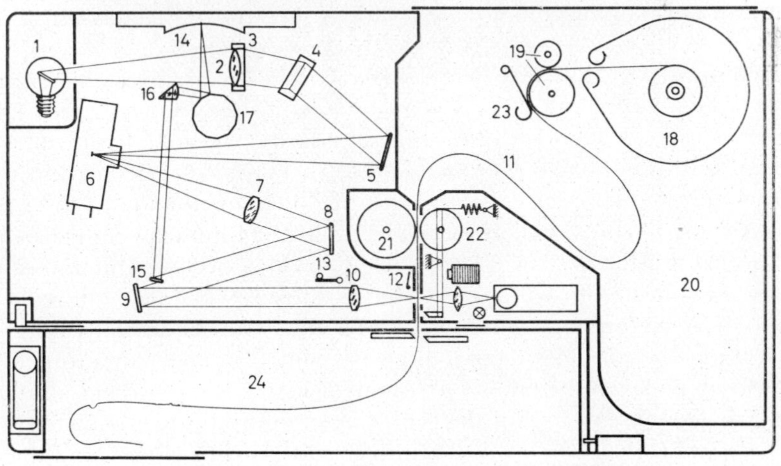

Fig. 3.64. Siemens & Halske Oscilloport recorder

slit diaphragms, to the galvanometers, and from them on via a deflecting system to the cylinder lens. This focuses the beams from all the galvanometers on the recording paper. Split-off portions of the beams are projected for observation by a motor-driven twelve-sided polygon mirror to a ground-glass screen. A hooded viewing mirror enables the screen to be seen by the operator from a sitting position.

The Oscillomat is supplied with a frequency generator, time marking vibrator, and optical system for recording a time reference. Paper consumption is indicated by a counter.

A four-digit counter photographically marks each exposure which is also indicated on the control panel. In addition, the end of each exposure is marked on the paper so that it may be cut before development.

THE SIEMENS & HALSKE OSCILLOPORT RECORDER. This portable oscillograph, shown in photo and functional layout form in Fig. 3.64, consists of two main sections: the recording section with the magazines and the paper drive system; and the optical section with the lamp and galvanometers.

In operation, paper in the feed magazine 18, is metered by the roller 19, through the paper drive-rollers 21 and 22, so that it travels past the exposure slit and piles up in loose folds in the take up magazine 24.

Instantaneous start is obtained by pinching the two feed rollers together with an electromagnet. Thus, the maximum "head-start" speed of 10 meters per second (390 inches per second) is attained in a few milliseconds. Automatic stopping and marking for a pre-set chart length is provided by an electrical interlock between the control and the pressure roller magnet.

Direct-feed, controlled by start-stop switches, is built in for speeds up to 3 meters per second (approximately 118 inches per second). Shift levers provide 8 speeds to be selected.

The feed magazine can accommodate rolls of 10 cm (approximately 3 15/16 inches) wide recording paper in lengths to 20 meters (approximately 66 feet). The takeup magazine can normally accommodate oscillograms up to 5 meters (16–1/2 feet) in length.

The Oscilloport has a time scale marker recorded by a galvanometer. This fifth channel may be used for other purposes if time marking is not required. The manufacturer states that the accuracy of the time standard is within ± 1 percent, with temperature fluctuations of 0 to 50 °C, and voltage variations of ± 20 percent.

The Oscilloport either can be triggered by the phenomenon that is to be recorded, or can trigger the phenomenon at the start. It has the following controls and provisions: 1) a push-button ON switch, the head of which becomes luminous when the equipment is connected to voltage; 2) a push-button two-stage voltage control for the motor; 3) an automatic recording counter that also sequentially numbers each record; 4) a manual-automatic lamp intensity control; and 5) an automatic cut-off for the exposed paper in the takeup magazine. The recorder also has a safety provision to prevent reoperation until the following conditions are completed: the numbering process must be completed, the stop circuit must be closed, and the start key must be back in its rest position.

The general purpose oscillograph recorders that have been described are often used for unusual applications as discussed in Chapter One. Some of these applications could classify the recorder as a special purpose instrument when used for a single purpose only. However, many single-purpose recorders also exist, the most significant being the fault-type recorder and the seismograph. These instruments are described in the following sections.

FAULT-TYPE OSCILLOGRAPH RECORDERS. The fault-type automatic oscillograph is a special recorder specifically designed to meet the needs of the electric power industry. The instruments can be used at various points on power systems to record magnitudes and duration of faults and transients. They can also be used for monitoring the operation of relays and circuit breakers, including staged testing. The records indicate the complete sequence of operations on all protective equipment. References[3.13, 3.14, 3.15] and [3.16] describe fault-type recorders and typical applications.

Typically, when a fault occurs, i.e., overcurrent, overvoltage, or undervoltage, a monitoring relay senses the variation and starts the automatic recorder within approximately 3 milliseconds of the occurrence.

As power faults occur, high frequency galvanometers record magnitude and duration of the fault and the operational sequence of protective equipment.

The month, day, hour, minute and second at which a record is taken is printed on the chart in digital form. Upon termination of the fault, the recorder automatically stops, resets, and returns to stand-by position to await a new abnormality. This type of record provides the electric utility industry with a detailed, photographic account of disturbances and their source. These are the most reliable records possible. Their value can be measured in terms of both dollars and system efficiency, and help to account for the reliability that has been developed in the electric utility industry.

THE HATHAWAY RS-9(30) AUTOMATIC OSCILLOGRAPH. Illustrated in Fig. 3.65 is the RS-9(30) oscillograph that automatically charts disturbances of electric utility service. The 12-inch paper, in any length up to 300 feet, records up to 32 independent channels at 3 or 12 inches per second chart speed. Automatic recording of either 12 or 125 volt dc models starts approximately 3 milliseconds after the operation of any of the 4 monitoring relays. Hathaway Group 155 galvanometers are recommended. These have a frequency response 800 cycles per second, 1.1 ohms resistance, and 2.2 mm/ma/m sensitivity. Other types are available for specialized recordings including watt galvanometers.

Centralized front panel controls include operations counter, viewing screen, in-use lamp indicator, ready light and all necessary controls for manual or automatic operation of the oscillograph. A flange is provided so that the recorder can be mounted on switchboard panels; the recorder projects approximately 8 inches from the front of the panel. The weight of the unit is 155 pounds.

A time stamp automatically indicates the month, day, and time in hours, minutes and to the nearest second, AM or PM. Space is provided for writing station identification and notes on the record.

Starting relays have a chart-speed selector switch. All records are made at a chart speed of 12 inches per second for the first 10 to 15 cycles (at 60 cycles per second). After this period, the chart will be driven at 3 or 12 inches per second as determined by the setting of the selector switch. The paper is driven by a positive gear train powered by a governor controlled dc series motor. After the initial starting period, chart speed is constant, with an accuracy of ± 3 percent.

THE HATHAWAY QS-12 AUTOMATIC OSCILLOGRAPH. The model QS-12 oscillograph, shown in Fig. 3.66, is similar in appearance to the RS-9(30) recorder. It weighs approximately 125 pounds, uses 12-inch-wide paper, and begins operation approximately 3 milliseconds after operation of an initiating relay. The recorder normally is supplied with from one to twelve Type OA, Group 155 galvanometers.

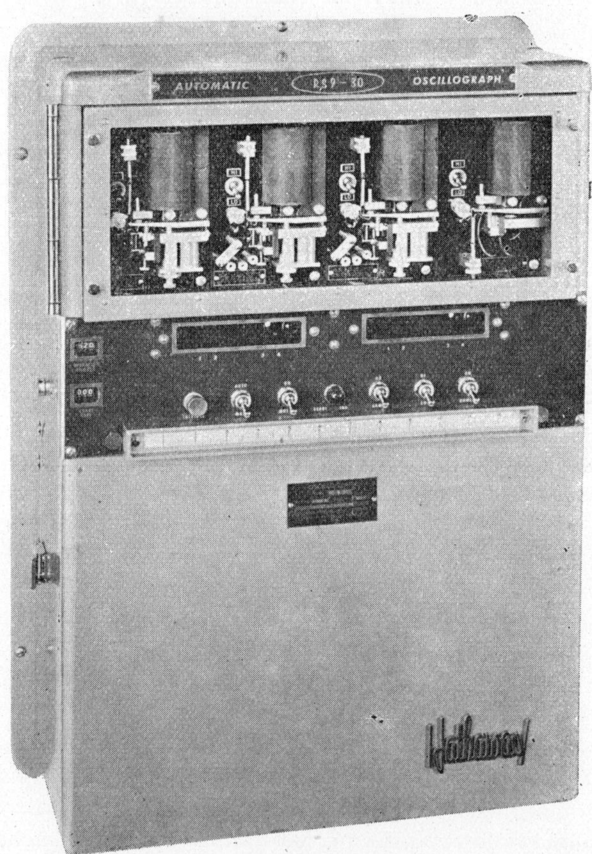

Fig. 3.65. Hathaway RS9(30) automatic fault-type oscillograph

However, other types of galvanometers including watt galvanometers can be installed without alterations.

The power requirements are handled by a power supply furnished with the unit, 115 ac or 125 volts dc and 12 volts dc.

The recorder is furnished with from one to four initiating relays (a maximum of four relays can be installed in the oscillograph.) Others may be installed external to the oscillograph unit. Controls include footage counter, operations counter, view-

Fig. 3.66. Hathaway QS-12 automatic fault-type oscillograph

ing screen, ready light, all centralized on the front panel. An automatic shut-off turns the oscillograph off when the paper supply is exhausted.

The recording paper is driven at 12 inches per second by a synchronous stepper ac motor through a positive gear train.

WESTERN ELECTRODYNAMICS OS–10 FAULT RECORDER. The OS–10 Automatic Fault Recording System, manufactured by Western Electrodynamics, Inc., Colorado Springs, Colorado, is a multichannel oscillograph for recording electric power system faults. Its features include: a fast-start mechanism that begins recording less than three milliseconds after the faults occur; a light source that burns only during recording; modular design that enables easy assembly of those units needed to meet particular requirements; high overload capacity switchboard type shunts capable of withstanding 300 amperes for 15 seconds and 100 amperes for 1 minute; all tapped resistances in the galvanometer control units; leak-proof galvanometer of single opening cell-box construction sealed with an "O" ring; interchangeable and adjustable starting relays; close speed regulation; and 50 times normal current overload capacity on the starting relays. Optional equipment available includes an Anticipator which produces a record of several milliseconds before the fault occurs; solid-state starting relay with closing times of one tenth of a millisecond; record calibrator which automatically calibrates each galvanometer through the galvano-meter control unit with an accuracy of $\pm\frac{1}{2}$ percent.

The OS-10 is designed to meet the electric power industry's preference for oscillo-graph channels to be in multiples of seven, and is usually ordered in 7, 14, 21 or 28 channels though any number may be supplied if desired. Up to four mechanical or solid-state starting relays are available. The OS-10 can be supplied to operate from either 125 volts dc or 115 volts ac. Because of its unique "building block" modular design, any desired individual requirements can be met.

Hub of the automatic fault recording system is a continuous drive multi-channel oscillograph, shown in Fig. 3.67. Triggered by the fault signal, a high speed clutch is engaged and powers the chart-drive roller from the continuous running electric motor. Paper transport speed is 10 inches per second. The roll of sensitized paper, 200 feet long and 10 inches wide, is loaded into the recording compartment through a hinged cover in the recorder panel. The time and date are superimposed at the end of each record.

As records are produced, the record receiver unit winds the exposed paper onto a spool in a cartridge. Whenever records are to be processed, the entire cartridge is removed to the darkroom.

Anticipator Unit. One of the more important features of the Automatic Fault Recording System is its ability to provide a permanent record of what happens a few milliseconds before a fault occurs! This is possible with the optional Type SA-1 Anticipator Unit.

This unit contains devices which cause a delay in the signal reaching the galvano-meters, therefore, the recorder can be started before the phenomenon that caused the start has begun to deflect the galvanometer.

OF-2 MINIATURE RECORDER. The Type OF Miniature Oscillograph is a very small recording oscillograph specially designed for those applications which require an instrument of small size and light weight. It is particularly suitable for installation in missiles, space devices, aircraft, torpedoes, in fact anywhere that size and weight is a consideration.

The OF-2 records on continuously-driven photosensitive chart, which is expos-ed by beams of light reflected from miniature pencil-type oscillograph galvanometer mirrors.

Fig. 3.67. Western Electrodynamics Model OS-10 fault recorder

The entire chart-transport system, containing the supply spool, the takeup spool, the drive motor and the gear train, are mounted in an aluminium case which serves as a cover to the main case whenever it is desired to remove the records or insert a fresh supply of chart. The main case contains the galvanometers, the galvanometer resistor bank, the light source and the optical system.

An unusual feature of the Western Electrodynamics OF Recorder is the provision for adjusting each galvanometer individually, not in banks as in other types. Each galvanometer element also has its own magnet.

The OF-2 has a viewing screen, either integral or removable, depending upon the model. The chart is quickly and easily loaded into the oscillograph. Up to 20 channels, and 100 feet of chart length are available, with chart speed from $\frac{1}{4}$ to 20 inches per second.

OS-30 FAULTOGRAPH. The Western Electrodynamics Model OS-30 Faultograph has been developed to meet the need of power companies for a light weight, fast starting, multi-channel fault recorder. Easily transportable from point to point in an electrical power system, it is especially designed for rugged field use. This entirely automatic fault recorder can be used manually for staged testing; as a portable survey oscillograph; or for unattended fully automatic operation.

Seven galvanometer channels record on a 6-inch strip chart 200 feet in length. Up to four starting relays can be used to start the oscillograph recording in under 5 milliseconds. Galvanometer controls are included, with shunt connections provided at the rear of the cabinet. Connection to power lines is easily and quickly made to a terminal strip on the rear of the cabinet.

The lamp which is used to supply light to the galvanometers is brought to full brilliance in less than 2 milliseconds.

A 115 volt ac motor is normally supplied.

Seismographic Equipment

This special class of instruments is especially designed for recording earth motions. The low-level, low-frequency signals are picked up by special transducers and recorded on high-density-storage recorders, some of which use photographic marking medium. Special purpose support equipment has been developed for seismological exploration, and includes van-mounted portable units and telemetering systems.

CENTURY MODEL 404 OSCILLOGRAPH. Chapter One described a method of geophysical recording utilizing equipment developed by Century Electronics and Instruments, Inc., Tulsa, Oklahoma. The recorder is described in the following paragraphs:

From the Century Model 404 Oscillograph the operator observes the phenomena as they occur. In addition to recording the data, the oscillograph automatically fires the explosive charge. The safety devices incorporated into this unit are so arranged that the utmost in safety precautions are assured. The unit is mounted on a portable daylight developing unit which allows immediate development of the record.

The portable, 25-pound Model 404 is furnished in a dustproof aluminium case measuring 7–1/2 by 9–1/2 by 16 inches. The magazine has a capacity of 100 feet of 3–5/8-inch paper. Paper is transported at speeds ranging from 12 to 16 inches per second, by a governor-controlled spring-wound paper drive. Full-width timing lines are recorded at 0.01 second with each 0.05 and 0.1 second line accented.

A matching amplifier control unit is available. Twelve Model 506 Amplifiers, for 6-volt battery operation, are incorporated, together with ac balancing controls, testing and sensitivity circuits.

GEOTECHNICAL CORPORATION EQUIPMENT. The Geotechnical Corporation, Dallas, Texas specializes in equipment and sevices for the geosciences. Their seismographs include all types for monitoring low-level motion, and for detecting earthquakes and explosions. Low-level signals down to 10^{-20} watts, 10^{-12} amperes and 10^{-9} volts are enhanced and handled; and earth motion down to 0.1 millimicron, 10^{-9} g, and 0.001 cycles per second is reputedly detected on Geotechnical units. The equipment includes seismometers, amplifiers, galvanometers, recorders, and Benioff instruments; telemetering devices for transmission of 1 to 7 channels of low-frequency data over one radio or telephone circuit; van-mounted seismic measurment systems; deep-well seismographs for operation 2 miles underground; auto-processing oscillographs; and film viewers. Portions of the following section are reprinted from Geotechnical publications.

Seismograms can be made by various recording media, including photographic systems. A block diagram for a complete photographic seismograph is shown in Fig. 3.68. The system, using Geotechnical Corporation equipment, has an optional direct-writing monitor. The monitor is the Geotechnical Model 2484 Helicorder, a direct-writing drum recorder that produces a 72-hour chart on 1 foot by 3 foot heat-sensitive paper. Up to three channels are available, and drum speed can be adjusted by changing gears.

The high-density storage capacity, suitable for recording data ranging from dc to 35 cycles per second, is achieved by recording on the revolving drum in compact helical form, just as a lathe cuts screw threads.

CALIBRATION PROCEDURE. Calibration of the system is performed from a remote position at operating gains as high as 2×10^6, without reducing gain or interrupting normal data recording. Fig. 3.69 is a block diagram illustrating the components of the calibration system.

When properly adjusted, the calibrators provide seismometer calibration analogous to weight-lift calibration. In addition, other response data can be obtained. These data include constant-amplitude response and constant-acceleration response.

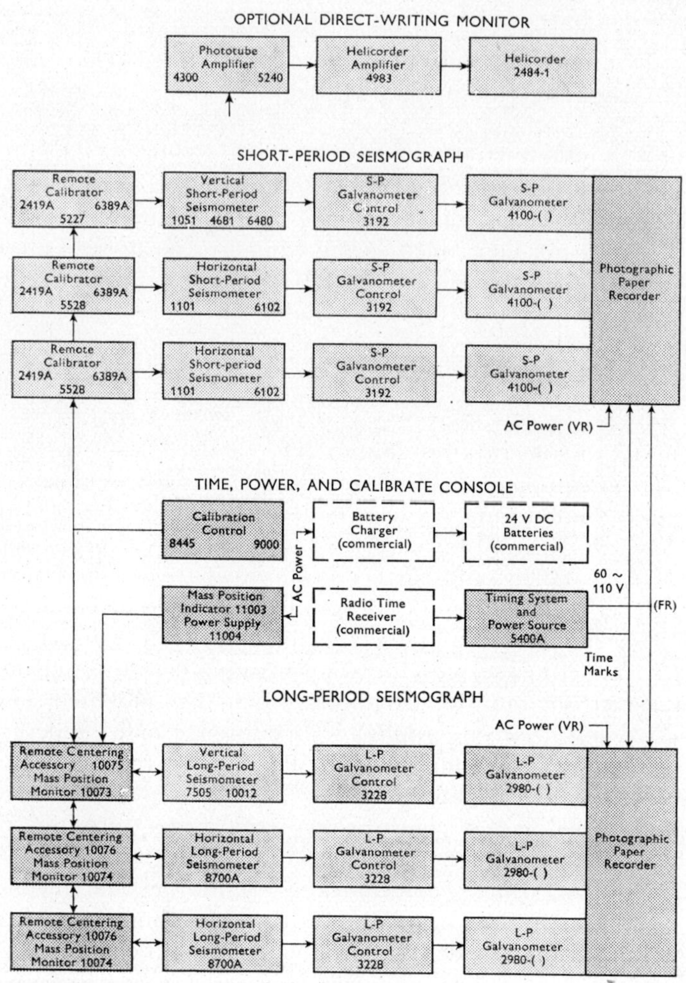

Fig. 3.68. Block diagram for complete photographic-paper seismograph using Geotechnical components

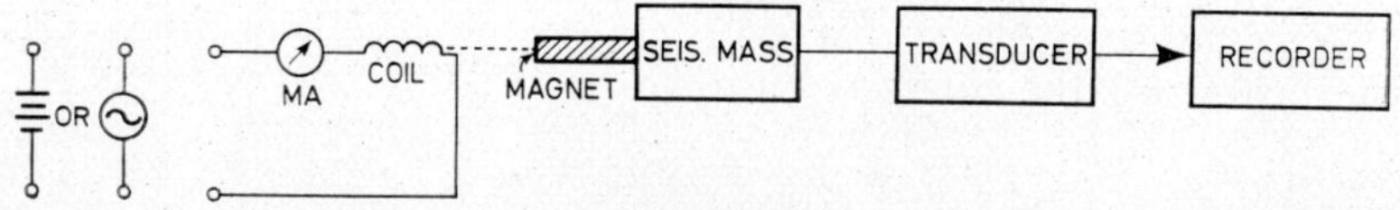

Fig. 3.69. Block diagram of recommended calibration procedure using Geotechnical electromagnetic calibrator kit

The following equipment is required for operation of the calibrator:

Current source to 10 milliamperes
Milliammeter
Function generator (for frequency calibration)

Adjustment of the calibrator is easily accomplished by determining the amplitude of a pulse necessary to produce the same recorder deflection as produced by a weight lift. The system megnification can be determined with an electromagnetic calibrator by dividing the recorded pulse amplitude or the peak-to-peak (sine wave) amplitude by the equivalent ground motion the calibrator induces in the seismometer. The equivalent ground motion is a function of the calibrator motor constant, the frequency at which the calibrator is driven, and the current through the calibrator coil. The motor constant is readily calculated using data obtained during the initial weight lift and pulse calibration. A typical calculation is shown below.

$$\text{System magnification} = \frac{\text{output}}{\text{input}} = \frac{A}{Y} \tag{1}$$

where

A = amplitude on recording (meters)
Y = equivalent earth motion (meters).

It can be shown that:

$$Y = K \, \frac{i_s}{f^2} \text{ meters} \tag{2}$$

where

$$K = \frac{G}{M4\pi^2}, \text{ a constant} \tag{3}$$

and

$$G, \text{ the motor constant,} = \frac{\dfrac{X_{1p}}{X_{1w}} 980 W_t}{i_p} \times 10^5 \frac{\text{newton}}{\text{ampere}} \tag{4}$$

Combining (1), (2), (3), and (4),

$$\text{system magnification} = \frac{A X_{1w} i_p \pi^2 M f^2}{245 \times i_p i_s W_t} \times 10^5 \tag{5}$$

where

X_{1w} = initial recording deflection from weight lift, meters
X_{1p} = initial recording deflection (pulse), meters
i_p = calibration coil current (pulse), amperes
i_s = calibration coil current (sine wave), amperes
M = seismometer suspended mass, kilograms
f = frequency of calibration coil current, cps
W_t = weight lifted, grams

Using values obtained with a Model 4681 seismometer, equation (5) showed a magnification of 2.98×10^4. Testing the same seismometer on an accurate shake table indicated the system magnification was 2.97×10^4.

THE GEOTECH FILM RECORDER. The Geotech Film Recorder, shown in Fig. 3.70, has provision for four separate recording traces employing four strips of 35 mm motion picture film. It employs the standard short-period galvanometer with a

Fig. 3.70. Geotechnical Model 1301 film recorder

natural period of 0.2 second and a sensitivity of 3×10^{-8} amp/mm at one meter. (Other galvanometers available on request.) For time comparison one trace may be equipped with a special galvanometer for recording radio time signals.

Mechanical Features. Each of the four drums is individually reset against a stop for uniform recordings. The unit is entirely enclosed so that it may be operated in a lighted room except when changing film. The rear cover may be removed and replaced with a loading bag. This allows the film to be changed without requiring the use of a darkroom. Fungus and moisture-repellent paint is used throughout the instrument. Bolts, nuts, and miscellaneous hardware are of stainless steel or plated brass. The optical system is completely enclosed. The recorded trace is a developed helix, 71 feet (21.6 m.) in length for a 24-hour run.

Electrical Features. Each trace of the recorder has an independent set of operational controls including recording lamp current, galvanometer damping, seismometer damping, and magnification. The magnification control is a calibrated attenuator with 2 db steps from 0 to 42 db with an "off" position. The setting of the attenuator in no way affects the damping of either the recorder or the seismometer.

The instrument measures 20 by 22 inches on the base, is 16 inches high, and weighs 140 pounds.

GEOTECH DEVELOCORDER. The Develocorder records on 16 mm film using either a cathode-ray tube or Geotech galvanometers. After exposure the film passes through the film-processing compartment where it is automatically developed, fixed, washed, and dried. The operator can then cause the film to either loop out of the instrument into a film bag, or to continue through the projector to the take-up reel. When projected, the data are magnified 10 times on the 6 inch by 17 inch viewscreen, as shown in Fig. 3.71.

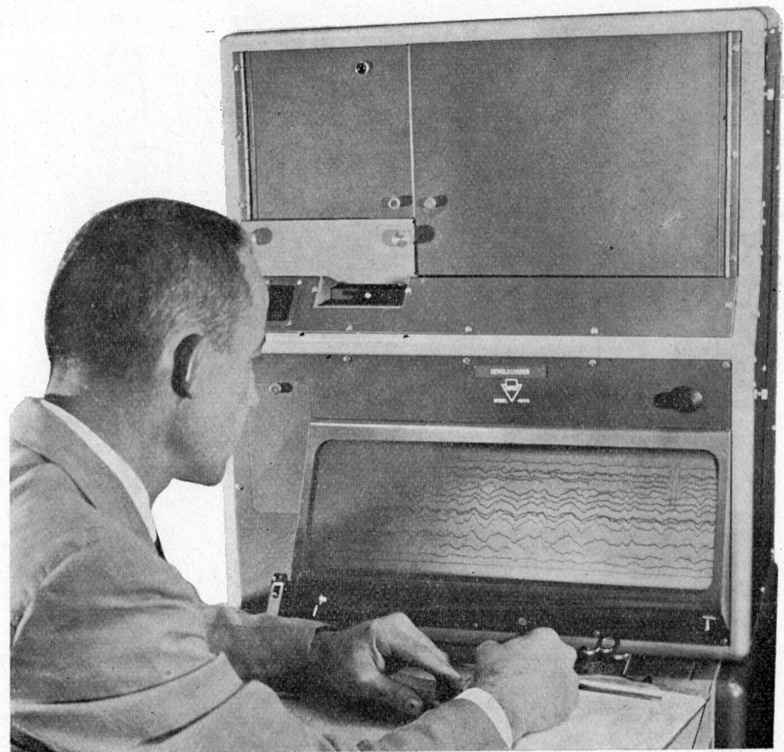

Fig. 3.71. The Geotech Develocorder

A variable-speed, two-directional drive facilitates data viewing and study.

The viewscreen, film-speed control, and film direction lever are conveniently located on the front panel.

Conventional developer and hypo are stored in small containers in the instrument. The diaphragm pump replenishes the developer and hypo, which are allowed to drain continuously from the processing compartment. Washwater is supplied from external plumbing.

The Develocorder is supplied in two types, one which records from galvanometers, and the other which records from a cathode-ray tube.

Model 4000C is furnished with Geotech Miniature Galvanometers to record and view up to 20 channels of low frequency analog data for long periods of time without interruption.

An accessory is available to record data-time numerals on the film. Attenuators, a trace coding unit, and a channel switching unit are also available. Time marks may be fed directly to the galvanometers.

Model 4000A can be furnished with a cathode-ray tube installed or it can be furnished adapted for use with the user's oscilloscope to record and view CRT data for very long periods of time without interruption.

Low-frequency data can be recorded along the film. High-frequency data can be recorded across the film in nested form.

The recording of analog data, on–off data, multi-gun, and frequency spectra suggest but few of the applications possible.

The Develocorder measures 23–3/8 inches wide, 18 inches deep, by 26–3/4 inches high, and weighs 175 pounds. A stand is available as an accessory.

Non-perforated, daylight-load film in 200-foot rolls is the standard length, and is contained within the unit. Other roll sizes are optional, but the 1000 foot roll must be mounted externally. With Recordak #1320 Film, Kodak Liquid X-ray Developer and Fixer are used for processing.

The standard film recording speed is 3 cm/min (approx. 1 3/16 inches/min), providing 17 data hours per 100 feet of film. The projector speed can be varied from 0–360 cm/min (0–142 inches/min).

GEOTECH BENIOFF HORIZONTAL SEISMOMETER. Motions as small as 1 millimicron at 1 cycle per second can be detected with the variable-reluctance Benioff Horizontal Seismometer, Model 1101, shown in Fig. 3.72. The instrument was designed by Dr. Hugo Benioff, Professor of Seismology at California Institute of Technology. Dr. Benioff is a leading designer of modern seismological instruments.

The seismometer is capable of magnification up to 300,000 at one second, and will operate up to 10 miles from the recorder without an amplifier. The unit, which measures 19–5/8 by 22 by 24–3/4 inches, and weighs 431 pounds, has eight separate coils to drive various types of recorders.

The frame of the seismometer consists of two sturdy triangular end plates held by three rigid horizontal frame members. Fixed to the frame members is the permanent magnet of the transducer on either side of which are two vertical laminated cores holding four coils each. These cores and coils are free to move horizontally with the two portions of the mass to which they are fixed. This inertial assembly is supported by lateral steel ribbons. The two cores complete the magnetic circuit except for opposing pairs of 2-mm air gaps on either side of the magnet. Any horizontal movement increase one pair of magnetic air gaps while decreasing the opposing pair. This changes the reluctance, and consequently the flux in the balanced, push-pull magnetic circuit. The changing flux produces emf in the eight separate coils, which are individually terminated.

GEOTECH BENIOFF VERTICAL SEISMOMETER. The Geotechnical Model 1051 Benioff High-Sensitivity Vertical Seismometer, illustrated in Fig. 3.73, has characteristics similar to the horizontal Model 1101 previously described. The base of the in-

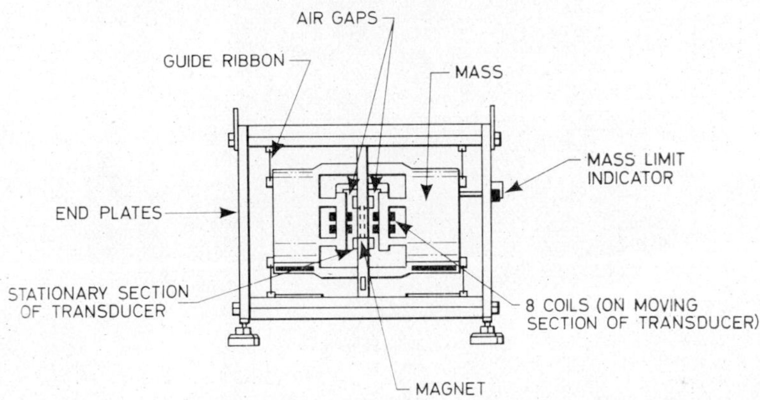

Fig. 3.72. Geotechnical Model 1101 Benioff high sensitivity horizontal seismometer

strument measures 20–3/8 inches on a side, and the overall height is 32–1/2 inches. It weighs 454 pounds.

The frame of the seismometer consists of sturdy top and bottom triangular steel plates held by three rigid vertical frame members. Fixed to the bottom plate is the permanent magnet of the transducers. Above and below the magnet are two horizontal laminated cores each holding four coils that are free to move vertically with the spring-suspended mass to which they are fixed. This inertial assembly is stabilized by lateral steel ribbons. The two cores complete the magnetic circuit except for a pair of 2-mm air gaps above the fixed magnet and an opposing pair beneath it. Any vertical movement increases one pair of magnetic air gaps while decreasing the opposing pair. This changes the reluctance, and consequently the flux in the balanced, push-pull magnetic circuit. The changing flux produces emf in the eight separate coils, which are individually terminated.

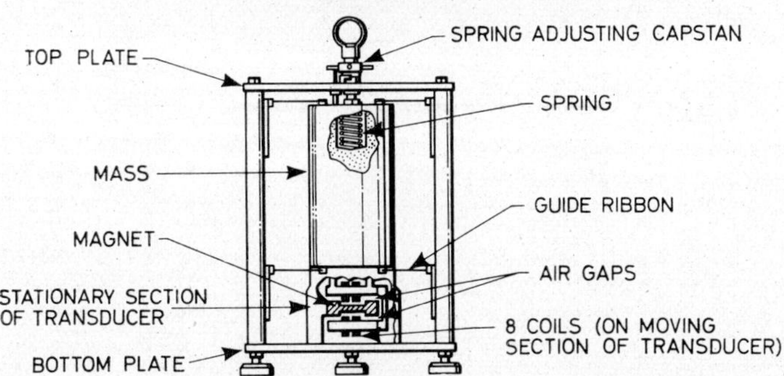

Fig. 3.73. Geotechnical Model 1051 Benioff high sensitivity vertical seismometer

LONG-PERIOD SEISMOMETERS. The vertical Model 7505A, and the horizontal Model 8700C, long-period seismometers produced by The Geotechnical Corporation are shown in Fig. 3.74.

The new, compact Long-Period Seismometers feature ruggedness, accuracy, and stability. A remote calibration coil is provided. Optional accessories for remote mass centering and mass position indication are available.

The seismometers have a cast-aluminum cover and a homogeneous, cast-Mechanite base for rigidity and stability. The alloys of other structural members are carefully chosen and heat treated to minimize thermal instability. Thermal insulation can be provided by covering the instruments with the thick polystyrene container. Both seismometers are the same size, 15–1/2 by 12 by 24 inches. Model 7505A weighs 139 pounds and Model 8700C weighs 120 pounds. Sturdy flexure hinges mounted in tension are used to eliminate friction, minimize hysteresis, and improve stability.

Fig. 3.74. Geotech long-period vertical (*left*) and horizontal (*right*) seismometers

The vertical instrument uses a short, stiff "zero-length" spring. The first mode of vibration of the spring is well above the operating range of the instrument, reducing parasitic noise.

The transducer magnet is fixed to the base of the instrument, and the transducer coil is fixed to the end of the boom. The mass is mounted on the boom. The coil is counter-balanced by a horizontal spring beneath it. Relative motion between the magnet and coil generates a voltage proportional to the velocity of earth motion. The use of Cardan hinges eliminates friction and increases lateral mass stability. Trouble-free electromagnetic damping is employed.

The Geotechnical Corporation also manufactures other seismometer models than the ones described here.

GEOTECH FM TELEMETERING SYSTEMS. Remote long-distance recording, computing, simulating and control are made possible on a practical and economical basis with a Geotech FM Telemetering System.

From 1–8 channels of low-frequency analog data can be transmitted over one voice circuit, coast-to-coast if required, via radio, microwave or telephone. Thus, real-time data-processing is accomplished without the delay and expense of A-D and D-A conversion. Overall system accuracy to ± 1 percent has been realized. Dependability is inherent; Geotech installations have operated with outage-time figures of less than 1 percent over year-round periods.

Transmitting. The system transmits 1 to 8 channels at from 0 to 35 cps with a dynamic range of over 45 db. IRIG subcarrier frequencies with ± 7–1/2 percent deviations are used. Station components consist of a Geotech frequency-division multiplexer and a power supply. Geotech voltage-controlled oscillators are mounted in the multiplexer, one for each channel of data to be transmitted. Each data-signal frequency-modulates the subcarrier output of an oscillator. The multiplexer then mixes the subcarriers and couples directly to a 600-ohm telephone line, radio transmitter, or microwave transmitter.

Receiving. At the receiving station, one Geotech sub-carrier frequency discriminator is required for each channel of data transmitted. The discriminators separate the individual subcarriers by selective filtering, demodulate the subcarriers, and amplify the original data to a level high enough to permit direct recording, without the need for further amplification.

Drift Compensation. If the transmission link introduces objectionable frequency translation, a correction circuit can be added to the system. This circuit uses an oscillator to transmit an unmodulated reference frequency subcarrier. The reference subcarrier is received by a discriminator whose output voltage is proportional to the error between the transmitted and received reference frequency. A translation corrector then chooses the correct proportion of the discriminator output voltage and applies it as a corrective voltage to each of the other discriminators.

Operational Features. A spare channel can be switched in immediately to replace another channel. The input or output of any data channel can be monitored without disturbing the operation of the channel. A built-in calibration system applies a standard calibration voltage to the input of any oscillator without interrupting service on other channels.

LITERATURE REFERENCES

[3.1] HORAK, R. J., "Oscillography," *CEC Recordings*, Vol. 17, No. 1, p19 (1963).

[3.2] *Galvanometer User's Handbook*, Bulletin 7300A, Consolidated Electrodynamics Corp. (360 Sierra Madre Villa, Pasadena, California).

[3.3] How to Select a Galvanometer, HC203, Honeywell Denver Division, (4800

East Dry Creek Road, Denver 10, Colorado), 1962.

[3.4] LEGETTE, M. A., *The Theory of Recording Galvanometers*, Bulletin 1582, Consolidated Electrodynamics Corp. (360 Sierra Madre Villa, Pasadena, California).

[3.5] SHIVER, R. A., "The Photographic Oscillograph," *Electronics World*, Vol. 68, No. 1, p52 (July 1962).

[3.6] STAUFFER, N. L., "Oscillograph Optical Systems,"*Instruments and Control Systems*, Vol. 38, No. 7, p88 (July 1965).

[3.7] JACOBS, J. H., "Rapid Access Methods for Oscillograph Recording," *Instruments and Control Systems*, Vol. 34, No. 7, p1236 (July 1961)

[3.8] JACOBS, J. H., "Very Rapid Processing of Oscillograph Records," *Journal of Photographic Science*, Vol. 6, No. 4, p120, (1958)

[3.9] JACOBS, J. H., "Flash Processing: A Rapid-Access Photographic Technique," *Photographic Science and Engineering*, Vol. 1, No. 4, p156 (1958).

[3.10] ERICKSON, R. D., *High Pressure Mercury Lamps used in Visicorder Oscillographs*, Honeywell Denver Division, (4800 E. Dry Creek Road, Denver 10, Colorado).

[3.11] *Compact Arc Lamps*, Hanovia Lamp Division, Engelhard Hanovia, Inc. (100 Chestnut Street, Newark, New Jersey).

[3.12] CARLSON, F. E. and CLARK, C. N., "Compact-Source Arc Lamps," *Applied Optics and Optical Engineering*, Vol. 1, p73, Academic Press (New York, New York) 1965.

[3.13] HORTON, L. S., *Experiences with an Oscillograph on a Transmission System*, Presented at NELPA Engineering and Operating Conference, Salt Lake City, Utah, April 23, 1959. Reprints available from Hathaway Instruments, Inc., (5800 E. Jewell Avenue, Denver 22, Colorado).

[3.14] FEIN, M. J. and CARTER, L. E., *A 32-Channel Automatic Oscillograph for Utility System Monitoring*, presented at conference on Protective Relaying, May 4–5, 1961, Georgia Institute of Technology, Atlanta, Georgia. Reprints available from Hathaway Instruments.

[3.15] O'BRIEN, R. J., *Résumé of the Applications of the Automatic Oscillograph, System Qualities Measured and Interpretation of Results*, Presented at the Electric Council of New England Relay Committee meeting, Waltham, Massachusetts, April 11, 1962. Reprints available from Hathaway Instruments, Inc.

[3.16] TURNER, J. A. and VANDERGRIFT, J., *Electrical South*, December 1962. (Reprints available from Hathaway Instruments Inc.)

IV ELECTRON-BEAM OSCILLOGRAPHY

The chief advantage of electron-beam recording over light-beam and other recording methods is that the electron displacement is virtually inertialess. For this reason, very high-frequency recording is possible, and performance is believed to be the highest of any existing recording method. Recording can take place either directly by exposure from electrons, or indirectly by photographing the fluorescent trace on the face of a cathode-ray tube. The latter method is used more commonly and is exemplified by photographic records of oscilloscope traces. In practice, the sensitized material is exposed by luminescent emission from the cathode-ray tube (CRT) when its phosphor coating is activated by electrons.

The oscilloscope, developed from the original cathode-ray tube used in the physics laboratory, has come into popular use as a visual indicating instrument. Although visual observation is sufficient for many purposes, measurement and evaluation of the visual presentation may be difficult or impossible. This is true because a trace cannot be compared with other traces that occur at a different time. A photographic record overcomes this drawback and enables the oscilloscope to be used for both indicating and recording purposes. In addition, an oscillogram of a CRT trace offers the following advantages:

1) A permanent record is obtained

2) The oscillogram can be examined at any time

3) It can be compared with other similar records taken at a different time under identical or completely different conditions

4) More precise quantitative data can be obtained, especially if the oscillogram is projected or enlarged prints are made

5) Transients may be recorded that have a duration too short to be observed or measured

6) Recording can be automatic in some cases, thus eliminating the need for constant monitoring.

Because of its universal use, the oscilloscope and the photographic recording of its CRT images have been well documented in both electronic and photographic literature. A review of references[4.1] through[4.14] will provide a good background into past and present techniques, and will supplement both the data presented and the equipment described in the following sections.

Cathode-ray Tube Principles

The cathode-ray tube, often called a "CRT", (or sometimes an "indicator tube" or "scope") is a special funnel-shaped vacuum tube that displays the interpretation of electrical phenomena on a screen. The principal components of the CRT are an electron gun, an electron-beam-deflecting system, and a phosphorescent screen, shown in Fig. 4.1.

The *electron gun*, axially-located within the neck of the tube, has an electrically-heated *cathode* as a source of electrons. The electrons are beamed in an axial stream at high velocity toward the screen. This is accomplished by a perforated-diaphragm

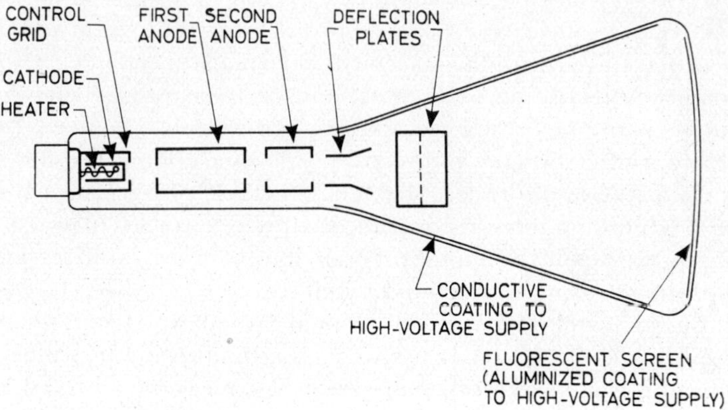

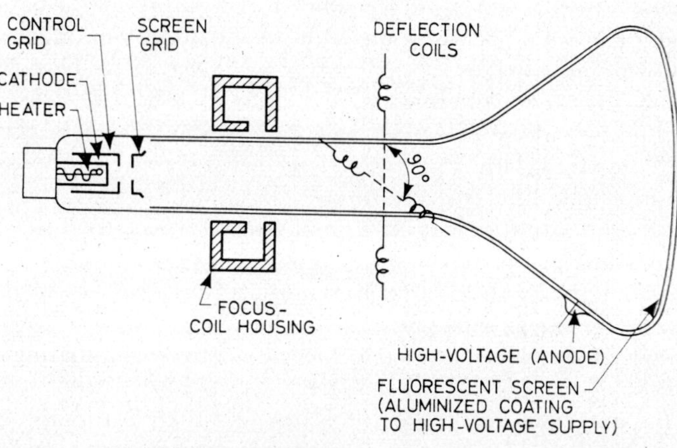

Fig. 4.1. Cathode-ray tube schematics: (*Top*) electrostatic deflection (*bottom*) magnetic deflection

negative *control grid*, located just in front of the cathode, and by accelerating *anodes* in front of the first grid. The final velocity of the electron beam is determined by the relative potentials between the cathode and the anodes.

The electron beam can be deflected either electrostatically or magnetically.

Electrostatic deflection is accomplished by a pair of deflector plates located inside the tube on either side of the beam. Voltage applied to these plates causes a deviation of the beam and consequent movement of the spot on the screen. To achieve movement of the spot in both X and Y coordinates, a similar pair of plates are oriented at right angles to the first pair.

Both *magnetic deflection* and focusing can be provided by magnetic means. Either a permanent magnet or an electromagnet can be used for focusing.

Vertical and horizontal deflection control is achieved by coils external to the tube as shown in Fig. 4.1. The coils are arranged in two pairs, perpendicular to the tube axis, spaced at 90 degree locations around the tube. Magnetic deflection may be used either with magnetic or electrostatic focusing.

Under unvarying electrical conditions, the electron beam produces a fixed spot of light on the luminescent screen at the end of the tube. However, with variations in the electrical inputs, the electron beam is deflected, and the spot moves across the screen producing a visible trace.

The size of the spot, controlled by the electron optics of the tube, and the energy of the cathode-ray beam, are the principal electronic factors that determine CRT sharpness and brightness. Spot diameters vary from a nominal 40μ in high -resolution tubes to 500μ or more on many oscilloscopes. The resolution capability of CRT screens coated with phosphors having particle sizes in the 2 to 20μ range greatly exceeds the spot size obtainable by the electron optics.

Indirect Exposure

PHOSPHOR-SCREEN CHARACTERISTICS. Cathode-ray tube screens may be divided into two types:

1) light-emitting screens, and 2) light-absorbing screens. The characteristics of light-emitting screens are best suited to photographic recording, and the recording requirement must be considered in determining the most effective phosphor to be used. Table 4.1 lists phosphor-screen characteristics as classified by JEDEC (Joint Electron Device Engineering Council). Table 4.2 identifies phosphor types chemically and indicates the approximate average particle size as furnished by one manufacturer.

The relative visual brightnesses of a number of different phosphors operating in a Tektronix oscilloscope are given in Table 4.3. The measurements, taken by Tektronix, Inc. with a Spectra Brightness Spot Meter which incorporates a CIE standard eye filter, are representative of 10 kv aluminized screens. Because the spectral response of films is different from that of the eye, films will respond differently to the various phosphors. Relative photographic efficiences are also compared in Table 4.3. (The latter information from[4.11] was obtained with other equipment under different conditions.) Individual films and the manner in which they are exposed will also affect the relative photographic efficiency. Therefore, the data given can be considered as an approximation only, and the user must be guided by his own tests. Further information on sensitized materials is given in Chapter 5. References[4.15] through[4.19] give further details on film sensitivity, detail retention, and exposure determination.

JEDEC Designation	JEDEC Persistence Class[1]	Decay Time[2]	Color of Fluorescence and Phosphorescence	Wavelength of Peak Radiant energy
P1	Medium	24.5ms	Yellowish-green	525mμ
P2	Medium-short	35 to 70μs	Yellowish-green	535mμ
P4 (sulfide)	Medium-short	22μs	Violet	455mμ
		60μs	Greenish-yellow	565mμ
P4 (silicate)	Medium-short	40μs	Violet	425mμ
	medium	12.5ms	Greenish-yellow	550mμ
P4 (silicate-sulfide)	Medium-short	40μs	Violet	425mμ
	Medium	12.5ms	Greenish-yellow	550mμ
	Medium-short	34μs	Blue	460mμ
P5	Medium-short	25μs	Blue	415mμ
P7	Medium-short	25 to 75μs	Violet	435mμ
	Long	400ms	Yellowish-green	555mμ
P11	Medium-short	34μs	Blue	460mμ
P12	Long	210ms	Orange	590mμ
P14	Medium-short	27μs	Violet	435mμ
	Medium	5ms	Yellowish-orange	600mμ
P15	Very-short	0.05μs	Near Ultraviolet	391 mμ
	Short	2.8μs	Green	510mμ
P16	Very-short	0.12μs	Near Ultraviolet	383mμ
P20	Medium to Medium-short	60μs	Yellowish-green	560mμ
P22	Medium-short	22μs	Violet	450mμ
		60μs	Green	515mμ
		60μs	Reddish orange	680mμ
P24	Short	1.5μs	Green	510mμ
P31	Medium-short	38μs	Green	522mμ

1. Joint Electron Device Engineering Council persistence classifications (based on time for phosphorescence to decay to 10% of initial brightness):
Very Long—1 second and over
Long—100 millisec to 1 sec
Medium—1 millisec to 100 millisec
Medium-short—10 microsec to 1 millisec
Short—1 microsec to 10 microsec
Very short—Less than 1 microsec
2. Based on the time required for the phosphorescence to decay to 10% of the initial brightness.

Usual Applications	Types of Film for Recording[3,5]
General-purpose oscilloscopes for visual observation of recurrent phenomena	Ortho & Pan
Oscilloscopes for visual observation of low-speed non-recurrent phenomena	Ortho & Pan
Direct-view black-and-white television picture tubes	Pan
Projection black-and-white kinescopes	Pan
Projection black-and-white kinescopes	Pan
Special oscilloscopes for photographic recording	Blue, Ortho & Pan*
Radar indicators and pulse-modulated applications	Pan$^*_4$
Special oscilloscopes for photographic recording, transcriber kinescopes, and image-converter cameras	Blue, Ortho & Pan*
Radar indicators	Pan
Oscilloscopes for observing high-frequency pulse-modulated phenomena	Pan*
Flying-spot scanners and special oscilloscopes for photographic recording	Ortho, Pan*
Flying-spot scanners	Blue, Ortho & Pan*
Direct-view display storage tubes.	Pan*
Direct-view color television picture tubes	Pan
Color flying-spot scanners	Ortho & Pan*
Oscilloscopes for visual observation of low-speed non-recurrent phenomena	Ortho & Pan

3. The types of films shown are necessary to record the entire spectral output of the phosphor and therefore achieve the highest possible photographic efficiency.
4. Only if a blue filter, such as the **KODAK WRATTEN** Filter No. 47B, is used to remove the long-persistence longer wavelength component. Alternately, a blue-sensitive film can be used without a filter.
5. All phosphor types are suitable for single-frame recording. Types marked* are suitable for moving-film photography.
Note: This table is a condensed version of a table appearing in **KODAK TECH BITS**, No. 1, 1965, and is reprinted with permission of Eastman Kodak Co.

TABLE 4.2

CATHODE-RAY TUBE PHOSPHORS SUPPLIED BY UNITED STATES RADIUM CORPORATION

JETEC No.	USRC No.	Phosphor Type	Approx. Avg. Particle Size (Microns)
P1	GS115	Zn_2SiO_4 : Mn	2
P2	G155F	ZnSCdS : Cu	18
	G180	ZnSCdS : Cu	18
	1559	ZnSCdS : Cu	13
	4121	ZnSCdS : Cu	10
P4		Blend $ZnS : Ag + ZnSCdS : Ag$	7-8
P5	T5	$CaWO_4$	4
P7	920B	ZnS : Ag	14
	3324	ZnS : Ag	4
	1036	ZnSCdS : Cu	17
P11	1039F	ZnS : Ag	11
	272B	ZnS : Ag	15
	3129	ZnS : Ag	2.5
	3324	ZnS : Ag	4.0
P14	920B	ZnS : Ag	14
	1038	ZnSCdS : Cu	20
	1541	ZnSCdS : Cu	12
P16	S1022	$CaMg(SiO_3)_2$: Ce	4.5
P20	S345-D12	ZnSCdS : Ag	7
	2544-860	ZnSCdS : Ag	4.5
P22	GR120	ZnSCdS : Ag	3.5
	RS480	ZnSCdS : Ag	6.5-7.5
	BL60	ZnS : Ag	4
P28	1036	ZnSCdS : Cu	17
P31	4737	ZnS : Cu	9
	3094	ZnS : Cu	8

Phosphor coatings consist of one or more layers of sulphides, silicates or other materials such as those listed in Table 4.2. Electron beam energy causes the coating(s) to luminesce, and a trace is formed as the spot moves. This result occurs because of the *persistence* of the phosphor. The light emitted during electron exitation

TABLE 4.3

**RELATIVE BRIGHTNESS AND PHOTOGRAPHIC EFFICIENCY
OF VARIOUS CRT PHOSPHORS**

JETEC No.	Relative Brightness	Approximate Relative Photographic Efficiency
P1	128	0.35–0.40
P2	238	0.40–0.60*
P4	165	–
P7	128	0,50
P11	100	1.0–1.4*
P15	32	0,90
P20	250	–
P31	284	0.58–0.9*

* Aluminized screen

is termed *fluorescence* and the afterglow is called *phosphorescence*. Persistence characteristics of various phosphors are classified in Table 4.1.

Conversion efficiencies of electrons (incident on the phosphor) to light are on the order of 3 to 4 per cent or less. If the conversion efficiency and the maximum power loading are exceeded, the screen may be burned, although there is considerable variation in burn resistance among screens. Burning darkens the screen and reduces the light output.

Electron dissipation from the screen must balance the incident electron current rate, or the screen will become negatively charged. Some screens are *aluminized*, i.e., an aluminum layer approximately 0.1 micron in thickness is coated over the phosphor. The coating prevents the screen from becoming negatively charged. In addition, the aluminized coating reflects toward the camera lens, light that is emitted by the phosphor; it also effectively filters out unwanted light produced by cathode glow. In using screens that are not aluminized, there is the possibility of fogging the film by radiation from the incandescent heater and cathode, which might pass through some phosphor screens.

Aluminized screens require higher voltage accelerating potentials to allow electrons to pass through the thin aluminum barrier. Since many CRO's already have higher voltage capability, selection of an aluminized tube will optimize conditions for photo-recording.

DARK-TRACE CATHODE-RAY TUBES. Generally, dark-trace screens are used for projection applications. They are principally applicable to radar, storage, and display purposes, rather than to conventional oscillographic practice. The origin and principles of dark-trace tubes are discussed in[4.20] pages 578–584.

12*

DUAL-TRACE TUBES AND MULTIPLE-TUBE CRO's. Simultaneous multiple-channel recording is useful for comparing two or more dependent or related signals. The traces, side by side on one piece of film, permit direct evaluation without matching the X axes of separate pieces of film. In continuous strip recording, multiple traces appear similar to those obtained in multiple-channel light-beam recording.

Multiple-channel CRT records are produced either by a special tube or by multiple tubes mounted adjacent to each other so that they can be photographed simultaneously. The special CRT instruments employ either a normal gun and a "splitter" plate, or alternately, a tube with multiple guns and deflecting systems. Multiple traces can also be formed by electronically switching from one input to another in fast sequence, to produce a time-shared display.

Multiple-tube CRO's, as exemplified by the Minirack and MUR recorders manufactured in England by Southern Instruments Limited, are described in a later section. The CEC 5-140, a 16-channel CRO recording instrument was also available until recently in the United States from Consolidated Electrodynamics Corporation.

OSCILLOSCOPE RECORDING TECHNIQUES. Today, the oscilloscope is used in nearly every laboratory and single-shot recording cameras are available to fit nearly every make of instrument. Modern oscilloscopes, coupled to improved cameras loaded with high-speed, high-resolution sensitized materials have popularized luminescent-screen cathode-ray oscilloscope photography.

The following basic techniques are used for making CRT oscillograms:

1) A "still camera" is employed and the photographic material is stationary during its exposure to the CRT

2) A camera loaded with continuously-moving film is utilized and the CRT sweeps in one direction only

3) Variations of 1 and 2.

It is easy to make photo-oscillograms with single-frame still cameras focused on the CRT screen. It is even easier to obtain high-quality results when special cameras are used that have been designed as integral accessories of the oscilloscope. This is especially true when the camera is loaded with high-speed sensitized material, for underexposure was previously a major problem in CRT photography. The still camera can record either continuous events in which the electron beam follows the same repetitive path for a period exceeding the exposure time; or transient events in which the beam makes one trace only during the exposure. In the first case, the camera exposure can be as long as necessary to obtain a satisfactory record. In the second case, making the exposure may be more complex. If the occurence time of the transient is known or in under the control of the operator, the oscilloscope time base can be started a little beforehand, or may be operated by the transient itself. The shutter of the recording camera is opened just before the trace occurs and is kept open for its duration. The single transient condition may necessitate a fairly wide lens aperture and fast film, particularly at high writing speeds. Today, Polaroid materials are available for oscilloscope photography that have speeds as high as 10 000

ASA equivalent, which greatly simplifies the exposure problems associated with slow
er materials.

To record transient events that occur at random, and/or which may be prolonged
and vary slowly but continuously, a different type of camera is required. The camera must move the film continuously at a constant speed thus providing the time
axis (just as paper is transported in the light-beam oscillograph). Therefore, the
time base of the oscilloscope is switched off so that the spot moves only in the amplitude direction.

Transient events can also be photographed in a similar manner with a drum camera. Cameras of this type are capable of very high film-travel speeds. However,
the length of the record is limited to the circumference of the drum. In turn, the
drum must be designed to be within practical size and weight limitations.

CRO RECORDING EQUIPMENT. The need for records of CRT traces led to photographing the tube face. Initially, this was done with conventional cameras. Because
setup and operation of such equipment was bothersome and time consuming, users
in need of more than an occasional record often built adapters to align cameras
with their oscilloscopes. Extraneous room light was excluded with tubes or bellows.
It was not until commercial units appeared that photographic records could be made
easily and good results could be obtained consistently with a greater degree of assur-
ance. Reference[4.11] describes the development of oscilloscope camera equipment by
one manufacturer, beginning in the 1940 to 1950 period. (pages 29 through 32).

The use of Polaroid Land equipment and materials added reliability to CRT photography by its high photographic speed and by its capability of allowing an immediate check of the result. By trial and error, the user learned rapidly how to obtain
successful results because he could see his record during the course of his experiment; usually he could make an improved record if conditions permitted.

The recording equipment in use today is elaborate by comparison with equipment of 20 years ago. The development of the recording equipment also led to improved oscilloscopes. As a result of these combined advances, plus the use of improved photo-recording materials, high-quality records can be made routinely without inconvenience or interference with laboratory procedures.

Many of today's sophisticated oscilloscope cameras include such advantages as
binocular viewing, built-in data-card recording, interchangeable and multiple-exposure backs, removable film supply and takeup magazines, exposed film cut-off, and
swing-away housings.

The *optical system* may be composed of one or several optical channels. One channel images the CRT trace on the film. In its simplest form it requires an objective
lens which, as in any camera, inverts the image. Depending upon the image-to-object ratio, the image may be the same size, or larger, or smaller than the original
trace. Generally, CRO recording cameras are designed for either actual size (unity
magnification) or reduced images.

It is desirable to have a well-corrected wide-aperture lens for CRO recording.
Characteristics of the phenomenon under observation, the phosphor persistence

and spectral output, and the recording material, could cause underexposure in certain instances if very fast lenses were not used. Lenses having apertures ranging from F/2.5 to F/0.87 are available that produce optimum performance at close conjugate distances over a broad spectral range.

In addition to being designed for close conjugate photography, a good CRO recording lens should be corrected for the various aberrations. Such image errors reduce sharpness, contrast, and quality of the image and are corrected by a lens design that incorporates several elements in the final lens assembly. The aberrations are: spherical aberration, coma, astigmatism, curvature of field, distortion, and chromatic aberration. Six-or-eight-lens elements, usually of two or more types of glass, effectively correct the aberrations in CRO recording lenses.

Reflection of light at the lens surfaces reduces transmission and causes a reduction in image brightness. In addition, multiple reflections cause nonimage-forming light, and this glare light, incident on the film further reduces image contrast and definition. Antireflection coatings are applied to the lens surfaces to minimize reflections at the glass-to-air surfaces. A coating is effective if the coating-glass surface is one-half wave length behind the air-coating surface reflections.[4·21]. Often the applied coating is for light having a radiant energy wavelength peak of approximately 550 mμ. Some users have reported transmission increases of up to three times when the original coating was removed and replaced with a coating optimum for the spectral output of the particular CRT being used.

WRITING SPEED. The transient-recording-speed limitation is dependent upon the following factors:

1) Oscilloscope characteristics (beam current, accelerating potential)
2) CRT phosphor (brightness, persistence)
3) Optics (lens aperture, image-to-object ratio, filters)
4) Photographic material and processing (sensitivity, spectral response, other sensitometric characteristics; development)

For any system containing a given set of the above components there is a maximum writing speed. This is usually expressed as the maximum spot velocity which produces a useable record. Such a record will have a specified trace density (usually at least 0.1 above base plus fog) on the photographic negative. Positives, in which the recorded trace appears as a white line against a black background, requires a slightly different definition. For instant diffusion-transfer prints, the maximum-writing speed is defined in[4·22] as "the maximum spot speed which produces a discernible trace against the black background."

The recording speed is usually expressed in centimeters per microsecond for CRO recording. Light-beam writing speeds are expressed in inches per second, since they are several orders of magnitude lower than CRO writing speeds. Corresponding damped sine-wave frequencies are given in megacycles for CRO instruments, and in cycles for light-beam recorders. The nomograph in Chapter 5 is used to determine writing speeds for light-beam recorders; it can also be used for determining writing

NOMOGRAPH RELATING AMPLITUDE, FREQUENCY, AND MAXIMUM WRITING SPEEDS FOR SINUSOIDAL TRACES

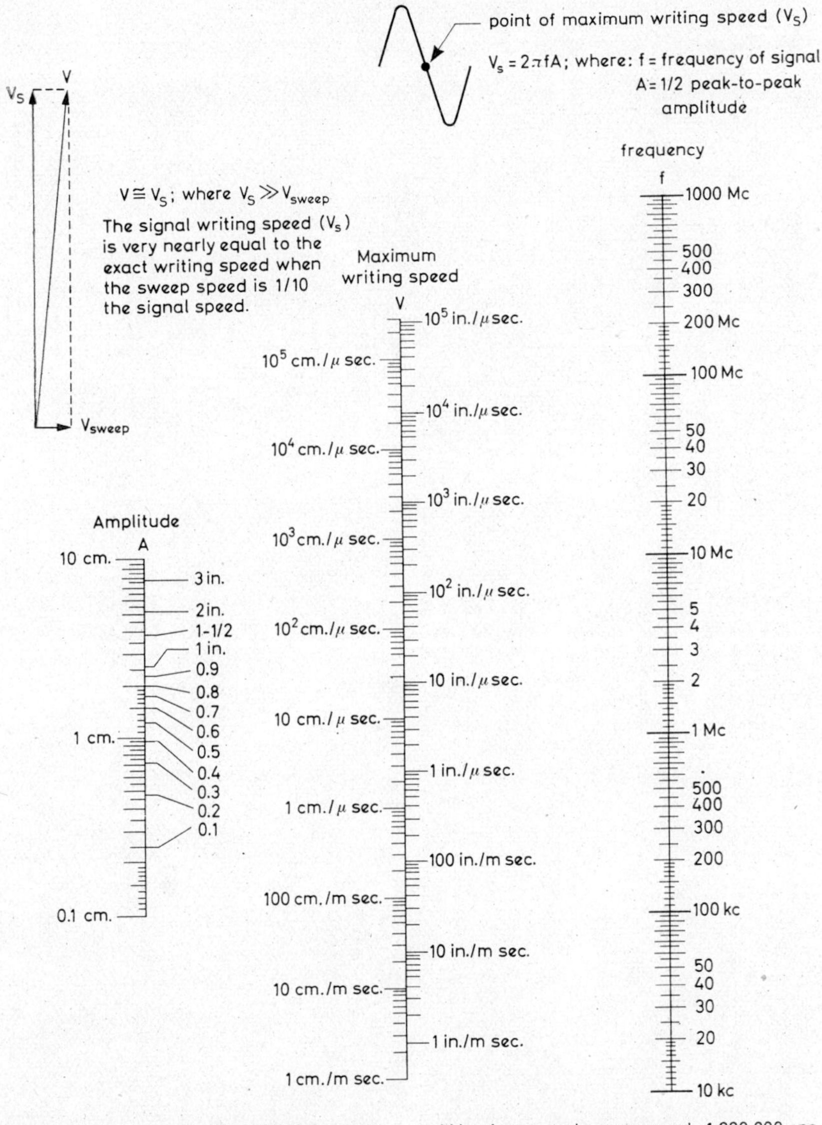

1 μ sec. = 1 microsecond = 1/1,000,000 second.
1 m sec. = 1 millisecond = 1/1000 second.

1 Mc = 1 megacycle per second = 1,000,000 cps.
1 kc = 1 kilocycle per second = 1000 cps.

Frequency range may be extended below 10 kc or above 1000 Mc by applying a suitable factor.

Fig. 4.2. Nomograph for determining sinusoidal writing speeds. (Courtesy Fairchild Instrumentation)

NOMOGRAPH FOR DETERMINING WRITING SPEED IN TERMS OF LENS APERTURE f, OBJECT-IMAGE RATIO, & MAXIMUM WRITING SPEED

$$V = \frac{4\,V_{max.}}{f^2\left(1+\frac{1}{M}\right)^2} \qquad\qquad V' = \frac{V_{max.}}{f^2}$$

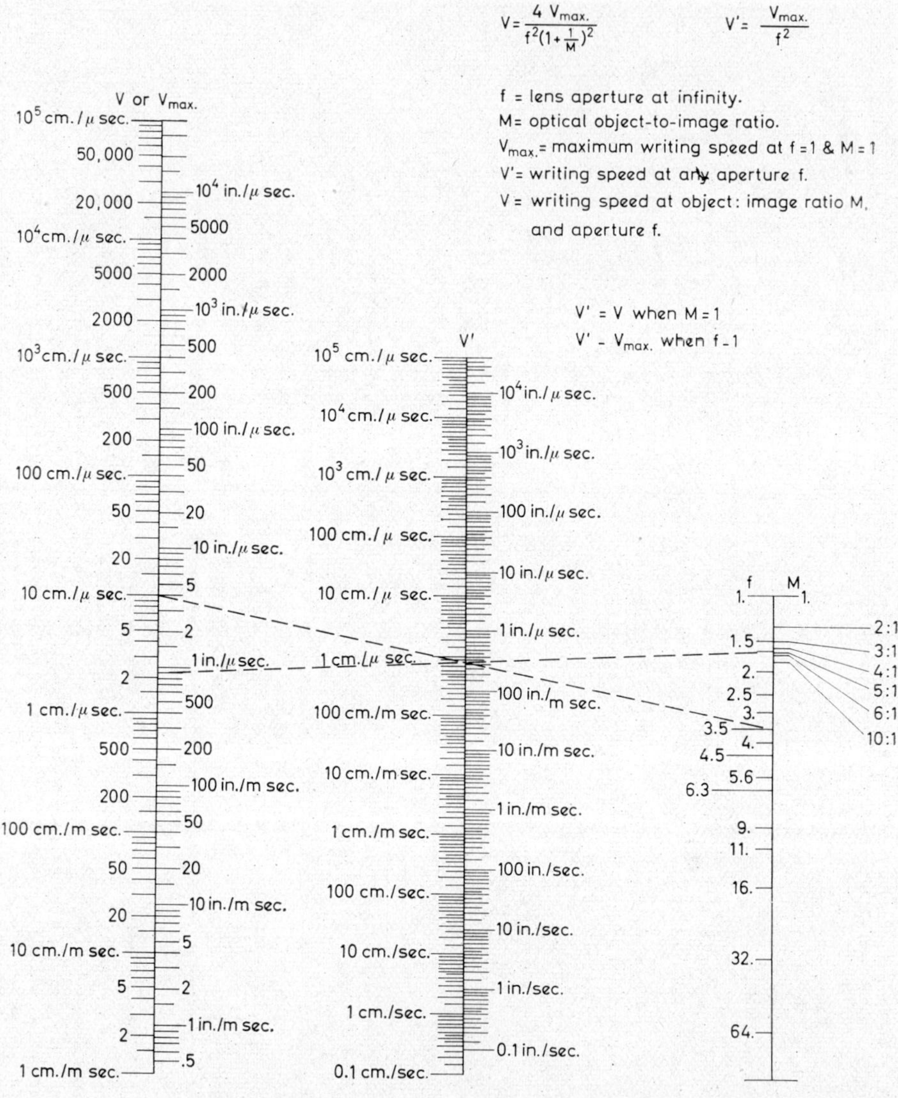

1 μ sec. = 1 microsecond = 1/1,000,000 second   1 m sec. = 1 millisecond = 1/1,000 second

Fig. 4.3. Nomograph for determining writing speed with reference to optical capabilities and adjustments. (Courtesy Fairchild Instrumentation)

TABLE 4.4

SINGLE FRAME RECORDING CRO CAMERAS

Manufacturer	Model	Viewing method	Film	Polaroid	Options, acessories, remarks
Agfa Aktiengesellschaft Camera Werk 8 Munchen 2, Germany	RCB	Adapter tube with eye-piece	35 mm		Manual (VB) and automatic (A) control units
Analab Instrument Corp. 30 Canfield Road Cedar Grove, New Jersey	3000 series	Binocular, dichroic mirror	35 mm 4×5	×	Interchangeable backs to accommodate various sizes/types photo-materials Manual and pulse models
Beattie-Coleman Products	K−5	Direct binocular	4×5	×	Interchangeable backs to accommodate various sizes/types photo-materials Manual sync. shutter, data chamber
Coleman Engineering Co., Inc. Santa Ana, California 92702	Mark II	Direct binocular	4×5	×	Same, but with electric shutter
	Mk II 565	Direct binocular	4×5	×	Same as Mark II but with 86 mm f/1.2 Oscillo Navitar lens
Fairchild Instrumentation Div. Fairchild Camera and	405 A −1 & −2	Direct binocular		×	Data recording facility; interchangeable lenses
Instrument Corp. Clifton, New Jersey 07015 750 Bloomfield Avenue	405 A −3 & −4	Direct binocular	4×5	×	Same
	405 A −5 & −6	Direct binocular	35 mm		Same, plus Robot 35 mm automatic advance recorder
	405 A−7	Direct binocular	35 mm		Automatic unattended recorder, with time, frame number and write-in data chamber Optional 30 ft. and 200 ft. magazines
General Atronics Corp. 1200 E. Mermaid Lane Philadelphia 18, Penn.	GA-300	Viewing port	×	×	Interchangeable backs for sheet film and roll or sheet Polaroid
	GA-309	Viewing port	×	×	
Hewlett-Packard	196-A	Direct binocular		×	
1501 Page Mill Road	196-B	Direct binocular		×	Has U.V. source to illuminate graticule
Palo Alto, California	197-A	Direct binocular		×	Has U.V. source to illuminate graticule. Graflok back is interchangeable with Polaroid back
N. V. Philips Eindhoven, Netherlands	PM 9300	Ground glass screen	55×55 mm	×	Uses Rolleicord Reflex Camera; extra counters and masks for other frame sizes, 35 mm adapter, Polaroid back. Camera can be used for normal photography
Tektronix, Inc. Beaverton, Oregon	C-12	Binocular, dichroic mirror	×	×	Polaroid and Graflok backs available; shutter actuator.
	C-27	Direct binocular	×	×	Same
	C-30	Swing away	×	×	Same

185

speeds for CRO recording by making the necessary conversion. However, the nomograph in Fig. 4.2 will read directly for CRO work.

The optical characteristics and object-to-image ratio are the other factors that determine writing speed. The nomograph in Figure 4–3 enables speed to be determined by relating object-to-image ratio to lens aperture and writing speed.

In continuous-motion recording (described in the next section) the film-transport rate can replace the CRO time sweep in determing the maximum writing-speed limitation.

Writing speed may be determined experimentally by utilizing the following technique. The film to be tested is used to record a single transient of a damped sine wave whose frequency fails to be recorded in the first cycle or two (see Figure 4–4). After being processed, the record is examined to find the first rapidly rising or falling portion of the damped sine wave that is photographed in its entirety. A measurement in centimeters is made of the peak-to-peak amplitude of the selected half cycle. A close approximation of the maximum writing speed, V_{max} can then be obtained from the following formula:

$$V_{max} = \pi F A$$

where F is the frequency of the damped wave in megacycles and A is the peak-to-peak amplitude in centimeters for an actual size image (1:1 or unity magnification) If the image size is other than 1:1 the following formula should be used:

$$V_{max} = \frac{\pi F A}{M}$$

in which M is the camera lens magnification.

SINGLE-FRAME CAMERAS. Single-frame CRO cameras are available from most oscilloscope manufacturers. They can also be obtained from camera and instrument manufacturers who furnish such recording devices as accessories for oscilloscopes manufactured by other firms. Cameras vary greatly in size, photographic material accepted, and special features. A partial list of such equipment appears in Table 4.4, and a representative group of recording instruments is illustrated in Fig. 4.5. Several types of equipment are also described in more detail in a later section. However, the varieties of recording cameras available, particularly those which fall into specialized classes, are too extensive to list or to describe in detail.

RECORDING ON MOVING FILM. The advantages of moving-film recording over single-frame recording are as follows:

1) Individual exposures can be made in sequence more rapidly than is possible with single-frame equipment
2) Records of transients can be made whose occurrence cannot be accurately predicted, and which cannot be triggered or otherwise synchronized with a single-frame camera

3) Recorded data can be analyzed that cannot be visually observed bacause they change relatively slowly over a long time span
4) Short-lived data can be recorded that cannot be shown as a "stationary" image with the conventional CRO time-base generator
5) Recordings can be made of high-frequency transients that cannot be resolved by light-beam or similar recorders. Such recorders sum or average the variations due to their inability to respond to such changes because of the mass of the moving element.

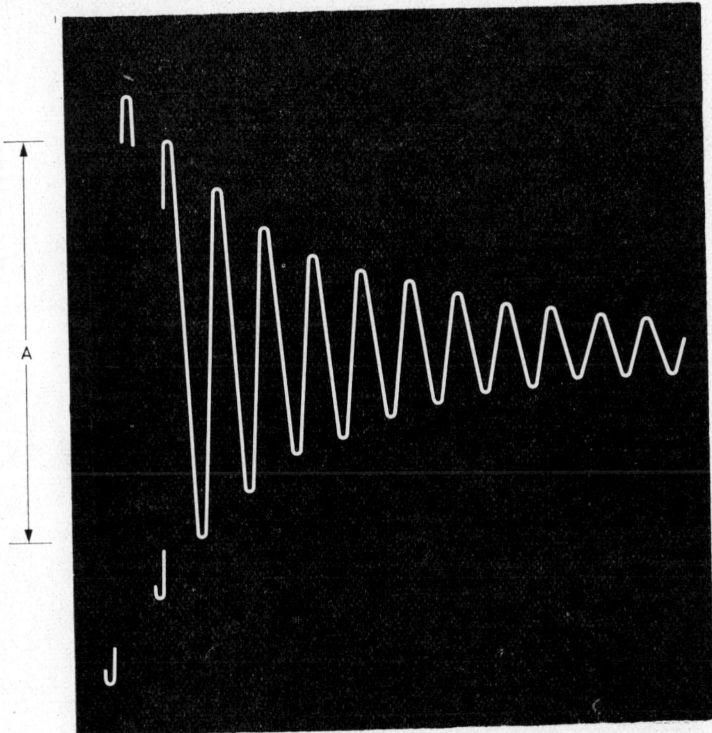

Fig. 4.4. Example of type of trace used to calculate writing speed

A comparison of continuous-recording methods is necessary to understand the use and capabilities of equipment designed for this application. A cursory review of the standard motion picture camera will show that cameras which produce discreet frames have limited application to CRO recording. The standard motion picture camera, in effect, produces a series of still pictures. The film moves and stops intermittently. The shutter is closed during film movement, and the exposure is made when the film is stopped. Good results are obtained with such equipment under conditions in which the "action" takes place continuously. In CRO recording, transients may not occur during the expusore period. If the transient cannot be

TABLE 4.5

MOVING FILM CRO RECORDING CAMERAS, CONTINUOUS AND/OR PULSE-OPERATED

Manufacturer	Model	Viewing method	Film size	Cont-inuous	Pulse or single film	Transport speed	Provisions and/or accessories
Analab Instrument Corp. 30 Canfield Road Cedar Grove, New Jersey	3020	Binocular, dichroic mirror	35 mm	×		1/16 to 8 ips linear in 8 speeds 0.016 to 16 ips with slight loss in speed linearity	Interchangeable pulse and conti-nuous magazines; data chamber; Integral Rapro-matic processor
	3030	Binocular, dichroic mirror	35 mm		×		
Beattie-Coleman Products Coleman Engineering, Co., Inc. Santa Ana, California 92702	KD-5hD	Binocular, dichroic mirror	35 mm	×		1/16 to 8 ips in 8 speeds	High speed maga-zine, CS-220-240 ips; data chamber timing light; Dual 35 mm Polaroid capability
	KD-5dT	Binocular, dichroic mirror	35 mm		×		
Cossor Instruments Ltd. Elizabeth way Harlow, Essex, England	1428	Hood. door	35 mm	×	×	0.05 to 25 ips in 9 speeds	Data chamber
	1458	Hood. door	35 mm	×	×	0.05 to 25 ips in 9 speeds	
Fairchild Instrumentation, 750 Bloomfield Avenue Clifton, New Jersey 07015	321-A	Binocular, dichroic mirror	35 mm	×	×	16 speeds, 0.013 to 60 ips. 90 and 180 ips speeds for short runs	Data chamber; timing light
General Atronics Corp. 1200 E. Mermaid Lane Philadelphia 18, Pennsyl-vania	SM-100	Binocular	35 mm	×	×	0.008 ips to 200 ips in three overlapping ranges 0.028 ips to 3.6 ips	Timing light Data chamber
Granger Associates 1601 California Avenue Palo Alto, California	541-B	Binocular, beam splitter mirrors	35 mm	×		Synchronously: 1.02 to 202 cm/s Asynchronously: 304 and 475 cm/s	Single frame records made by using hand crank.
N.V. Philips Eindhoven, Netherlands	PP1014	Eyepiece on adapter cone		×	×	Continuous: 15 speeds, 0.005 to 380 cm/s	Zeiss recording camera drum cas-sette, controls, adapters, etc. available
	PP 1021	Eyepiece on adapter cone	35 mm	×	×	Drum: 10 speeds, 3.5 to 5100 cm/s	

Manufacturer	Model	Viewing	Film width			Speed	Accessories
Southern Instruments Camberley, Surrey, England	M 731	Available on some models	35/70 mm	×	×	Continuous: 0.4 to 100 ips Drum: 4 to 1200 ips	Slow speed attachment for 40 to 1 reduction
Vought Camera Division Computer Equipment Corporation 19 18 North Central El Monte, California	NP-1	Boresight to focus and align camera	16 mm 35 mm	* × * ×	× ×	Pulses on demands in excess of 20 pps. (*Optional continuous motion transport avaiļable)	Frame-void light, data chamber, coding for automatic retrieval accessories
Warrick Electronics Division McDonnell Aircraft Corp. St. Louis, Missouri, 63166	F-8	Focusing: 10× microscope	35 mm	×		10 ips to 2220 ips	Two marker lamps
J.A. Maurer, Inc 37 — 01 31st Street Long Island City, New York 11101	KD-3	Periscope	16 mm		×	6.6 frames per second max.	Data chamber with card,24-hour clock 4 coding lights, & frame counter
	283	Operator sees front of CRT; camera records inside face of CRT through "picture window"	16 mm		×	Shutter speeds from 0.15 to several seconds	Data chamber
Photo-Sonics, Inc. 820 South Mariposa Street Burbank, California	CFA	Reflex view through camera, limited to aperture slit area	70 mm *	×		60 to 1440 inches per second	Timing lights, control box, boresight, magazines.* 35 mm adapter accessory available.
	CFC	Reflex view through camera of full 35 mm standard format	35 mm	×		0.83 to 166 inches per second	Timing lights, control box, magazines
Robot-Foto GmbH Dusseldorf, Germany	Robot Recorder F	Viewing port in hood	35 mm	×	×	2 ranges from separate motors: 1 to 6 inches per second and 6 to 20 inches per second	30 and 200 foot magazines, control set, speed marker lamps, variable-width gate

Fig. 4.5. Representative single-frame CRO recording cameras

a. Agfa 35 mm recording camera with 17 m (55ft). Magazine for 450 Exposures; b. Beattie-Coleman K-5 Oscillotron recording camera with Polaroid Roll Back; c. Fairchild Instrumentation Type 450A oscilloscope trace recording camera; d. Fairchild Instrumentation Type 450A-7 Automatic Unattended Camera System for recording oscilloscope single sweep transients; e. General Atronics Model GA-309 oscilloscope camera with swing-away feature; f. Philips Type PM 9300 oscilloscope camera (Rolleicord special, Polaroid back, and swing-away adapter) mounted on a Philips Type CM5601 oscilloscope triggered by a sync-pulse for the camera shutter in its open position, a blank film record might be obtained.

Similar results could occur from rotating prism cameras which produce discrete pictures at high frame rates. The film in the camera moves continuously and the image is optically displaced by the rotating prism so that image motion is equivalent to film movement thereby making the image "stand still" during exposure.

Although discrete-frame motion pictures can be taken when relatively long-persistent CRT screens are used, such recordings have limited value. Except for reference purposes, it is doubtful if more information is obtained from such records than can be perceived visually.

To achieve the greatest benefit from moving-film CRO recording, it is necessary to use a camera that moves the film continuously rather than intermittently. To prevent image blur on continuously-moving film, a CRT having a short-persistence screen is necessary. Usually a P11 tube phosphor is used, with an aluminized screen to increase the light output. Film motion up to approximately 120 inches per second will produce no blur. The P5 and special order P15 phosphor screen CRT's are also used.

The maximum writing-speed limitations previously given also apply to continuous-motion recording. The upper frequency limit at any given film transport-speed may be considered to be the resolution limit of the system.

There are several techniques for continuous-motion strip recording the least complex of which is to transport the film in the proper direction to obtain a time axis. More elaborate techniques can be used to record very high frequencies or narrow pulses. These techniques may involve changing film direction and/or changes to the CRO. Pages 69-73 of[4.11] explain several methods that are suitable for various applications.

Examples of cameras used during the 1940 to 1950 era include a 35 mm continuous-motion camera produced in the United States by the General Radio Company, and several models of continuous motion strip and drum cameras manufactured by Avimo Limited, Taunton, Somerset, England. High speed cameras of all types have constantly improved over the ensuing years and a representative list of continuous motion cameras that are currently available is given in Table 4.5. In addition, several models are illustrated in Fig. 4.6, or are described and shown in a later section. Reference [4.23] gives further details on the application of one of these instruments.

Some high-speed cameras that are normally used for instrumentation purposes (other than oscillography) utilize a rotating prism to produce discrete frames on continuously-moving film. Waddell[4.24] describes the adaptation of such cameras for oscillography; this is accomplished by removing the rotating prism so that discret frames are not produced.

A

B

C

D
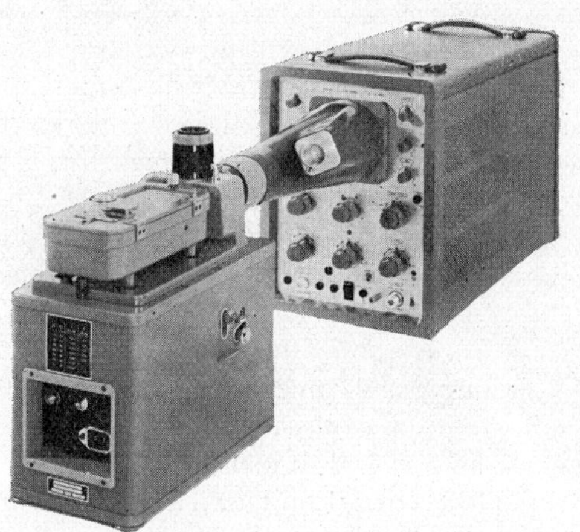

E

F

SOUTHERN INSTRUMENTS EQUIPMENT. Southern Instruments Limited, Frimley Road, Camberley, Surrey, makes electrical gages and pickup devices; FM pre-amplifiers, high-gain direct-coupled pre-amplifiers and driver amplifiers; single and multi-channel oscilloscopes and oscillographs; high speed continuous-feed and drum cameras; electrical transient recorders and electronic engine indicators, direct writing pen and galvanometer oscillographs. Southern Instruments has developed a modular 14-inch wide miniature rack-panel system of components, distributed under the trademarked name "Minirack." Minirack equipment covers a comprehensive range of single and multi-channel oscillographs, which may be broadly divided into two types: the transportable or box type and the trolley-mounted type (the former intended for table mounting). The trolley-mounted units offer the advantage of easy mobility when recording at more than one location.

SOUTHERN INSTRUMENTS OSCILLOGRAPHS, GENERAL DESCRIPTION. Two of Southern Instruments general-purpose oscillographs are pictured in Fig. 4.7. The M978 is a two-channel oscilloscope for visually indicating two variables plus one time-base reference on a 5-inch tube and flasher tube located in the camera hood. Data can be recorded by adding the M1227 70 mm camera. The MUR 12 will record from 6 channels plus one time-base reference, to 12 channels plus two time-base references. A number of transportable and trolley-mounted oscillographs, having various numbers of channels, lie between the channel capabilities of the M978 and the MUR 12 instruments. For example, two-, three-, four-, five-, six-, and eight-channel instruments are also manufactured.

The MUR Recorders. The outstanding features of this system are as follows: frequency response 0–20 KC/s; 2 to 12 channels plus time reference; interchangeable recording cameras of still, continuous-feed and drum types; incorporated monitoring facilities; and optional availability trolley or box-mounted units.

The main units are the recorder and camera, and the electronic console containing amplifiers, time base generator, time marker, and power supplies. Other types of multi-channel cathode-ray oscillographs are available capable of recording frequencies to 1 M/C.

Fig. 4.6. Representative moving film CRO recording cameras

a. Beattie-Coleman 35mm KD-5 rapid sequence and continuous-flow recording camera; b. Type 321A continuous motion camera manufactured by Fairchild Instrumentation; c. General Atronics Model SM-100 single-frame or moving film oscilloscope recording camera; d. Philips photographic recording camera Type PP 1014 mounted on Philips GM 5602 oscilloscope; e. Zeiss recording camera PP 1021 mounted on Philips GM 5603 oscilloscope. Single-frame/continuous motion cassettes, and a drum cassette are interchangeable; f. Step-servo NP-1 pulse camera manufactured by Vought Camera Division, Computer Equipment Corporation

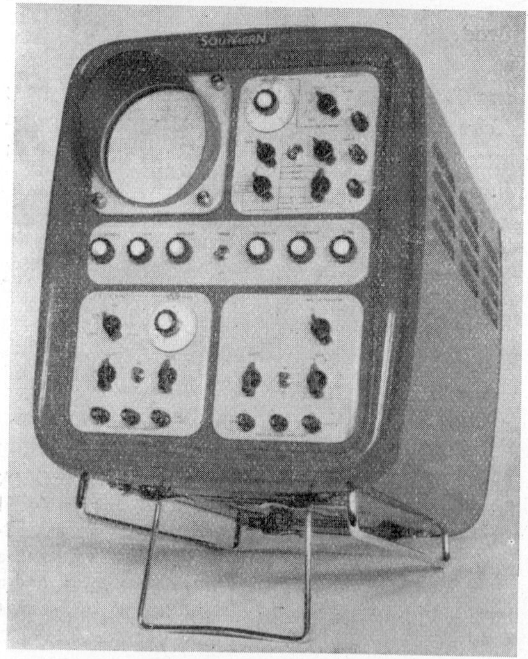

Fig. 4.7. Oscilloscope and recording equipment manufactured by Southern Instruments Ltd. *Top*: MUR 12, a 12-channel universal oscillograph. *Bottom*: M978 oscilloscope

Engine Indicators. Southern Instruments has specialized in equipment for engine indicating utilizing a cathode-ray tube as the display element. Indicator measurements of internal-combustion engines include the pressure cycle both in the cylinder and at different points in the induction and exhaust manifolds; mechanical events such as timing and extent of needle lift in a fuel injection valve; torsional vibrations and other movements of a camshaft or crankshaft; and variations in engine fuel, timing, load and other operating conditions.

The indicator diagram shown on the cathode-ray screen consists of two simultaneous movements of the spot, the vertical movement corresponding to the physical quantity being measured, and the horizontal movement corresponding to time or crank angle. Synchronization of crank angle to successive vertical spot movements can be achieved with magnetic pick-ups, contacts, or a photocell device. The Southern Instruments Type M. 738 Photocell Sweep Unit, described in Chapter 2, can be used for this purpose.

A permanent photographic record is desirable for measurement and analysis of indicator diagrams. Two alternative methods are possible: 1) A still photograph of the complete diagram; or 2) A moving-film record of only the vertical spot displacement on the cathode-ray tube. Southern Instruments manufactures eight types of cameras which can be used with their general-purpose oscillographs and engine indicators.

Southern Instruments M731 Camera. The M731 Universal Camera, shown in Fig. 4.8, can be used for taking either stills, continuous-feed, or drum records. The 50 foot magazine will accommodate either 35 mm or 70 mm, perforated or unperforated film or paper.

The case and the principal parts of the camera are made of aluminum-alloy castings. The driving motor operates on 6 to 12 volt power and has a wide speed range and constant torque. It has interchangeable gears and two-step pulleys with a belt drive that enables the film speed to be varied over very wide limits. The gear box forms the mounting for the 6-inch diameter drum. It contains the manually-tripped timing contacts that ensure that the exposure is limited to one drum revolution by operating the oscillograph's beam trigger. Contacts for initiating the event to be recorded are also fitted. A continuous-feed attachment can be fixed in the place of the drum. This draws the film uniformly through the exposure gate from the magazine and re-spools any length up to 50 feet automatically. Single-frame exposures can also be made at a rate of about one every three seconds. The exposed piece can be cut off by a guillotine knife, and a footage indicator and record marker are provided. There is a simple sliding shutter and a 3 inch focal length F: 1.9 lens in a sliding mount. Daylight loading is effected through black cloth sleeves fitted to a frame at the top of the camera.

The continuous speeds of the camera range from 0.4 to 100 inches per second. A slow-speed attachment for an additional 40 to 1 reduction ratio is available as an accessory. Drum speeds are 4 to 1200 inches per second. Dimensions are 13 by 12 by 13 inches, and the weight is 50 pounds.

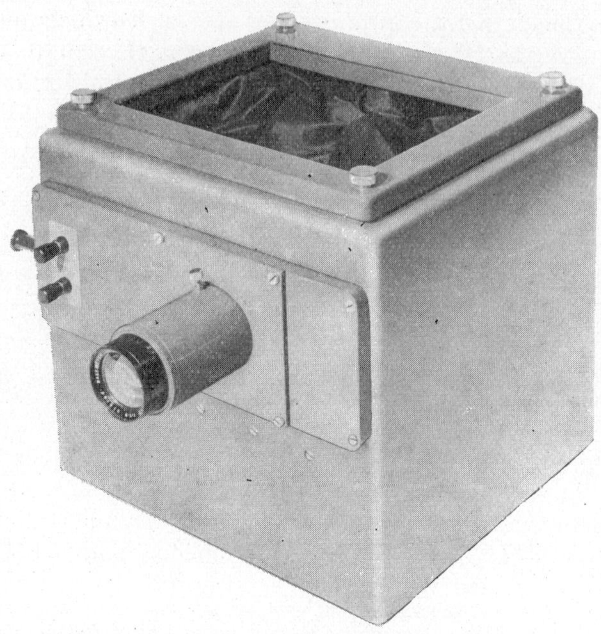

Fig. 4.8. Southern Instruments M731 Universal Camera

Southern Instruments M800 Camera. The M800 camera is intended for very fast single-stroke transient recording. It uses a Wray F:1.0 lens that produces a 4 to 1 reduction ratio. The camera has been designed primarily for use with the TR10 or TR12 transient recorders, but may be adapted to other oscillograph equipment.

The body is an aluminum alloy Y-shaped casting, as shown in Figure 4.9. The lens is in a screw-focussing mount with means of access to the lens aperture adjustment. The film holder, secured by knurled screws for quick removal, contains loading and re-wind spools and a one inch-square exposure gate. It accommodates 5 feet of 35 mm film which is sufficient for 50 exposures though any short exposed length can be cut off if required. A sliding shutter makes it light-tight between exposures or while being removed to the darkroom for loading or unloading. The viewing hood fits snugly to the head and enables the operator to see the transient while it is being recorded. A simple flap shutter closes the viewing aperture when not in use.

The overall length is 12 inches. The film holder measures 7 by 3 by 2 inches. Total weight is 15 pounds.

MODEL 35P OSCILLOSCOPE RECORDING CAMERA (CAI). The Model 35 P recording camera manufactured by Chicago Aerial Industries can be easily mounted or adapted to any 5-inch oscilloscope. It records oscilloscope traces, exposure time, exposure number and specific test data on each exposure.

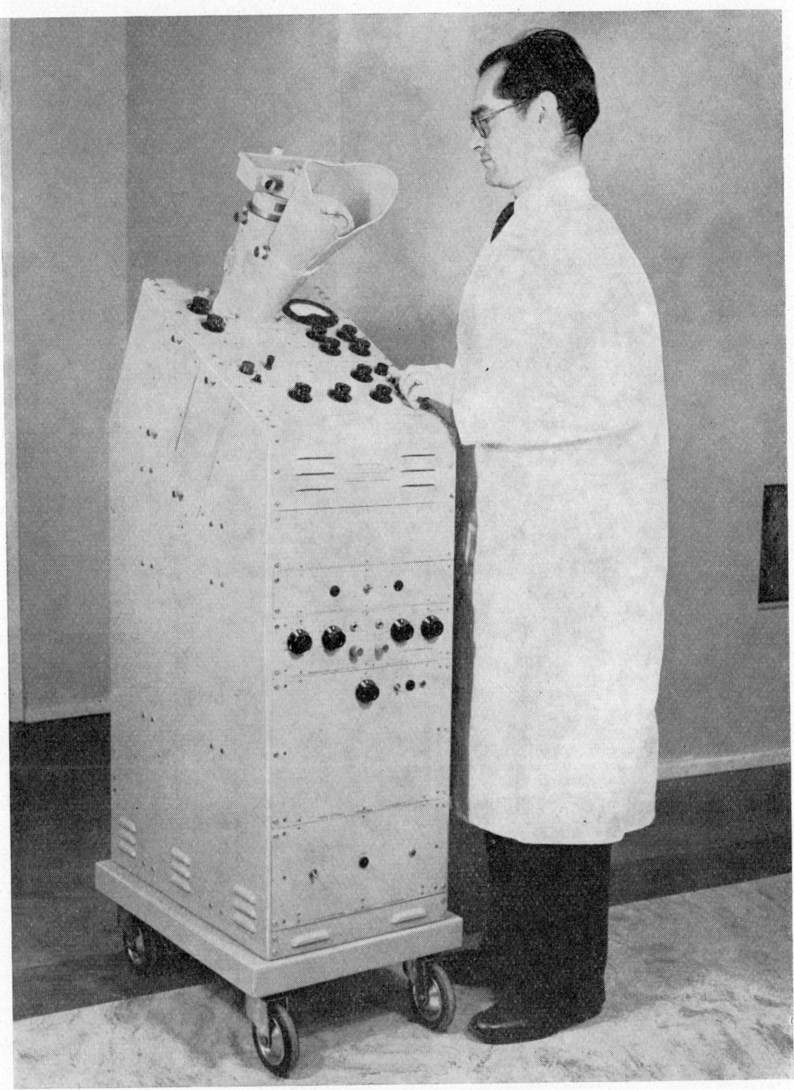

Fig. 4.9. Southern Instruments TR 12 transient recorder with Model M 800 camera

The camera portion of the unit is a 35 mm still camera equipped with a fixed focus, f/1.8 anastigmatic lens, an iris diaphragm calibrated in f stops from f/1.9 to f/16 and a leaf-type shutter with speeds of 1/5, 1/10, 1/25, 1/50, 1/100 and bulb. The lens is prefocused on an object plane one inch (± 0.015 inch) in front of the camera mating surface, to cover both the data chamber and reflected image of the oscilloscope. Diaphragm and shutter speed controls are readily accessible even when the camera is mounted.

The 35P is equipped with a 28 volt dc motor and speed reduction gearing which advances the film one frame after each exposure. The film is mechanically flattened to guarantee image acuity. An exposure counter, graduated every five frames, shows the number of exposures remaining. The counter dial is easy to read and can be set for 70 or 35 frames.

The data chamber contains a 24-hour clock with sweep second hand; removable data card; an electrically-operated three-digit additive counter showing exposure number; an edge-lighted illumination panel; and three coding lights. A time limit relay located in the camera cuts off illumination in the data chamber after 1/10 second during bulb exposures. All operating controls are located on a panel on the upper front face of the data chamber.

A scope adapter faceplate, supplied with the camera, is used for installation. The automatic shutter reset and film advance simplify operation. The camera measures 10–1/4 by 9–1/4 by 13–1/2 inches (with hood) and weighs 14 pounds. It is available from Chicago Aerial Industries, Inc., 550 West Northwest Highway, Barrington, Illinois.

CAMERZ INSTRUMENTATION CAMERAS. The Photo-Control Corporation, 5225 Hanson Court, Minneapolis, Minnesota, manufactures completely automatic, electrically-operated cameras which can be used for instrumentation purposes. The housings are light-weight aluminum construction, which accept a variety of interchangeable lenses and film magazines. Double extension bellows and rackover base permit extreme closeups. Matched viewing and taking lenses are available in 2, 3, 4, and 5 inch focal lengths. From 300 to 3000 exposures can be made per roll; the camera accepts 16, 35, 46, and 70 mm Camerz magazines.

Components include automatic trip and wind control, negative numbering assembly, lenses, and magazines.

Medical Equipment

ELECTROMYOGRAPH. The Electro-Medical Laboratory, Inc., South Woodstock, Vermont, has developed an electromyograph to supply the need for an instrument with extreme simplicity of operation for use in the neuromuscular field of electro-physiological work. The instrument is illustrated in Fig. 4.10.

Clinical studies in which the use of the electromyograph is recognized include testing of nerve sutures; location of the Gasserian ganglion for alcohol injection or cautery in neuralgia; study of nerve and muscle action currents in tic and stammering; study of deficiencies in the myoneural junction in myesthenia gravis; investigation of the mechanism in cases of myotonia congenita and dystrophia myotonica, and investigations in other forms of paralysis.

Routine laboratory use of the electromyograph includes demonstration to students of the Weaver and Bray effect in the auditory system, and investigation of action current in nerve and muscle preparations.

198

The first amplifier of comparable sensitivity to be operated without batteries, this instrument is capable of recording the action currents of peripheral nerves and skeletal muscles. It has been designed with regard to maximum simplicity of operation for clinical investigations and for demonstration purposes. Oscillograms are recorded by cathode-ray oscillograph and camera. Action currents are simultaneously made audible in a loud speaker.

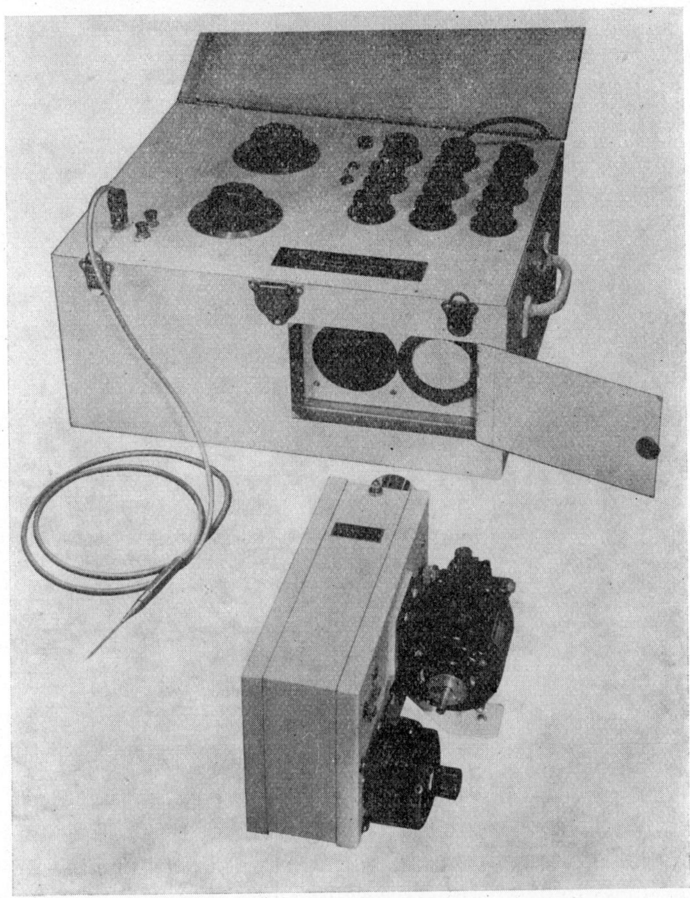

Fig. 4,10. Electromyograph developed by the Electro-Medical Laboratory, Inc.

Design. The electromyograph consists of a single cabinet containing an amplifier of great sensitivity, 3 inch cathode-ray oscillograph, 5 inch loud speaker, and a fixed-focus spring-driven camera.

The oscillograph is equipped with a self-luminous timing device. The cathode ray face is photographed by the camera which records simultaneously traces of time, electrical phenomena. Frequencies of the action currents lie within the audible

199

range, and, with the use of the loud speaker, can be recognized by ear as well as by inspection of the oscillograph.

The only controls necessary are those for gain of the amplifier, speed of sweep circuit, and volume of loud speaker. Focus and intensity of the cathode ray spot need not be adjusted.

Specifications. The electromyograph is housed in a portable steel case, 11 inches high by 19 inches wide by 15 inches deep.

Amplifier: 2 microvolts give 1 millimeter deflection on cathode ray tube; audible response in speaker.

Brilliance: cathode ray spot can be seen and photographed in moderately lighted room.

Camera: the camera body matches the cabinet and measures 8 inches high by 6 inches deep by 3 inches wide. Lens aperture: F: 1.9; standard 35 mm motion picture film in 100 foot reels is exposed at transport speeds of 1/2 inch to 5 inches per second.

Power supply: 115 volts 60 cycles, 150 watts.

Shielded sterilizable Adrian-Bronk electrodes, any size and length specified, and exploring and injecting electrodes for the Gasserian ganglion, are available as accessories.

SIMULTRACE RESEARCH RECORDERS. The SIMULTRACE equipment has been designed primarily for physiological monitoring and recording. The PR-4 and PR-7 models, which have four and seven channels respectively, use 12 cm recording paper. The DR-8 records eight channels on 18 cm paper. Recording is accomplished by a dual gun cathode-ray tube with a P11 phospher in the camera section of the basic unit. The camera can make both loop and scalar tracings. A circuit is provided for addition and subtraction of traces. The recorders feature absolute correlation between monitoring and recording, full screen deflection for all traces, and simple superimposition of tracings for simultaneous pressure and gradient measurement.

A large variety of amplifier channels are available, and special amplifiers can be made for problems not handled by standard ones. A common arrangement for cardiovascular studies is two ECG/EEG/Phono amplifiers and two or three pressure amplifiers. For respiratory studies, the addition of an integrator amplifier makes possible the measurement of lung compliance and airway resistance.

Other available channels include the QTD-1 for synchronizing other devices with the ECG and PHD-2 for polarographic and potentiometric studies. Special channels include multipliers and other computing amplifiers, FM tape recorders, and multiplexing amplifiers to permit recording of more than one signal with a single channel. This makes possible the display of more than eight traces on a DR-8 or PR-7.

The accessory Rapid Writer Attachment for the DR-8 Recorder processes dry, stable records in as little as 4 seconds. Good contrast is obtained with the single-chemical process. An alternate method, conventional photo processing, can be used if desired, to obtain maximum contrast and/or archival quality.

This equipment is manufactured by Electronics for Medicine, Inc., 30 Virginia Road, White Plains, New York.

SCHWARTZER 10–CHANNEL LARGE-SCREEN SCOPE. The large-screen scope is a multi-channel unit capable of displaying up to 10 different phenomena simultaneously on a viewing screen, 220×300 mm (17 inch tube). The 10-channel oscilloscope has a special long-persistance screen. Frequency response is 15,000 cycles per second. The device has been used in medicine, industry, as a measuring tool in research, as well as for monitoring. A camera attachment for Polaroid, Robot, Leica, Contaflex, and other cameras is available as an accessory.

Space is provided to accommodate up to ten preamplifiers (5 inch twin units), dc or ac coupled, designed as plug-in components and available in various types. Further details are available from the Schwartzer Company, 46 Salmi Road, Framington, Massachusettes.

Direct Exposure

COMPARISON OF DIRECT AND INDIRECT EXPOSURE METHODS. In the conventional CRT system there is a tremendous energy loss from the electron-to-light conversion, and added losses within the optical system of the recording camera. Evans and Tarnowski have pointed out[4.25] that an electron accelerated to a 27 kv potential has as much as 10,000 photons of energy (at a wavelength of 460 mμ corresponding to a P11 screen). However, only a small fraction of the electron energy is converted to light. Further light losses in the camera lens reduce the overall system efficiency to only a few tenths of one percent.

Direct electron-exposure cathode-ray oscillographs, circuitry, and applications were described by Wilson in[4.26]. The need for direct-exposure equipment in the 1940 era was stimulated perhaps by the difficulty of obtaining good results with phosphor-screen CRO equipment, cameras and lenses, and sensitized materials available at that time. The direct-exposure equipment, in effect, increased film speed, improved resolution, and eliminated the phosphor grain and other inherent phosphor characteristics. As conventional phosphor-screen CRO's, cameras, optics, and sensitized recording materials were improved, it became commonplace to obtain high-quality results without the necessity for direct-exposure.

A renewed interest in direct-exposure recording has occurred in recent years, and may be attributed to television recording, and to recording techniques suited to reconnaissance and computor memories. Electron-beam sensitometry has been advanced by these studies; in turn, films with improved electron sensitivity have become available.

DIRECT-EXPOSURE OSCILLOGRAPH. A direct-exposure oscillograph for the observation and recording of very fast transient electrical phenomena is manufactured by Trüb, Taüber & Company, A. G., Zürich 37, Switzerland.

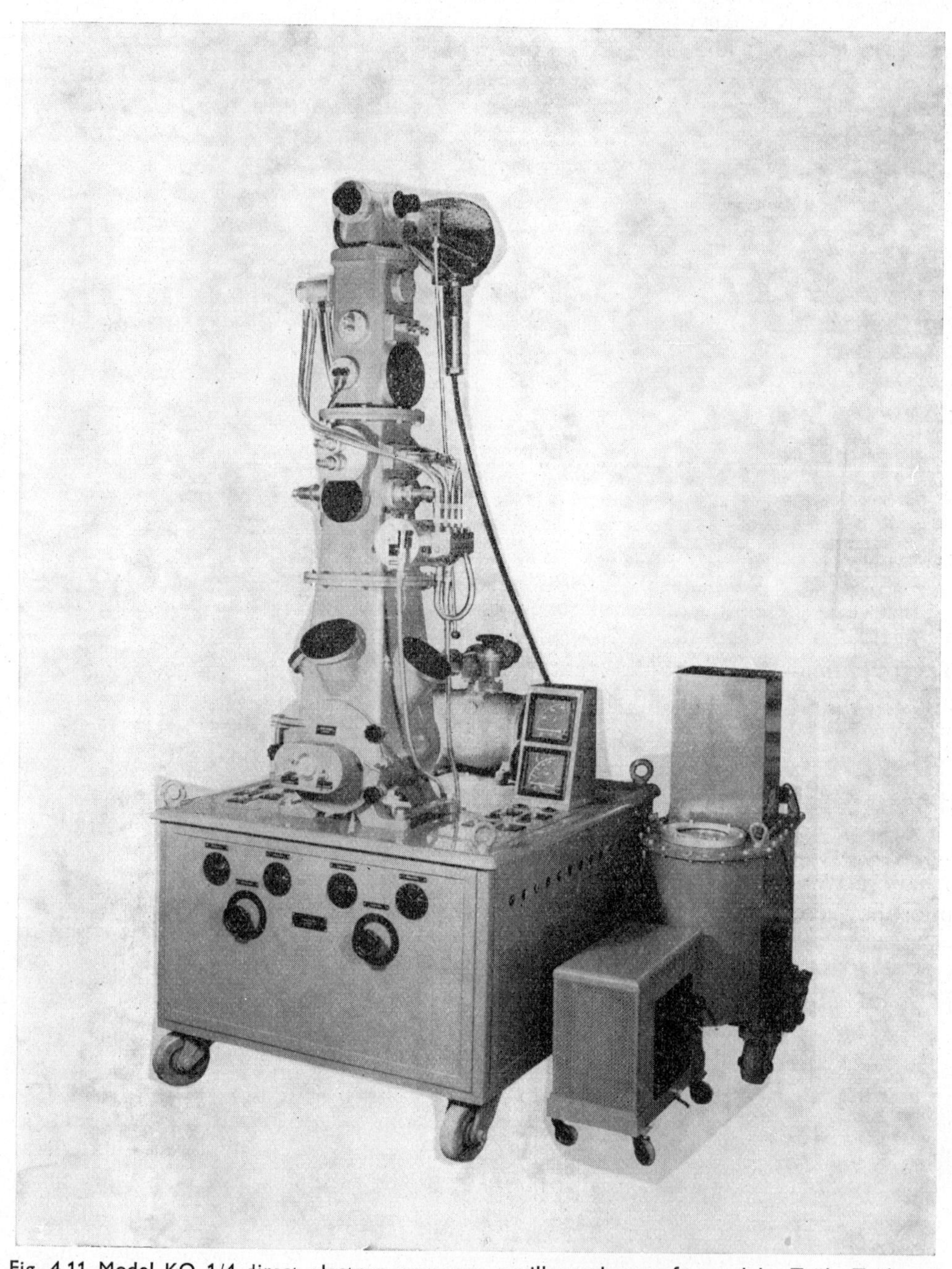

Fig. 4.11. Model KO 1/4 direct electron-exposure oscillograph manufactured by Trüb, Taüber & Company, AG

The oscillograph, shown in Fig. 4.11, and described in[4.27], uses a cold cathode electron source, and has the following components:

Discharge Chamber. The cold cathode, the anode, anode apertures, and an adjustable air inlet to control beam current are housed in this chamber. Four simultaneous beams can be produced from the cathode.

Beam-Trapping Chamber. To prevent fogging of the film four traps are used which allow recording to begin automatically by a change in trap bias. Focusing coils, combined with the trapping devices, enable the operator to properly orient and focus the image.

Deflection Chamber. Five pairs of deflection plates and two time-base plates are mounted in this section.

Sweep Initiating Relay. The relay, triggered by the phenomenon itself, initiates the recording, and again traps the beam after the transient has been recorded.

Recording Chamber. Three viewing ports permit observation of the data on a fluorescent plate. The data may be recorded either on stationary film or on film attached to a rotating drum.

Vacuum Supply. A two-stage oil pump and a molecular pump provide the necessary working vacuum.

Power Supply. A high-voltage supply, a low-voltage supply, a low-voltage rectifier, and an isolating transformer are normally used.

Accessories. The following accessories are available for calibration and operation of the oscillograph: mobile beam-supply unit, film holder with rotating film drum, delay line (approx. 1 microsecond delay), time base calibrating generator, calibration voltage rectifier, and relaxation oscillator.

The general operating and performance characteristics are as follows: cathode voltage, 35 to 50 kv; frequency range, 0 to 50 mc/s; switch-on time lag for beam and time base sweep, 0.2 microsecond; time base for smallest sweep, 2.5 microseconds; and measuring accuracy at constant beam voltage, ± 1.0 percent of maximum deflection.

Chemical Processing of Exposed Film

High-speed Processing. Access to recorded CRO data is usually required immediately, just as it is for light-beam recordings. For single-frame records this need can usually be met with Polaroid materials. Still or continuous-motion recordings on film are generally developed by the conventional processing methods described in Chapters 6, 7, and 8. However, some films can be processed with specialized techniques which usually require special chemistry, elevated temperatures, and perhaps special processors. Further informaton on two high-speed methods of processing CRO records is given in Chapter 8, reference[8.7] which describes the Raproroll web, and reference[8.8] which describes the Rapidata airborne camera processor. Other high speed methods are also available.

Processing Underexposed Records. The potential danger of underexposure at the high-writing speeds encountered (especially in CRT photography) sometimes dictates that unconventional photographic techniques be used. Any one or a combination of two or more of the following procedures may be used as an aid in obtaining a satisfactory negative under unfavorable exposure conditions:

1) Hypersensitizing (See Chapter 5)
2) The use of fresh developer
3) The use of more energetic developer
4) Development at elevated temperature
5) Prolonged development
6) After treatment of the processed film (intensification).

LITERATURE REFERENCES

[4.1] FELDT, R., "Photographing Patterns on Cathode-Ray Tubes," *Electronics*, February (1944). Reprinted as "Maximum Photographic Writing Rates of Commercial Cathode-Ray Tubes," Allen B. DuMont Laboratories. Contains 25 references on same subject covering 1934 to 1942 period.

[4.2] GARVIN, E. L., *Bibliography on High-Speed Photography*, Research Library, Eastman Kodak Company, (Rochester, N. Y.) 1960. Contains approxiamately 100 references to high-speed oscillography.

[4.3] PARR, G., *The Cathode-Ray Tube and its Applications*, Chapman and Hall, (London) 1941

[4.4] HOPKINSON, "The Photography of Cathode-Ray Tube Traces," *Journal of the IEE*, 93Pt. IIIA (1946)

[4.5] REYNER, J. H., *Cathode-Ray Oscillographs*, Putman (London) 1947

[4.6] HERCOCK, R. J., *The Photographic Recording of Cathode-Ray Traces*, Ilford Limited, (Ilford, London) 1947

[4.7] *Photographic Abstract Cumulative Index, 1931–1940*. Twelve references. Royal Photographic Society (London)

[4.8] *Photographic Abstract Cumulative Index, 1941–1950*. Royal Photographic Society (London)

[4.9] SCULLY, L. M., "Oscilloscope Recording," *International TV Technical Review*, January to December (1962) p502–506

[4.10] TYLER, R. W., and STRAUB, C., "Theory and Photometric Results," (Part I) and "Photometric Techniques and Instrumentation," (Part II). EISEN, F. C., VEAL, T. G., and TYLER, R. W., "Photography and Photometry of Cathode-Ray Tube Displays," (Part III). *Photographic Science and Engineering*, Vol. 7, No. 5 (1963)

[4.11] TAILEUR, M. C., Editor, *Techniques of Photo-Recording from Cathode-Ray Tubes*. Fairchild Instrumentation, a Division of Fairchild Camera and Instrument Corp. (formerly DuMont Division) 730 Bloomfield Avenue (Clifton, N. J.)1963

[4.12] BEISER, L., "A Unified Approach to Photographic Recording from the Cathode-Ray Tube," *Photographic Science and Engineering*, Vol. 7, No. 3 (1963)

[4.13] CLARK, G. L., "High Speed Data Recording," *Photographic Science and Engineering*, Vol. 8, No. 1, p44 (1964)

[4.14] DENNY, E. W., "Multiple-Exposure CRO Photography," *Electronic Instrument Digest*, Vol. 1, No. 2, p 18–21 (1965)

[4.15] NITKA, H. F., "Sensitivity and Detail Retention in the Recording of Light Images and Electron Beams," *Photographic Science and Engineering*, Vol. 7, No. 3, p 188 (1963)

[4.16] LEHMANN, M., "Development of a Light Meter for Cathode-Ray Tube Photography." Paper by M. Lehmann, Stanford Electronics Laboratories, Stanford University, Palo Alto, California, presented at SMPTE Conference, October 26, 1962.

[4.17] TYLER, R. W., and EISEN, F. C., "Emulsion Sensitivity for the Photography of Cathode-Ray Tubes," *Journal of the SMPTE*, Vol. 88, No. 3, pp222–225 (1959)

[4.18] *Kodak Films for Cathode-Ray Tube Recording*, P–37, Eastman Kodak Company (Rochester, N. Y.) 1962

[4.19] BEISER, LEO"Energy Transfer from CRT to Photosensitive Media," *Information Display*, Vol. 2, No. 5, p 32 (1965)

[4.20] SOLLER, T., STARR, M., and VALLEY, G. E., *Cathode-Ray Tube Displays*. McGraw-Hill Book Company, Inc. (New York, N. Y.) 1948

[4.21] NEBLETTE, C. B., *Photography, Its Principles and Practice*, 6 th Edition, p 64, H. Van Nostrand Company, Inc. (New York, N. Y.) 1962

[4.22] PIERCE, R. T. B., "High-Writing-Speed Film for CRO Recording," *Electronic Products Magazine*, p40, Sept. 1961

[4.23] "Automation in Recording Oscillographs," N. V. Philips' Gloeilampenfabrieken, (Eindhoven, Nederland) 1963

[4.24] WADDELL, J. H., "Oscillographic Moving Film Photography," *Research/Development*, Vol. 16, No. 3, pp33–34 (1965)

[4.25] EVANS, C. H., and TARNOWSKI, A. A., "Photographic Data Recording by Direct Exposure with Electrons," *Journal of the SMPTE*, Vol. 71, No. 10, p765 (1962)

[4.26] WILSON, W., *The Cathode-Ray Oscillograph in Industry*, Chapman and Hall (London) 1943

[4.27] BIRD, R. T. H., "A New Swiss Quadruple Beam High Voltage Cathode-Ray Oscillograph," *Instrument Practice*, August, p764 (1955)

V PHOTOGRAPHIC RECORDING MATERIALS

PAPERS AND FILMS. The previous chapters have discussed in detail the methods and equipment used for making oscillograms. In this chapter, the product—the oscillogram itself—is considered. The characteristics of the materials, both physical and sensitometric, are analyzed, and recommendations are given for matching the material to the recording conditions. Because sensitometry plays such an important role in fitting the material to the function, the fundamentals of the D-log E curve are presented, together with other characteristics such as spectral sensitivity, halation, fog, and reciprocity failure. The properties of both print-out and develop-out materials are discussed.

THE EMULSION SUPPORT. The light-sensitive emulsion of photo-oscillographic materials is located either on a transparent, flexible plastic sheet to make film, or on a roll of high quality paper. The use of roll film in general purpose galvanometer recorders has dwindled to a point where it is seldom used for this purpose. However, wide film rolls (8–7/8 to 40 inches) are still used in seismic and electrical well-logging equipment. Also, film is used extensively in oscilloscope recording.

Roll paper, now used almost exclusively in general purpose galvanometer recorders, may be divided into two basic types: print-out paper and develop-out paper (sometimes abbreviated POP and DOP). Each type has a paper base ranging in thickness from 0.0025 to 0.0055 inch with a silver-halide gelatin emulsion coated on its surface.

The print-out material is relatively slow, and its slow speed demands a high-intensity light source for recording, as discussed in a previous chapter. It produces a visible image without chemical processing, as opposed to the development-fixation treatment that must be applied to developing-out materials to render them readable. The principal advantages of developing-out papers are that they can be used at high writing speeds, and that they produce a high-quality record after processing. They are also unaffected by subsequent exposure to light, and are therefore capable of reproduction by the high-intensity light sources used in certain copying machines.

Paper is an excellent and durable base on which to coat the light-sensitive emulsion. Paper made by the early Chinese nearly 2,000 years ago still survives. The life of paper is chiefly dependent on manufacturing and storing conditions: bleaching and sizing operations during manufacture, and storage in air polluted with sulfur fumes may cause rapid deterioration of the paper fibers.

In its manufacture, there is a progression in the quality of paper that extends from a low-grade pulp to a superior-grade ledger stock. Legal records require the top-grade ledger material for they must withstand the ravages of time and use for generations. In the photographic materials for oscillograph recording, the paper-base stock must have the durability of ledger paper, plus a purity not found in standard ledger stock. The special purity is necessary to prevent contamination that might change the emulsion characteristics during manufacture, and to prevent chemical fogging, at the time of sensitizing or afterwards.

Desirable characteristics of the paper on which the emulsion is coated are strength (both dry and wet); lack of brittleness; uniformity of texture, thickness and color; resistance to the chemical action and staining during processing and to scorching; ability to be coated evenly; good writing surface for pencil or ink notations; good density on thin stock (for more footage per roll and reproduction capability); absence of impurities that could cause degradation of the emulsion, its life, or subsequent image; and elimination of characteristics or additives that could adversely affect the reproduction of the resulting image.

The paper base is made of either rag stock or alpha-cellulose fibers (other fibers are also used). Of the two materials, the long fibers of the rag stock are stronger, and typical strengths are shown in Table 5.1.

TABLE 5.1

TYPICAL STRENGTHS, RAG-BASE PAPER

Thickness	Burst Strength	Tear Strength With Grain	Across Grain	Folding Endurance (359 degree fold)
0.0045″	37 dry lbs/sq. in.	80 grams	104 grams	47
0.0025″	30 dry lbs/sq. in.	42 grams	42 grams	300

The ability to manufacture high-quality-base paper, and to have it in a pure form begins with the quality of the original rag content. At one time this was no problem because rags could easily be separated into wool, cotton and linen. Today, as the result of the ever-expanding synthetic-fiber industry, rags cannot be easily identified. The materials may even be made of mixtures of synthetics, and/or synthetic and natural fibers; therefore, separation and control have become impossible. Photographic paper manufacturers have resolved this problem in two ways: one is to develop a control method that extends from growing natural fibers and using them as the only source of fibers that go into "rag" base papers. The other is to use alpha-cellulose (wood pulp) and to improve those characteristics found lacking for use in the photographic process. Lack of strength, particularly wet strength, is one shortcoming of alpha-cellulose paper when it is compared with rag base paper. This deficiency has been minimized by impregnating the base stock with a plastic that renders it water-resistant. Recording papers so treated may not equal 100 percent rag stock papers in every respect, but they have advantages over both regular alpha-cellulose and rag stock material: they hold size better, and they do not readily absorb processing solutions. Because they do not get wet, they have greater "wet" strength, they can be processed faster, and they dry more rapidly. Papers made of manufactured fibers also absorb very little water.

A desirable characteristic, not found in paper, is dimensional stability. This may be defined as the retention of original size, so that a trace amplitude may be measured accurately. Film, of course, has much better dimensional stability than paper. However, papers made of manufactured fibers will change size only about one-fourth as much as either alpha-cellulose or rag-base paper. In addition, correct processing and drying procedures can minimize the size change in paper. When the paper, exposed in its dry state, becomes wet from the processing solutions and is subsequently dried, it may not return to its original size, especially if it is dried under tension. Cycling of drier drum temperature may also cause size changes, as described in Chapter 8.

The influence of ambient humidity, both before exposure and after processing, is apparently a linear relationship. Average figures are 0.1 percent elongation change for each 10 percent of RH increase. Upon completely wetting a typical paper, it will expand approximately 0.4 percent in length, and approximately 2.1 percent in width. Dry rupture will occur at 3 percent elongation; wet rupture will occur at 5 percent elongation.

THE EMULSION. The light-sensitive coating applied to the paper or film is called the *emulsion*. It is composed of a dispersion of minute and somewhat imperfect crystals of silver halide in gelatin (Fig. 5.1). Slow emulsions contain silver halide in the form of silver chloride, and fast emulsions contain silver bromide with a small amount of silver iodide. These crystals are called grains, and are approximately one 0.001 millimeter in diameter. The thickness of the coating of emulsion on the base is typically less than 0.0005 inch.

The formation of a trace on a photo-oscillographic recording occurs because the silver halide is sensitive to light. The amount of silver formed by exposure is too little to be seen, and is referred to as the *latent image*. The latent image can be made visible by *development*, which broadens the latent-image centers, and converts them to dense metallic silver deposits that stand out clearly against the unexposed background. Development may be considered as an amplifying process; in a fast DOP emulsion, chemical development can extend the amount of silver in the latent image by a factor of 10^9.

PRINT-OUT PAPERS. The discovery of photographic printing is usually credited to Fox-Talbot (Chapter 1), who employed what today is termed the salted-paper process. Improvements to this early method produced a printing-out material which photographers used for printing portraits and other pictorial images. With the advent of better quality developing-out papers and printing equipment, the print-out material became obsolete (except for "proofs"). The search for a direct-recording photographic medium led to printing-out paper as a possible avenue for investi-

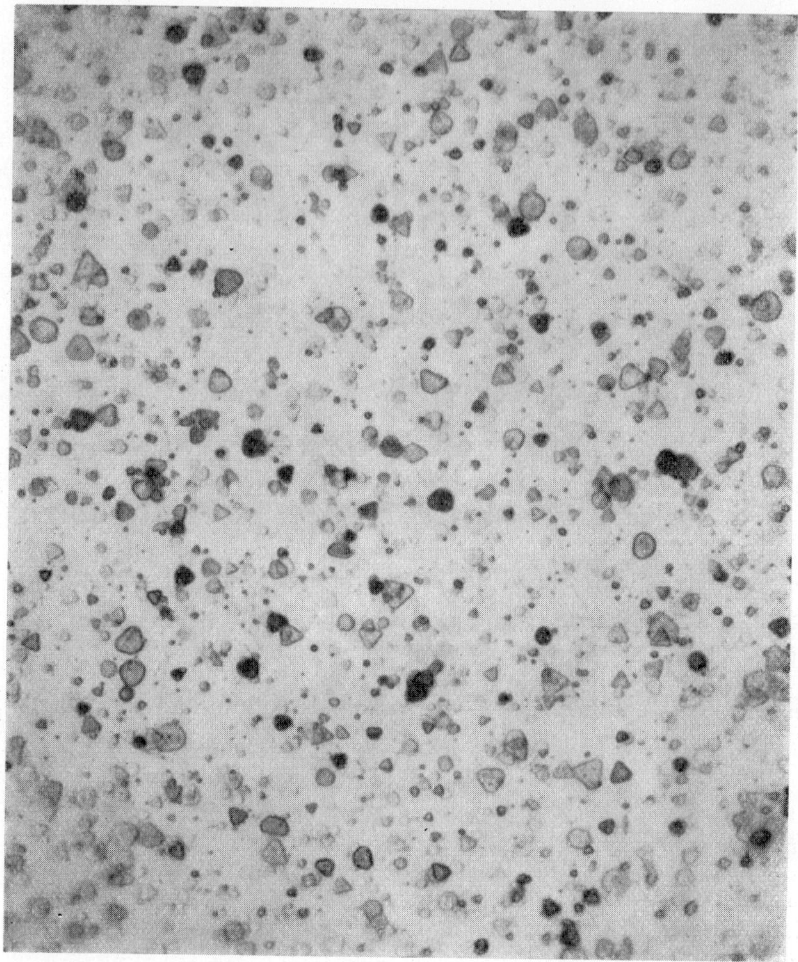

Fig. 5—1. Photomicrograph (2500×) of silver halide dispersion in gelatin

gation. In technical reports by Hunt[5.1] and by Jacobs[5.2] and also by Jacobs and McClure[5.3] research is described that fostered improvements in print-out recording techniques. The research has enabled better papers to be made, the latent image to be amplified at higher speeds and without degradation, and better post-exposure desensitization to be achieved.

The generic term, *direct-print* paper, refers to a specialized print-out paper used in photo-oscillography which produces a visible image by exposure to light rather than by chemical development. The galvanometer recorder must be equipped with a high-intensity light-source. Usually this is a miniature, high-pressure mercury or xenon lamp of the type described in Chapter 3. It is used with print-out paper which is chiefly sensitive to ultraviolet radiation and blue light. However, recent develop-

210

ments have produced print-out papers having sensitivities to longer wavelengths, and these materials can be used in recorders equipped with special high-intensity incandescent lamps.

Print-out paper is very slow when compared with conventional develop-out recording papers. Although the two materials bear no direct resemblance to one another, Fig. 5.2 shows (in a relative way on the same graph) a family of D-log E curves for develop-out papers and a curve for print-out material. While there is no direct speed relationship, and an exact exposure-index value cannot be assigned, various authorities report direct-print material to be capable of writing speeds ranging from 40,000 to 75,000 inches per second[5.4].

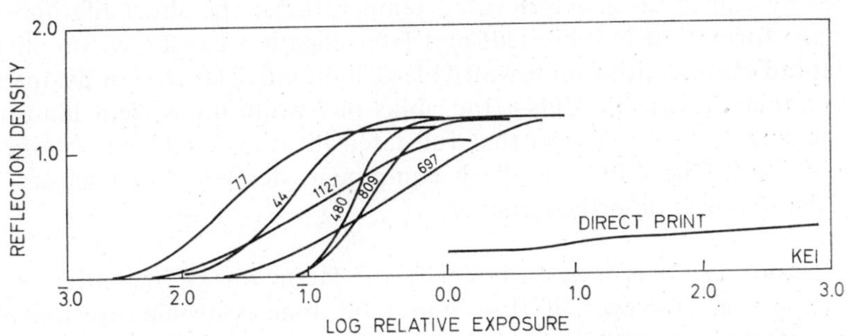

Fig. 5.2. Relative comparison of develop-out and print-out materials (Eastman Kodak Papers)

Two exposures are required: a specific exposure made by the galvanometer, and a general over-all exposure. The second exposure, of an amplifying nature, may be thought of as development by light, and is termed *photo-development, latensification, photo-activation,* or *photolysis.* A typical latent-image exposure trace made by the galvanometer recorder on papers currently available will appear in 15 to 30 seconds when latensified with approximately 30-foot candles of fluorescent light. The trace color, usually blue against a tan or greenish-gray background, will have a maximum density of 0.40 to 0.60. The peak density will be reached in 3 to about 30 minutes of exposure to a fluorescent source (approximately 5 minutes is optimum for some papers). Changes in image color then begin, and image loss will occur if exposure is continued.

In the latensification process, light reduces the remainder of the exposed silver-halide grain to metallic silver and amplifies the trace as much as 500 to 1000 times. This is a function of the low-intensity reciprocity law failure of print-out materials. The action is catalyzed by the latent image which probably is already metallic silver. Therefore, care should be used in handling and loading the paper to prevent accidental exposure to low-intensity actinic light.

There are two other significant factors in print-out materials: spectral response and reaction to temperature. To speed up amplification (or photodevelopment, as it is sometimes called), experiments were conducted by Jacobs (again see reference [5.2]) using special lamps and elevated temperatures, which resulted in very rapid latensification. A spectrographic study revealed that, while a wide band of radiation produces latensification, a narrow section of this band is responsible for background fog. A search for a light source that did not emit radiation in the fog band resulted in the use of a low-pressure discharge tube containing zinc, with a conventional argon starter gas. However, a lamp of this type that had sufficient output, together with its power supplies, proved to be impractical from the cost and size standpoint. An investigation into ways of eliminating the fog band (rather than of avoiding it) resulted in elevating the temperature of the paper during the latensification phase. It is believed that during latensification at room temperatures, unstable latent-image centers in the background begin to grow before they dissociate, thereby causing fog. At elevated temperatures, the short lifetime phase of latent image formation is inhibited, and latensification can be accomplished with ultraviolet radiation. With four 8-watt "black-light blue" (BLB) lamps, manufactured by Sylvania Electric Products, Inc., plus one white fluorescent lamp to aid in visual assessment, Jacobs found that full latensification could be obtained in less than one second. The CEC Dataflash equipment, described in Chapter 3, was a result of the research described above.

TRACE STABILITY. It is wise to latensify the latent image immediately after the primary exposure. However, if this cannot be done conveniently, the record, if properly protected, may be latensified as long as six months later with no appreciable loss of image quality.

After latensification to maximum density, the trace life will depend chiefly upon the way the record is handled. If protected from light, it will last indefinately; if exposed to artifical light, the image may be degraded after 8 to 24 hours. If the record is important, and is to be read carefully or is to be duplicated, it should be given a chemical treatment to insure that it will not be damaged. The treatment, which changes the blue trace to black and the background to gray, can be carried out under normal room light in a stabilization or other processor. If duplicates are desired, the stabilized record can be run through a diazo machine without damage, and although a light-background record is desirable to read, a gray background makes excellent reproductions.

Sensitometry

The choice of sensitized recording material is dependent upon the spectral output of the oscillograph light source, its intensity, and the speed of the motion of the light spot from the galvanometer. Certain qualities other than the physical aspects of the sensitized material play an important part in the type of material that should be used under a given set of conditions. The properties are called the sensitometric characteristics of the material.[5.5] This phase of oscillography is often misunderstood even though the photographic process is employed in recording data. The sensitized-material selection can often be made from information in tabular and graph form furnished by the manufacturer. This published data is determined by sensitometry, and a review of its principles will aid in interpreting the characteristic curves of paper and films.

Initially, sensitometric methods were developed by the manufacturer for control purposes. However, where exacting and repeatable results must be obtained by users of material, sensitometry offers a precise control method. The need for such control in the motion picture industry, arising from the exact processing required for sound films, resulted in the general use of sensitometric control to maintain the quality of both picture and sound. Other users of sensitometry are engaged in research and industry; they depend on sensitometric principles and practice to obtain consistent results and high quality even under unfavorable conditions. Therefore, it is important that an extension be made of this knowledge into oscillography; it is much easier to compare data published by the manufacturer and then make a few tests than to run innumerable tests on most of the available materials.

The D-log E Curve. Several important factors concerning choice of material are relative sensitivity, contrast, and spectral sensitivity. The sensitometric characteristics may be graphically represented by plotting the response (silver density) against the logarithm of the exposure. The curve obtained when a line is drawn through the plotted points is known as the "characteristic" or "D-log E" curve. Such curves are also referred to as "H and D curves" after Hurter and Driffield who originated this graphical method. The curve that is obtained is a combination of the emulsion characteristics, the conditions under which it was exposed, and the processing conditions.

Plotting the Curve. A 21-step sensitometric strip is shown in Fig. 5.3 together with a typical instrument used to expose such a scale. The instrument is named a sensitometer and it produces a series of carefully controlled graduated exposures on the material to be tested. Whatever method is used to expose the sensitized material must meet the following minimum conditions: the light source must be constant and have a known intensity and spectral output; a series of known exposures must be produced on the sensitized material. After exposure, the sensitometric strip is processed under precisely standardized conditions of development; exact formulation, temperature, time, and agitation are required. After being fixed, washed, and dried, the finished strip is "read" on a densitometer, and the densities are plotted on graph paper to produce the finished D-log E curve. Either reflection or transmission density may be measured, depending on whether the material is paper or film. American Standards (PH 2.17–1958 and PH 2.19–1959) describe the conditions for diffuse reflection and diffuse transmission densities. The densitometer may be the photoelectric type, an example of which is shown in Fig. 5.4. The densitometer must be capable of determining the least visible density under standardized and repeatable conditions.

213

Fig. 5.3. *Top*: Herrnfeld sensitometer; *Bottom*: Sensitometric strip

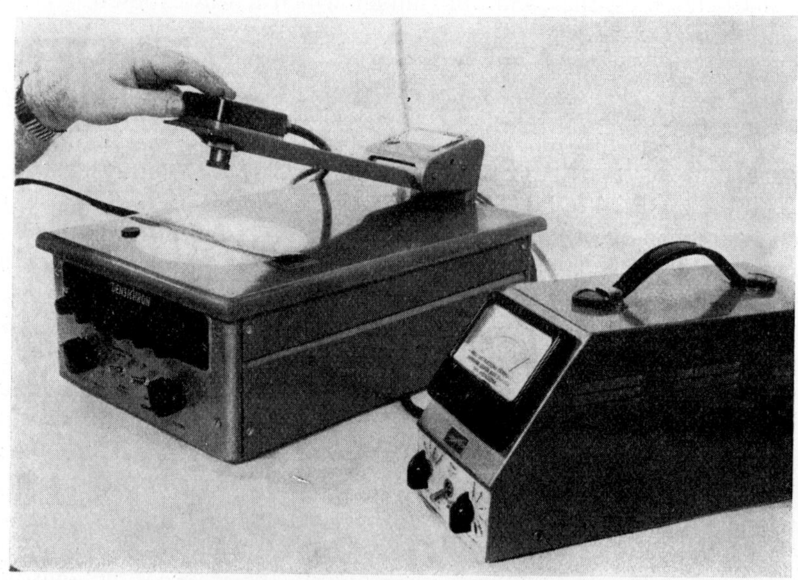

Fig. 5.4. Densichron densitometer

A typical D-log E curve, illustrated in Fig. 5.5, represents the "gray scale" or "step tablet" plotted as density changes against the logarithm of the exposure. The lower portion of the curve, at the left, is called the "toe", the upward sweep is the "straight-line portion" and the curved line at the top is the "shoulder". At the bottom of the graph is a logarithmic scale (to the base 10). A logarithmic rather than a linear scale is used because the large range of exposure values cannot be compressed sufficiently on a linear scale.

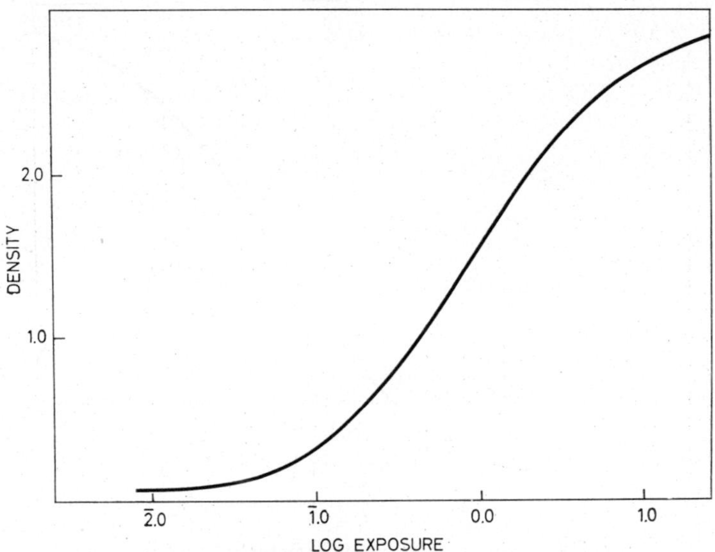

Fig. 5.5. D-log E curve plotted from density measurements read from sensitometric strip of Kodak Linagraph Recording Film

When the exact light level of the sensitometric exposure is unknown, an arbitrary logarithmic scale is used, and this is known as a "relative log-E" scale.

Negative materials are represented by a curve that begins at the bottom left and progresses to the top right. Positive materials and reversal materials have a slope that begins at the bottom right extends to the upper left. Therefore, for films and papers used in the conventional oscillograph, a negative-type D-log E curve is drawn. Similarly, a positive type D-log E curve is used for oscilloscope materials which use reversal films, or which deliver a print having a white trace against a dark background. In oscillography, confusion sometimes results on these two points, because dark backgrounds are commonly associated with negatives, and white backgrounds with positives.

INTERPRETING THE CURVE. Blackening occurs to the sensitized material only after a certain minimum exposure has been given. This point is known as the threshold exposure and is represented by B on the curve in Fig. 5.6. The region of no exposure (AB) is the fog level of the material. The region of under exposure, the toe, is nonlinear and is represented by the area B to C. The most significant part of the curve is linear; it is the straight line portion or region of constant gradient between C and D. It is the region of correct exposure in which density is proportional to $\log_{10} E$. The area from D to E is the shoulder, and is the region of overexposure. Additional exposure will produce reversal in which less density is produced.

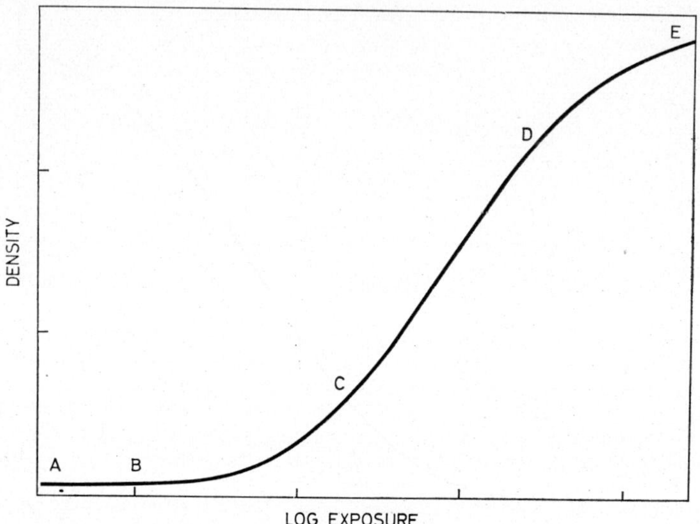

Fig. 5.6. Interpretation of characteristic curve

CONTRAST. The separation of tones in the step wedge is referred to as contrast. Contrast is dependent on two factors: the material and its development.

1) Material. Various materials differ in the manner in which they reproduce minimum exposures and maximum exposures. Sensitized films or papers that produce a sharp density difference as a result of a slight increase in exposure are said to have high contrast. This is shown by a steep characteristic curve since the increase in vertical distance (density) is great for a given horizontal increase in log E. Therefore, when sensitometric properties are judged by reference to graphic data, the high-contrast materials can be quickly spotted by the steepness of their curves.

2) Development. Contrast of a given material may be increased by an increase in development (time, temperature, agitation). When development is increased, the density differences on the straightline portion of the characteristic curve also increase, resulting in a steeper slope. The tangent of the angle formed by the straight-line portion of the D-log E curve and the E axis is a measure of these density diffe-

216

rences, and is therefore a measure of the amount of development. This contrast ration in the straight-line portion is termed *gamma*, and is represented by the Greek letter γ. It is expressed in the following equation:

$$\gamma = \frac{D_2 - D_1}{\log E_2 - \log E_1}$$

Gamma is usually determined with a transparent scale, known as a gammeter (Fig. 5.7). Gamma may be read directly from the scale; the arrow is placed at the

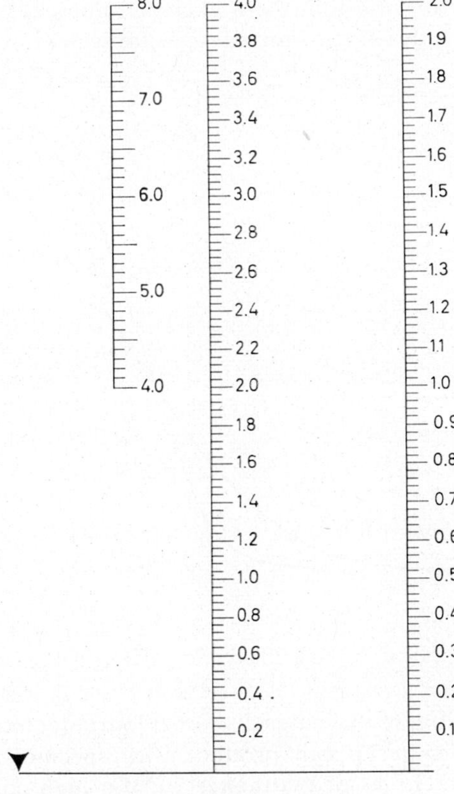

Fig. 5.7. Typical gammeter

point of intersection of the straight-line portion of the D-log E curve, extended to the base line (log E axis). The point of intersection of the straight-line portion of the curve with the gammeter scale indicates the gamma. Gamma reaches a maximum with prolonged development, and will then decline as fog (undesired density) builds up.

Fog. Fog may be due to two causes: that caused by unwanted exposure, and that inherent in the development process. Exposure fog may be due to light leaks in the magazine or in the oscillograph; or from incorrectly adjusted oscillographs. It may also be caused by light leaks in the darkroom, or from incorrect or improperly used safelights. Development fog is held to a safe level by restrainers in the developing solution. However, with prolonged development, or with solutions that are not fresh, the fog level may be higher than normal.

Latitude. The range of exposures in the straight-line portion is termed the *latitude* of a sensitive material. It is not a constant of the emulsion, but varies inversely with the amount of development or gamma, i.e., the latitude decreases as the development increases. In addition, the higher the inherent contrast in the sensitized material, the less latitude (or the shorter exposure scale) it will have.

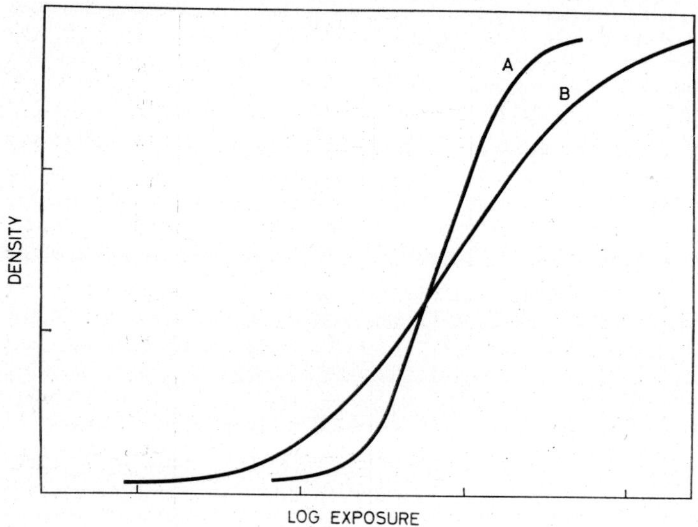

Fig. 5.8. Comparison of two D-log E curves

The latitude of oscillographic material is of importance because it indicates the range of exposure the material can produce with specific development. Fig. 5.8 shows the characteristic curves of two materials,[5.6] a high contrast paper *A* and a low contrast paper *B*. *A* may appear faster than, slower than, or equal in speed to *B*, depending on the density at which the comparison is made.

When it is necessary to record at high writing speeds, *B* will be better than *A*, because it requires less exposure to obtain sufficient density for a visible record. It is possible to see traces which are about 0.02 more dense than the background, but for good visibility it is better to have a higher density difference than this. It is becoming common practice to give writing speeds at a density level of 0.1.

218

 The sensitivity of oscillograph material, usually called *speed*, may be defined as the degree of blackening of the emulsion upon exposure to light. Sensitivity is probably the most important factor governing a choice of recording material.[5,7] Linked with the emulsion speed is *spectral sensitivity*, that portion of the spectrum to which the material is most sensitive. All photographic

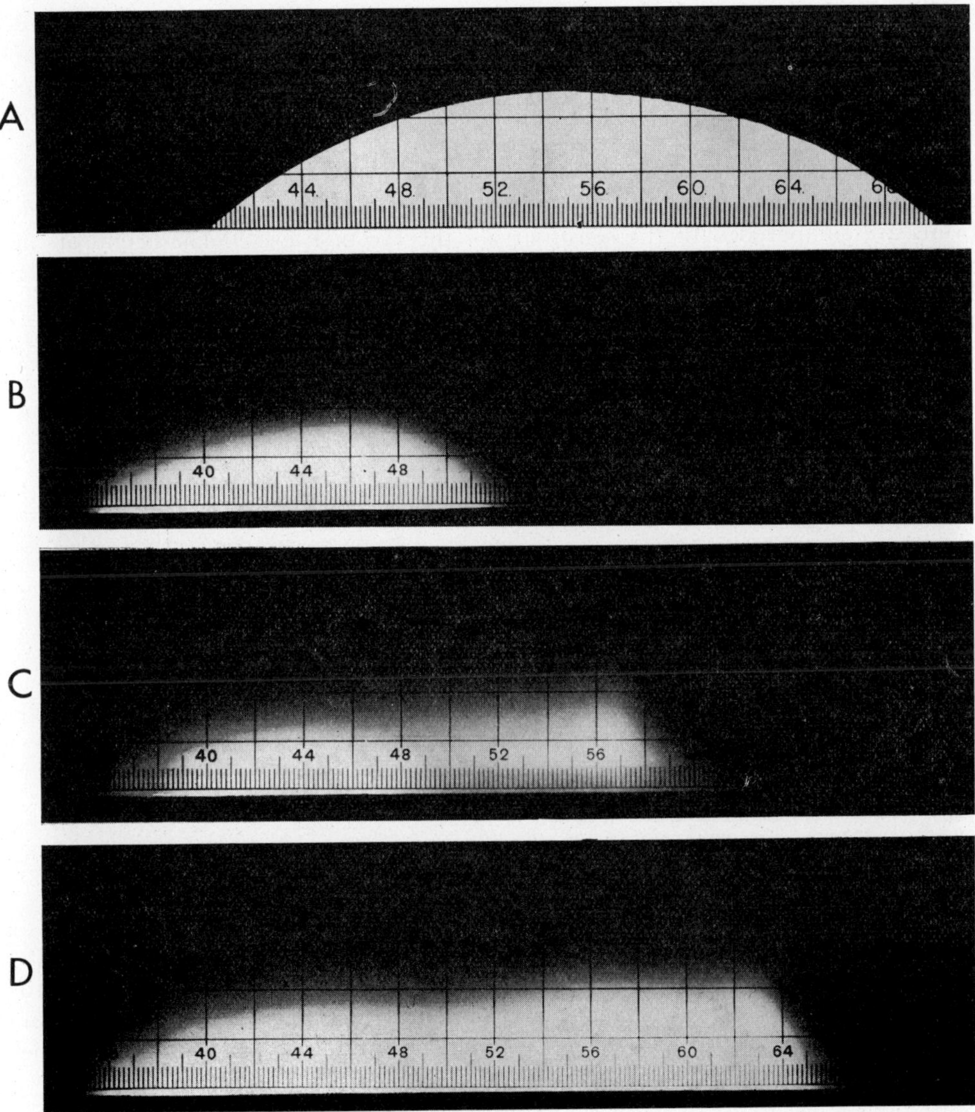

Fig. 5.9. Comparison of spectral sensitivity of the eye and photographic materials
A. Eye; B. Blue-sensitive film; C. Orthochromatic film; D. Panchromatic film

emulsions are sensitive to blue, violet, and invisible ultraviolet light. However, the light from the recorder contains other wavelengths as well, and to utilize as much of the output as possible, most recording materials are made sensitive to green as well as blue light. Such materials are referred to as *orthochromatic*. The sensitivity to longer wavelengths of the spectrum is achieved by the addition of a dye to the emulsion during the manufacturing process. Dyes used for this purpose may render the unprocessed material a pink shade, but the color from the sensitizing dye is removed by a "decolorizing" action during processing. If stabilization rather than conventional processing is used, the sensitizing dye may not be completely decolorized. Therefore, stabilized records may have a slight tint, although recent improvements in optical sensitization have greatly improved materials in this respect.

The color sensitivity of the material is indicated by a *wedge spectrogram*. These are positive prints made from original spectrograms, and typical spectrographic prints are compared with the sensitivity of the eye in Fig. 5.9. The height of the white area indicates the relative sensitivity of the material, and the scale at the bottom indicates the wavelength. A typical spectral sensitivity graph is shown in Fig. 5.10.

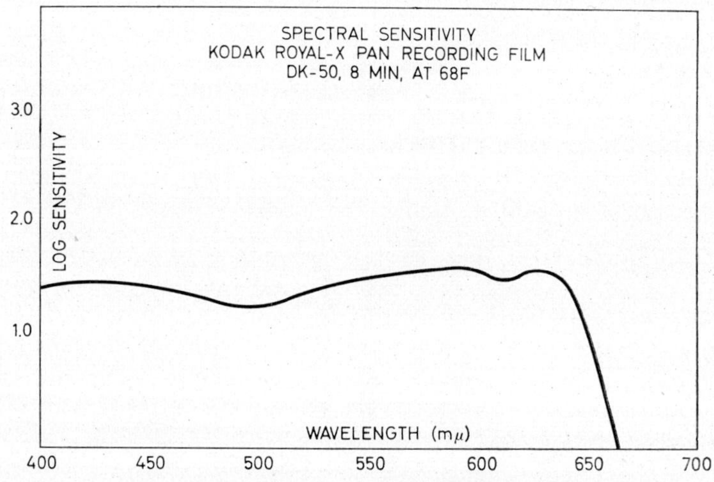

Fig. 5.10. Typical spectral sensitivity graph

In addition to blue-sensitive and orthochromatic materials, a third type of emulsion sensitivity is available. This is the *panchromatic* emulsion which is sensitive to all colors. Develop-out panchromatic emulsions are available on both film and paper bases. High speed panchromatic materials must be handled in total darkness. Some print-out panchromatic papers are also available.

Reciprocity Effect. In 1876, Bunsen and Roscoe found that the density resulting from exposure is the product of time and light intensity. This *reciprocity law*, which holds for most photochemical reactions, means that exposure time may be decreased if the exposing light is increased proportionally. Reciprocity-law failure occurs under conditions of extremely long or ultra short exposure. A typical reciprocity-failure curve is shown in Figure 5-11. If the curve were a straight horizontal line,

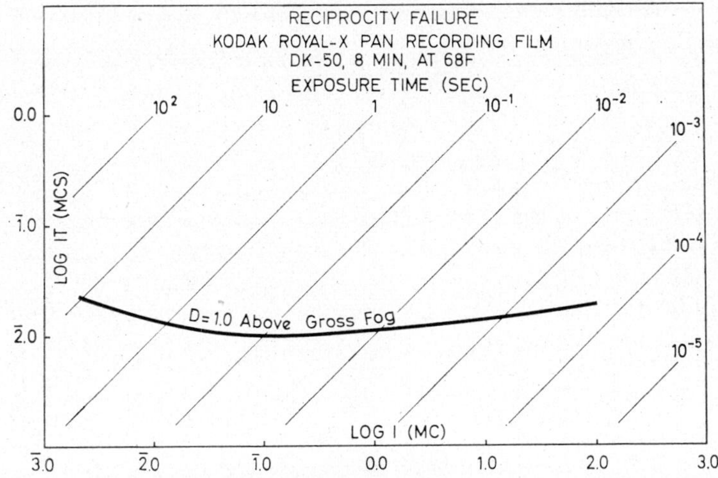

Fig. 5.11. Typical curve illustrating reciprocity-law failure

the material would exhibit no reciprocity failure. However, for the material illustrated in the graph, the upward bend of the curve indicates that disproportionally more exposure is necessary to achieve a given density.

When the oscillographer exposes material at high writing speeds, the resulting oscillogram is subject to reciprocity failure. He may find that sensitometric curves, produced under "normal" conditions of time, intensity, spectral output of the exposure source, etc., are inaccurate for his application. This is true because the exposure of a given point by the galvanometer spot may be on the order of microseconds or less, and subject therefore to reciprocity failure.

Because the oscillographer has such a problem in determining exposure, he must: 1) determine the maximum velocity of the writing spot of light which exposes the trace; and 2) select a sensitized material that has a writing-speed capability greater than the determined spot velocity. The nomograph in Fig. 5.12, relating maximum spot velocity, amplitude, and frequency, is exact only for pure sinusoidial waveforms. However, it will give close approximations as long as the harmonic content is low in the case of complex harmonic waves.

RESOLUTION. Resolving power is a measure of the ability of an optical device or a photographic material to produce distinguishable, fine detail. It is used as an index of this quality in optical components or systems, or in photographic materials. Resolving power is usually measured by photographing a parallel-line test chart at a greatly reduced scale (See Fig. 5.13). The chart lines are grouped according to width and are spaced apart by a distance equal to the width of the individual lines. When examined with the aid of a microscope, the group of lines that is just recognizable as individual lines is taken as the criterion for the resolving-power determination. By reference to a table, the resolution may be stated in terms of the number of lines per millimeter. Narrower, more closely spaced lines will appear as a gray area, not as individual lines.

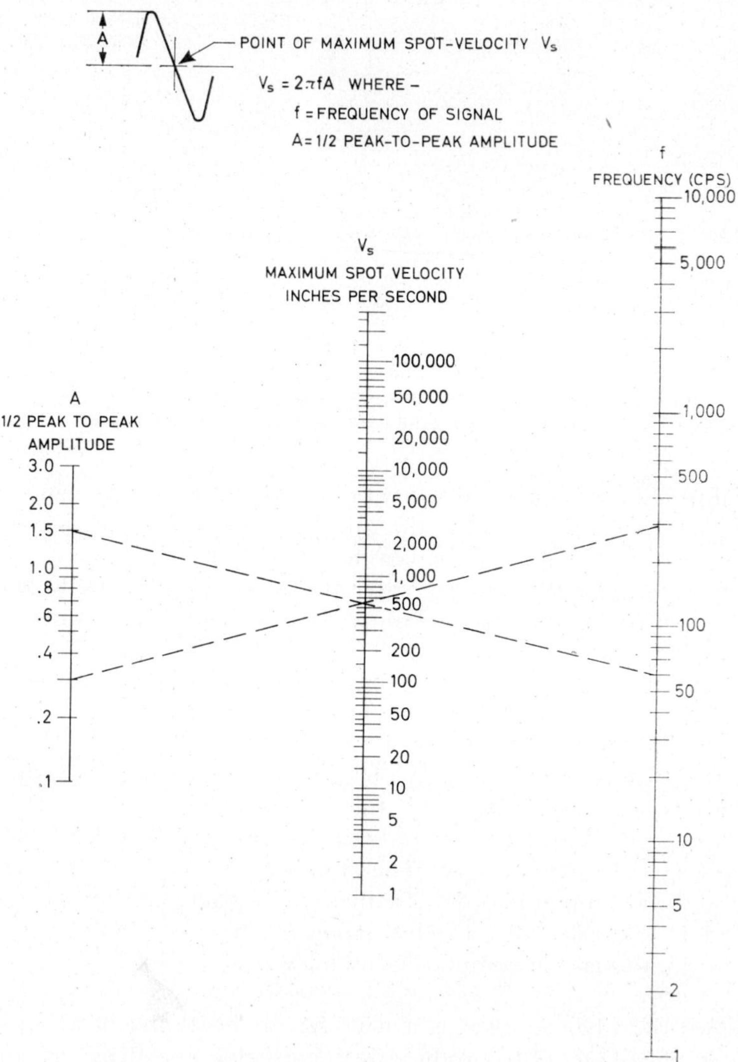

Fig. 5.12. Nomograph relating amplitude, frequency, and maximum spot-velocity for sinusoidal traces

Although the type of developer and time of development in general have little effect on resolution, some fine-grain developers can improve resolution. Two factors that greatly influence resolution are granularity of the emulsion, and exposure. When exposure is extreme, either under or over, the resolution will be less than when the material is correctly exposed.

Resolution of the photographic material is only of casual interest for full-size records, but may have significant value when recording on 35mm film with wide-aperture lenses.

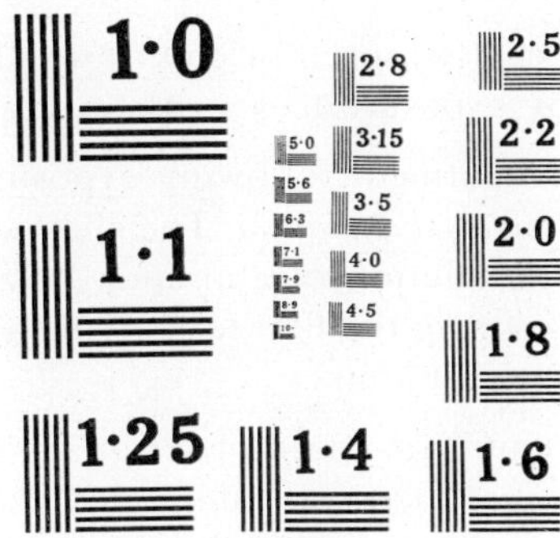

Fig. 5.13. Resolution test target

CHOOSING THE MATERIAL. After the probable writing speed has been determined, the sensitized material may be selected from data published in the oscillograph instruction manual (also see [5.8] and [5.9]). In addition, a review of the information published by the sensitized-material manufacturer will be helpful if it is considered on the basis of characteristics discussed in this chapter. If there is any doubt about the speed of the material that should be used, a test should be made prior to recording actual data. Generally, this can be done by recording a single trace at a fixed frequency, and then varying the amplitude to produce the required range of spot velocities. The calculation of the spot velocity is made at the point in the record where the zero-crossing just disappears. This value of "X" inches per second writing speed is assigned to the toe of the D log-E curve. The test, and the *subsequent recordings* should be processed under identical conditions of time (or processor speed) and temperature, and fresh chemicals should always be used.

Regular-speed emulsions produce better quality records than high-speed emulsions. Therefore, regular speed materials should always be used unless it is not possible to obtain the required writing speed at normal filament voltage.

The following characteristics should be considered when analyzing the test record: 1) consistent sensitivity for the writing speeds; 2) good exposure latitude; 3) good trace-to-background contrast; and 4) low background fog. Other qualities such as light fog, streaks, mottle, and image color should be considered.

In addition to the photographic characteristics of the emulsion, the user will want to choose a material having the most desirable physical characteristics. If record length is important, thin-base material will increase the magazine capacity by approximately 80 percent. Thin-base stock dries faster when it is processed, and drying speed is sometimes the limiting factor in obtaining maximum processing speed. If duplicate copies of the record will be required, thin-base material will permit reproductions to be made by ozalid and similar processes.

Generally it is desirable to identify the record and to annotate it with supplementary data. Therefore, the paper surface should readily accept pencil, ink, and other markings.

HYPERSENSITIZING. The effective speed of emulsions can sometimes be increased by chemical treatment or by exposure to a weak source of light.

Latensification. Chemical treatment following exposure or post image exposure to a weak light source is called *latensification.* The chemical treatment is accomplished by bathing the film in a solution of ammonia or ammoniacal silver chloride for several minutes or by exposing the film to mercury vapor in a closed container at room temperature for 20 to 30 hours.

Prefogging. The effective emulsion speed of sensitized materials can be increased by a general controlled exposure to a weak source of light before the image-forming exposure is made. Various emulsions respond differently to this treatment, and the speed of some materials (Polariod, for example) may be increased as much as two to three times by the prefogging exposure.

DESCRIPTION OF WINDING SPECIFICATIONS. When an order is placed for oscillograph recording materials, the following minimum information should be included on the order: 1) number of rolls; 2) roll size — width and length; 3) paper or film name (and number if there is one); and 4) specification number.

> NOTE: Do *not* confuse sensitized material numbers with specification numbers. For example, Linowrit 4, will *also* have a specification number that must be on the order. Specification numbers are identified in the following paragraphs and tables.

Specification Numbers. These numbers are used as part of the nomenclature to identify a particular spool or core, the winding (emulsion in or out), leader, perforation and other particulars. The use of "spec" or "sp" numbers reduces the possibility of error, and insures delivery of material that will function in the recorder. In addition, the simple one, two, or three-digit number eliminates a long verbal description concerning the winding data. Table 5.2 lists the most common windings. Table 5.3 relates the specification numbers to recorders by make and model number.

KODAK LIMITED WINDING SPECIFICATIONS. Typical recording paper sizes, supplied by Kodak, Limited, London, are given in Table 5.4.

Kodak Limited recording paper is normally wound emulsion inwards on cardboard cores of internal diameter 1/2 inch (approximately 13 mm) for widths up to

Table 5.2

WINDING SPECIFICATIONS

Spec. No.	Wound On	Dia. Inches	Emulsion Winding	Additional Description
1	Strawboard Core	1–1/8 ID	IN	
2	Strawboard Core	1–1/8 ID	OUT	
4	Strawboard Core	1–1/8 ID	IN	With 4 ft. black paper leader attached to paper.
12	Strawboard Core	2 ID	IN	Core has black ends.
16	Plastic Type U Core	2 OD	IN	Keyway through core center hole.
28	Strawboard Core	2 ID	IN	
39	No. 10 Spool	—	IN	Perforated, 8 ft. perforated spliced-on- red-and-black duplex paper leader and trailer.
42	Strawboard Core	1–1/8 ID	OUT	4 ft. attached black paper leader and trailer.
44	Daylight Load Spool	—	IN	5 ft. attached red-and-black paper leader and trailer.
79	Daylight Load Spool	—	IN	Unperforated; spliced-on 20 in. red-and-black duplex paper leader and trailer.
87	Strawboard Core	3 ID	IN	
109	Strawboard Core	2–1/4 ID	IN	
111	Strawboard Core	3/4 ID	IN	
119	Daylight Load Spool	—	IN	4 ft. attached red-and-black paper leader and trailer.
124	Strawboard Core	1–1/8 ID	IN	6 ft. attached black paper leader and trailer.
130	Strawboard Core	1–1/8 ID	IN	Perforated, one side only.
174	Strawboard Core	1–1/8 ID	IN	Same as Sp. 1, except paper attached to core.
195	Strawboard Core	3 ID	IN	
207	Daylight Load Spool	—	IN	5-ft., 2–3/4 in. attached red-and-black paper leader and trailer.
218	Daylight Load Spool	—	IN	Perforated, no splices.
219	Daylight Load Spool	—	IN	No leader or trailer.
287	Core 14	3 ID	OUT	Punch mark 10 feet from core end. Paper not attached to core.

TABLE 5.3

RECORDER WINDING SPECIFICATION REQUIREMENTS

BRUSH INSTRUMENT CO.

Series or Model No.	Roll Width	Standard length roll	Extra thin length roll	Specification
Direct Writing				
2300	6″	100′	200′	1

CENTURY ELECTRONICS AND INSTRUMENTS

Series or Model No.	Roll Width	Standard length roll	Extra thin length roll	Specification
Conventional				
440 E5	12″	– –	400′	1
408	8″	150′	– –	1
409				
Mag. 409 D6	3 5/8″	50′	75′	1
409 D9	3 5/8″	100′	150′	1
409 D11	3 5/8″	50′	75′	1
409 D12	3 5/8″	100′	150′	
409 D280	3 5/8″	100′	150′	1
409 E349	3 5/8″	300′	600′	1
409 E487	3 5/8″	300′	600′	1
414	3 5/8″	50′	75′	1
439(100′mag)	3 5/8″	100′	150′	1
(500′mag)	3 5/8″	350′	500′	1
Direct Writing				
440 E4	12″	– –	400′	1
444	3 5/8″	– –	150′	1
447	8″	250′	475′	1
445	3 5/8″	– –	150′	1
1000	6″	– –	400′	

HONEYWELL

Series or Model No.	Roll Width	Standard length roll	Extra Thin length roll	Specification
Direct Writing				
906	6″	100′	200′	2
1406	6″	100′	200′	2
1108	8″	200′	400′	2
1012	12″	200′	400′	2
1508	8″	100′	200′	2
1612	12″	200′	400′	2

CONSOLIDATED ELECTRODYNAMICS CORP.

Series or Model No.	Roll Width	Standard length roll	Extra thin length roll	Specification
Conventional				
5–114	7″	100′	225′	1
5–119	12″	250′	475′	28
Datarite Magazine				
5–114 5–047	7″	– –	400′	28
5–119 5–036	12″	– –	400′	28
Direct writing				
5–124	7″	100′	200′	111
5–119V	12″	250′	475′	28
5–123	12″	250″	475′	1
5–133	12″	250′	475′	1

MIDWESTERN INSTRUMENTS, INC.

Series or Model No.	Roll Width	Standard length roll	Extra thin length roll	Specification.
Conventional				
560	3 5/8″	50′	75′	1
561	3 5/8″	50′	75′	1
581	3 5/8″	100′	150′	1
581 (high capacity magazine)	3 5/8″	250′	350′	1
591	12″	250′	425′	1
590	7″	100′	225′	1
Direct writing				
615	5″	100′	150′	28
621	6″	100′	200′	28
616	7″	200′	300′	28
616	7″	100′	175′	28
602	12″	200′	350′	28
603	12″	200′	350′	28
606	12″	200′	350′	28
607	12″	200′	350′	28
601	7″	200′	350′	28
800	12″	250′	475′	1

SANBORN COMPANY

Series or Model No	Roll Width	Standard length roll	Extra thin length roll	Specification
Direct Writing				
650	8″	200′	350′	1

and including 70mm, and 1 inch (approximately 25mm) for widths over 70mm. Alternatively, it may be supplied on cores of internal diameter 1 1/8 inch (approximately 28.5mm), or 1 3/16 inch (approximately 30mm), to special order.

Paper widths up to and including 70mm may be supplied perforated in both edges, with perforations according to British Standard 1193: 1961, or not perforated. Paper wider than 70mm is supplied only without perforations.

Certain other items may be supplied to special order; full details of perforations, core, etc., are required.

CALCULATING ROLL DIAMETERS. The external diameter of rolls of film or paper may be calculated approximately from the following formula furnished by Kodak, Limited:

$$D = \sqrt{(1.27\ Lt + d^2)}$$

where D = diameter of roll of film or paper

L = total length of film or paper

t = thickness of film or paper

d = external diameter of core

(N. B. The dimensions must be in the same units. Factors t and d should be measured in each specific case.)

Designers of recording instruments should bear in mind that the diameter of the roll may vary to some extent, owing to variations in tension during winding, and that there should be enough room for the operator's fingers when loading or unloading the equipment.

TABLE 5.4

NOMINAL WIDTHS AND LENGTHS OF RECORD-
ING PAPER SUPPLIED BY KODAK, LIMITED

Nominal width	Nominal length	Nominal width	Nominal length
35mm	7·5m (24·6 ft) 15·0m (49·2 ft) 30·0m (98·4 ft)	90mm	30·0m (98·4 ft) 60·0m (196·8 ft)
		120mm	30·0m (98·4 ft)
45mm	15·0m (49·2 ft)	4 in.	60·0m (196·8 ft)
60mm	15·0m (49·2 ft) 30·0m (98·4 ft)	6 in. 8 in.	60·0m (196·8 ft) 60·0m (196·8 ft)
70mm	7·5m (24·6 ft) 15·0m (49·2 ft) 30·0m (98·4 ft)	10 in. 12 in.	60·0m (196·8 ft) 60·0m (196·8 ft)

The quantity of sensitized material remaining on a partially used roll may be calculated from the following formula furnished by CEC:

If D is outside diameter (inches)

C = core diameter (inches)

T = paper thickness in inches

$$\text{Number of wraps} \frac{D-C}{2T}$$

$$\text{Average diameter} \frac{D+C}{2}$$

$$\text{Length} = \pi \times \text{Ave. diam.} \times \text{No. of wraps}$$

$$= \frac{\pi(D-C)(D+C)}{4T}$$

$$= \frac{\pi(D^2-C^2)}{4T} \text{ inches}$$

$$\text{or } \frac{\pi(D^2-C^2)}{48T} \text{ feet}$$

STANDARD RECORDING MATERIALS. Since there are frequent changes in sensitized products, such as physical and sensitometric quantities, replacement with improved materials, etc., the user is urged to contact his supplier for current availability and characteristics. A list of standard sensitized recording materials now furnished by various manufacturers is printed in Table 5.5.

Special materials

EXTENDED RANGE FILM. Extended Range (XR) film is available from Edgerton, Germeshausen and Grier, Inc., 160 Brookline Avenue, Boston 15, Massachusetts. It consists of a standard film base coated successively with three emulsion layers having differing degrees of sensitivity. All layers are panchromatic (responsive to the entire visible spectrum), and each layer simultaneously records that brightness level which is within its particular response range. Fig. 5.14 shows conventional D-Log E characteristics represented by three separate D-log E curves, one for each emulsion layer. A more useful curve is also shown, which represents the total or summation of densities. The speed range is 400/7° to 0.004/0.2°.

After exposure, XR film is processed in Kodak Color Processing Kit, Process C-22. During development, silver is replaced by dye in direct proportion to exposure. Separate colored images form in each layer, each containing all the gradations of the silver image. The images will be yellow, in the top layer; magenta, in the

TABLE 5.5 A

PRINT-OUT RECORDING PAPERS

Manufacturer	Designation	Spectral sensitivity	Thickness*		Remarks
			Std.	Thin	
Agfa-Gevaert	Oscilloscript d	Ortho	×	×	Writing-speed sensitivity to 100,000 ips claimed.
Consolidated Electrodynamics (CEC)	Dataflash 55	Blue	×	×	In excess of 65,000 ips
	Dataflash 57	Blue	×	×	Similar to Dataflash 55, with develop-out option when run closed magazine.
	Dataflash 58	Pan	×	×	Extended spectral sensitivity for high intensity sources, plus extended unstabilized life. Speeds in excess of: 10,000 ips tungsten. 65,000 ips mercury.
Du Pont	Lino-Writ 5	Blue	×	×	Ortho sensitive materials can be used in high-intensity incandescent recorders.
	Lino-Writ 7	Ortho	×	×	
	Seismo-Writ Dry writing	Ortho	×	×	
Eastman Kodak	Linagraph Direct Print, Type 1843	Pan	×		80,000 ips, mercury
	Type 1855	Pan		×	20,000 ips, tungsten
	Linagraph Direct Print	Blue	×	×	18,000 ips, xenon
	Linagraph Direct Print, Type 1799	Blue		×	80,000 ips, mercury
	1801	Blue	×		70,000 ips, mercury
GAF-ANSCO	Anscotrace	Blue	×	×	80,000 ips, mercury
Kodak Limited	Linagraph Direct Print	Blue	×	×	In excess of 30,000 ips, with mercury—vapor lamp

*Note:

Approx. thickness range for standard weight papers is 0.0045 to 0.0055 inch, and for thin papers, 0.0025 to 0.0030 inch, depending on manufacturer.

TABLE 5.5 B
DEVELOP-OUT RECORDING PAPERS

Manufacturer	Designation	Spectral sensitivity	Relative speed	Approx thickness (inch)	Safe-light	Remarks
Black and white Agfa-Gevaert	Agfa ARP-1	Blue	6	—	Agfa 104 107 or 117L	
	Agfa ARP-2	Blue	6	—	Agfa 107	
	Oscilloscript 1	Ortho	6,000	0.003	Gevinac X601 L652	
	Oscilloscript 2	Blue	20,000	0.005		
	Oscilloscript 3	Ortho	40,000	0.003	Gevinac L652	
Consolidated Electrodynamics (CEC)	Datarite 22	Ortho	*	0.0030	2	* In Datarite 5–036: 2500 inches/sec; Closed magazine: 5000 inches/sec
	Datarite 33	Ortho	**	0.0030	2	** In Datarite 5–036: 5000 inches/sec; Closed magazine: 20,000 inches/sec Closed magazine speeds for records processed in CEC 23–109 processor using CEC's chemical kits nos. 156 455 or 217 050
Du Pont	Lino-Writ 1	Ortho	3,000	B = 0.0045 W = 0.0030	2	
	Lino-Writ 2	Ortho	6,000	Same	2	For conventional and stabilization processing
	Lino-Writ 3	Ortho	30,000	B = 0.0045 W = 0.0025	2	
	Lino-Writ 4	Ortho	50,000	0.0025	2	
	Lino-Writ 201	Ortho	3,000	Thin weight only	2	May be processed conventionally and in E.K. Ektaline 200, CEC 23–109 processors, or CEC Datarite magazing
	Lino-Writ 202	Ortho	6,000		2	
	Lino-Writ 203	Ortho	30,000		2	

Manufacturer	Designation	Spectral sensitivity	Relative speed	Approx thickness (inch)	Safe-light	Remarks
Du Pont (*cont*)	Seismo-Writ	Ortho	10	B = 0.0045 W = 0.0025	1	For conventional processin 9
Eastman Kodak	Ektaline 12, Type 1762	Ortho	5,000	0.003	2	For Kodak Ektaline 200 processor with Kodak Ektaline 200 chemicals. May also be processed in CEC 23–109 processor using Kodak Ektaline processing kit. May be loaded (not processed) in DIM room light.
	Ektaline 16	Ortho	20,000	0.003	2	
	Ektaline 18	Ortho	75,000	0.003	2	
	Ektaline Type 1729	Blue	50,000 (mercury)	0.003	1	
	Linagraph 480	Ortho	5,000	0.005	2	Conventional processing with Kodak Lina-graph developer.
	Linagraph 483	Ortho	5,000	0 003	2	
	Linagraph 809	Ortho	5,000	0.005	2	Conventional processing with Kodak Lina-graph developer or Kodak Linagraph 1000 processing kit. In CEC 23–109 processor.
	Linagraph 697	Blue	7,500	0.005	2	
	Linagraph 622	Ortho	20,000	0.005	2	
	Electrocardiograph 533	Blue	—	S.W.	OC or 6B	Conventional processing
	Electrocardiograph 797	Blue	—	S.W.		
GAF-ANSCO	Linatrace A-3	Ortho	15,000	0.0028	2	May be processed in CEC 23–109 processor using Linatrace chemical kit no. 1.
	Linatrace A-4	Ortho	15,000	0.0035	2	
	Linatrace N-3	Ortho	50,000	0.0028	2	
	Linatrace R-4	Blue	5,000	0.0035	1	Conventional, stabilization or rapid access (Datarite) processing.
Kodak Limited	Linagraph RP.30	Blue	60***	0.0068	6BR or 1	*** Relative speeds for 3200°K light
	Linagraph RP.35	Blue	60***	0.0045	6BR or 1	
	Linagraph 480	Ortho	4***	0.0050	2	
Xerox	Astroprint II	Ortho	13	B = 0.0046 W = 0.0026	2	Conventional and stabilization processing.

	Astroprint 2567	Ortho	55	$B = 0.0046$ $W = 0.0026$	2	Conventional processing.
	Astroprint 2701	Ortho	12	0.0026	1	Conventional, stabilization, and rapid access processing. (2802 Baryta coated 100 per cent Alpha sulphate)
	Astroprint 2703	Ortho	50	0.0026	2	
	Astroprint 2802	Blue	25	0.0064	1	
	Geoprint IIB	Ortho	11	0.0046	1	Conventional and stabilization processing
	Geoprint IIIB	Ortho	13	0.0046	1	
	Geoprint IIW	Ortho	11	0.0026	1	
	Geoprint IIIJ	Ortho	13	0.0052	1	
Color Consolidated Electrodynamics (CEC)	Datacolor 88	Blue & Green	2700-red 1200-cyan	0.003	2	Develop, stop, bleach, stabilize in CEC 23–109 Processor.
Eastman Kodak	Linagraph 705	Blue & Green	2700-red 1200-cyan	0.003	2	Develop, stop, bleach, stabilize in 4-tank stabilization processor.
GAF-ANSCO	Colortrace	Blue & Green	2000	0.003	2	Prebath, develop, 2 tanks stabilizer in 4-tank stabilization processor.

Note ①: Relative speeds having large values are manufacturers' rated maximum speeds in inches per second. Small values are arbitrary relative speeds by individual manufacturers for their product lines: consequently, evaluation between speeds of different manufacturers may not be possible.

Note ②: Safelight filters 1, 2, OC, 6, 6B and 6 BR are Wratten numbers.

TABLE 5.5 C

RECORDING FILMS

Manufacturer	Designation	Spectral sensitivity	Relative speed
Agfa-Gevaert	Fluorapid X-ray	Ortho	—
	Isochrome 22	Ortho	125 ASA
	Isopan 15	Pan	25 ASA
	Isopan 21	Pan	100 ASA
	Isopan 24	Pan	200 ASA
	Isopan 27	Pan	400 ASA
	Isopan Record	Pan	800 ASA
	Scientia Scopix B	Blue	High
	Scientia Scopix G(IS & HDS)	Ortho	High
	Scientia 37 C 50 MML	Blue	Medium
	Scientia 50 B 65	Pan	Medium
	Scientia 67 Λ 56 K	Ortho	High
Du Pont	Type 131 Cronar Rapid Processing	Pan	Neg 64 ASA(I) / Rev. 125 ASA(I)
	Negative (up to 125 °F) Type 136 Cronar Fine Grain Superior 2	Pan	100 ASA (I)
	Type 140 Cronar High Contrast, Fine Grain Negative	Pan	80 ASA (I)
	Superior 2	Pan	100 ASA (I)
	Superior 3	Pan	200 ASA (I)
	Superior 4	Pan	250 ASA (I)
	Lino-Flex 1	Blue	10
	Seismo-Flex 1	Ortho	4
	Seismo-Flex 2	Blue	30
	Cronar Recording Film	Blue	30
Eastman Kodak	Linagraph Drift Survey	Ortho	6
	Linagraph Ortho	Ortho	400
	Linagraph Shellburst	Ext. red. Pan	250
	Linagraph Recording	Blue	32
	Linagraph Recording (thin base)	Blue	32
	Verichrome Pan	Pan	250
	Timing Negative	Pan	100
	2475 Recording Film	Ext. red. Pan	1250
	2490 RAR	Blue	64
	2492 RAR	Ortho	160
	2479 RAR	Ext. red. Pan	—
	2496 RAR reversal	Ext. red	160
	2496 RAR negative	Pan	125
	5498 RAR (reversal or neg)	Pan	250
	High speed positive, Type 5305	Blue	—
	Television recording, Types 5374 and 7374	Blue	100

Manufacturer	Designation	Spectral sensitivity	Relative speed
GAF-ANSCO	Hyscan	Blue	50
	Hyscan pan	Pan	100
Ilford	8B21 Recording film	Blue	Slow
	5B11 Recording film	Blue	Slow
	5B52 Recording film	Blue	Medium
	BY Negative recording film	Blue	Medium
	BU Neg. or rev. recording film	Blue	Medium
	5G91 Recording film	Ortho	Very fast
	RE Neg. or rev. recording film	Pan	Medium
	RX Neg. or rev. recording film	Pan	Medium
	5R61 Recording film	Pan	Medium
	5R81 Recording film	Pan	Fast
	BR 101 & 5R101	Pan	Very fast
	3R 111	Pan	Ext fast
Kodak Ltd.	Linagraph Ortho	Ortho	400
	Linagraph Pan R.60	Pan	100

Notes:

① Material speeds are given in ASA values where available. Other values are relative speeds for products by given manufacturers, determined by their own criteria.

② The symbol (I) after speed refers to the ASA incandescent speed. ASA daylight speeds are higher for those materials.

ASA ratings without (I) are daylight speeds.

middle layer; and cyan, in the bottom layer. Each layer may be individually assessed through the use of light or filters of appropriate colors. This optical discrimination is summarized in Table 5.6.

TABLE 5.6

TYPICAL FILTER GUIDE FOR EXAMINING AND PRINTING XR NEGATIVES

Predominant Color of Image	Use Wratten Tri-Color Filter (or equivalent)		
	To Examine Negative	To Print onto Enlarging Paper	To Print onto Panchromatic Film or Paper
Yellow to orange	# 47 Blue	No filter needed	# 47 Blue
Orange to red	# 58 Green	No filter needed	# 58 Green
Red to purple	# 25 Red	Not printable	# 25 Red

Because each dye transmits light freely at certain wavelengths, the image will be invisible if it is inspected with light of those wavelengths. On the other hand, the

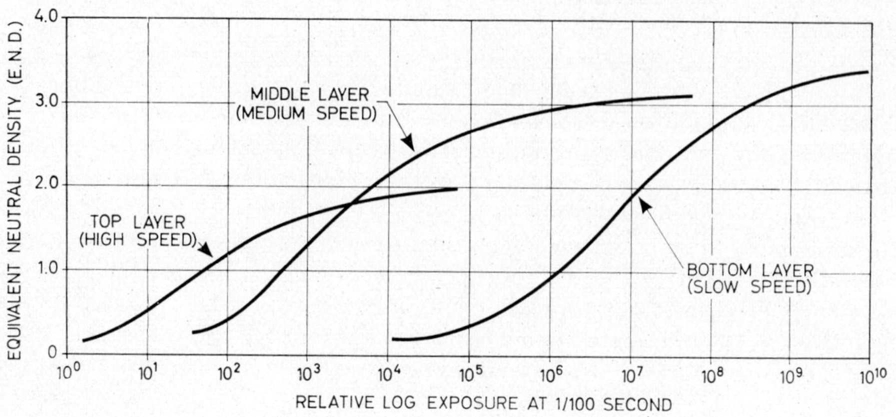

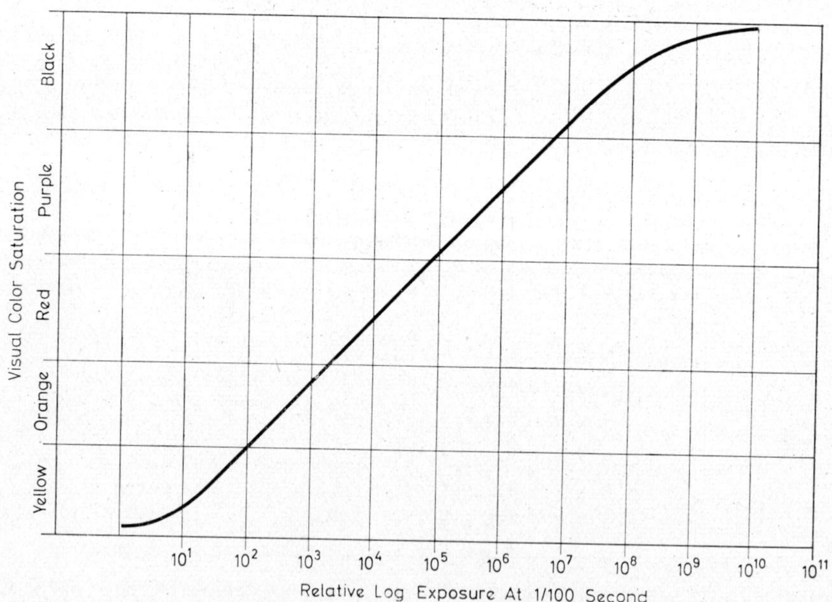

Fig. 5.14. Characteristic D-log E curves of extended range film

image will be visible in graded tones if it is inspected with filters that absorb light of those wavelengths.

Where the layer is completely overexposed it simply becomes a colored filter, which is completely transparent when examined with the appropriate filter.

After filter examination, black-and-white prints, including enlargements, may be produced by conventional printing methods except that the appropriate filter must be placed over the printer or enlarger lens (i.e., between the XR film negative and the photosensitive print material).

Printing paper must be selected according to the predominant color of the image on the XR film negative and the appropriate printing filter. Where the image color ranges from yellow through orange, prints may be made when necessary, on any standard printing or enlarging paper using the appropriate contrast grade. Printing filters may not be necessary with standard papers. However, where the image ranges from red to purple, necessitating a red series of filters, the prints should be made using Kodak Panalure paper (or its equivalent).

The application of this technique to oscillography is illustrated by the oscillograms shown in reference[5.10].

POLAROID EQUIPMENT AND MATERIALS. Quick access still records of oscilloscope traces may be obtained with the Polaroid Land process, which produces prints of cathode-ray traces in ten seconds by the diffusion-transfer reversal method. The Polaroid materials can be used in oscilloscope cameras equipped with a Polaroid Film Adapter.

A schematic view of the Polaroid camera is shown in Fig. 5.15. The camera is loaded with Polaroid film of the selected type. The roll is a tandum roll of paper, one strip containing a high speed orthochromatic or panchromatic-sensitive emulsion, and the other strip composed of a non-sensitive positive paper containing precipitating nuclei. The film is loaded into the camera so that the roll of sensitized

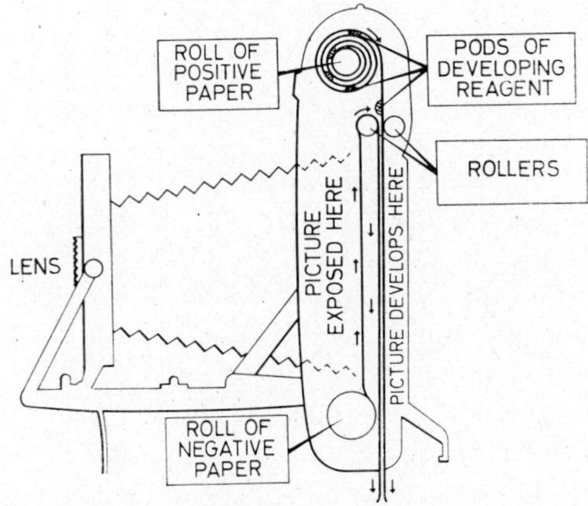

Fig. 5.15. Schematic of Polaroid Camera

material is at one end and the roll of positive paper at the other end. Attached to the non-sensitive roll are pods of a viscous processing reagent. There is one pod per picture, affixed to the paper immediately ahead of the picture area. After exposure of the light-sensitive roll, the sensitive and non-sensitive rolls are advanced between stainless-steel rollers into a developing chamber. The rollers exert enough pressure on the pod to cause it to rupture, and to spread the jelly-like developing fluid evenly between the sensitive and non-sensitive papers, now in intimate contact. The viscous developer reduces the exposed grains of silver halide to metallic silver. The unexposed grains are dissolved, diffuse out of the negative emulsion, and then migrate to the non-sensitive receiving paper.

The nuclei in the non-sensitive paper cause the migrating silver halide grains to precipitate silver in the form of a positive image on the paper surface. For current Polaroid materials used at normal room temperature the process reaches equilibrium in approximately ten seconds. At this point, the operator extracts the print from the camera through a door on the back. The print is weakly held to the negative by its edges, but can be peeled away by tearing it along pre-deckled perforations. The print is nearly dry because most of the viscous processing fluid adhers to the negative. The negative is pulled out of the camera and discarded.

The well-known Polaroid Land process makes finished pictures in the camera seconds after the shutter is snapped. For this reason, and because it can be used successfully without special training in photography, it is especially suited to standard oscilloscope test routines. Polaroid materials enable oscilloscope traces to be recorded, processed, and checked before continuing with the next successive step. The wide selection of materials cover a speed range of from 50 to 10,000 ASA equivalent—sufficient to record any condition from continuous static conditions to subnanosecond rise-time pulses.

Polaroid films are available in rolls for most oscilloscope cameras. In addition, single-shot packets are available which fit the Polaroid Land Film Holder; the holder fits some oscilloscope cameras and most 4 by 5 inch press and view cameras.

Polaroid Land roll film is made in two sizes: 2–1/4 by 3–1/4 and 3–1/4 by 4–1/4 inches in paper print material. Transparency materials are also available in the larger size only. Packets are 4 by 5 inch sheet only. They are furnished in three speeds of paper print including 10,000 ASA and in a combination paper-print film- and negative material.

Polaroid also makes high speed recording films. Polaroid PolaScope Type 410 and Type 510 films are specifically designed for oscilloscope and other high-speed recording purposes. Its 10,000 ASA equivalent for middle gray-scale tones provides the highest speed rating of any photographic material. PolaScope Film Type 410 has high contrast, resolution equivalent to regular Polaroid Land prints, and can be processed to completion in the standard development time of 10 seconds. The film may be fully developed outside the camera after an initial 2 seconds of processing in the camera, thus permitting recording in rapid-sequence situations. In this manner an entire roll of eight pictures can be exposed and processed in less than 20 seconds.

238

Projection Film. Polaroid Land Projection Film Types 46 and 46L produce a virtually grainless (resolution approximately 42 lines per mm) continuous-tone black-and-white positive transparency two minutes after the picture is taken. The speed of the film is 800 (ASA equivalent).

Polaroid Land Projection film produces a positive transparency from the back of the camera in much the same way a positive paper print is produced. However, the handling and the use of the film is quite different from that of the paper positive.

Standard developing time is two minutes, but as with paper prints, one can remove them from the camera in less than the standard time for most CRO work. One minute after the picture is taken a transparency can be pulled out of the back of the camera with only a slight loss in the density of the background.

The image layer of the transparency is wet and delicate before it is placed in the Dippit, a small, hand-held tank of hardening solution. While still soft it provides a good surface for writing information directly with a sharp instrument such as a pen point or pencil. The transparency is then treated in a Dippit as usual to harden the surface and to protect it from fading.

Accurate measurements of time intervals, amplitudes, and slopes can be made from a projected transparency more readily than they can be made from a print. Another technique is to make projection enlargements from a transparency onto matte surface photographic paper. Measurements again are traced from this enlargement.

One of the most desirable features of the transparencies is the ease with which they can be used for illustrating a small number of copies of report papers. Any diazotype machine (such as Ozalid, Technifax, Bruning, Pease, etc.) can be used.

By attaching a transparency directly to the translucent original of a report page for diazotype reproduction, both the printed information as well as the illustration can be produced in one pass through the machine.

Here's how it is done: First, information is typed onto a high quality vellum tracing paper, producing a master sheet. (Some manufacturers of diazo materials provide papers ideal for this purpose.) For best results one should use an electric typewriter. Unless an acetate ribbon is in the typewriter to give deep, black, sharp letters, the master sheet should be backed with a soft carbon paper, or "orange" carbon paper.

The Polaroid transparency is then taped on the back of the master sheet in the open area where the illustration is to appear. For a "right-reading" reproduction, the transparency must go emulsion side down. (Or, if the oscilloscope trace is reversed by the camera, one should place the emulsion side up to obtain a "right-reading" trace.) Minnesota Mining and Manufacturing Company's Tape No. 810 can be used for taping the transparency down to eliminate dark borders.

The master sheet is placed face up on top of the diazo-sensitized paper and fed into the machine. Any inexpensive standard blue-line or black-line diazo-sensitized paper can be used.

An example of another method of printing Polaroid transparencies is illustrated in Fig. 5.16. Polaroid transparencies of CRT screens are positives, but they have

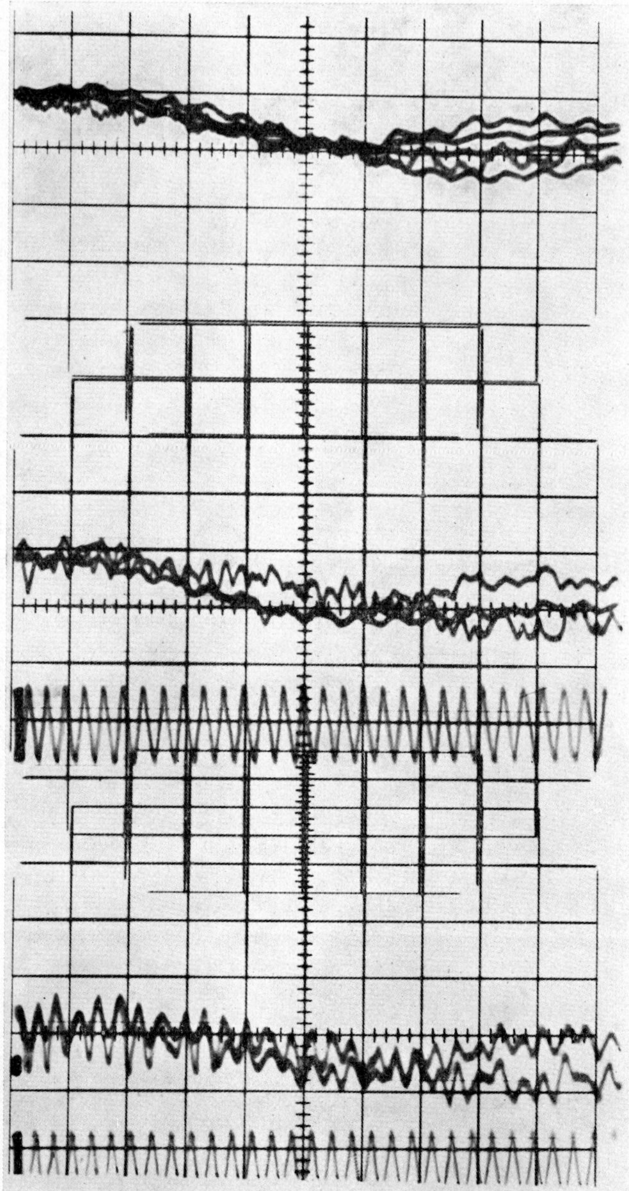

Fig. 5.16. Multiple-exposure print from Polaroid transparencies

dense backgrounds and transparent image areas. This characteristic permits several transparencies, representing different test conditions, to be exposed in sequence in a photographic enlarger. Masking the area around the transparency enables the resulting enlarged paper print to retain a white background; it is a negative, similar in appearance to light-beam records. The white background and comparative images in prints made by this method make them ideal for test reports.

MATERIALS FOR COLOR OSCILLOGRAPHY. Oscillogram traces in multiple colors instead of black can be made on either Eastman Kodak Linagraph 705 Recording Paper[5.11, 5.12] or on GAF-Ansco Colortrace Color Recording Paper. The papers are used in conjunction with colored filters that are inserted in the instrument. Color oscillograms can be produced in tungsten oscillographs, and also in print-out oscillographs if they are run "closed magazine."

The placement of the filters will depend on the individual instrument. They should be located between the source and the galvanometer, or fitted over the galvanometer lens. Placing the filter on the face of the galvanometer will cause double filtration, with additional loss of total energy. Regardless of the mounting method, care should be exercised in the installation of the filters. The user should take steps to insure that the filters will remain in the proper position during recording. If not properly installed, filters may move during tests in which the instrument is subjected to vibration, acceleration, or shock.

A snap-in filter bracket is available for the CEC 5–119 Oscillograph (either tungsten or mercury light source models). Once installed, the filter bracket can be instantly flipped in or out of the light path so that color and black-and-white recording can be done interchangeably. The galvanometers are not disturbed by the addition of this bracket, so there can be no damage to them. Similar filter brackets may be made for various other oscillographs.

Either brand (Eastman or Ansco) of color recording material produces colored traces which are easy to follow even when overlap occurs for long distances. Although only 3 or 4 distinct colors are claimed by the material manufacturers, some users have obtained as many as 6 to 10 distinguishable traces by using variations of the recommended filters.

While the color recording paper is higher in price than black-and-white papers, it reduces reading time and labor. This can result in a lower cost per data point. The cost analysis assumes that an analogous black-and-white oscillogram would have many traces that crossover, and that consequently it would be difficult to follow even with the aid of trace interruptions and numbering.

The use of a two-color emulsion results in a lower cost material than a three-color recording paper would cost, due to simplification of physical and sensitometric problems.

Color oscillogram materials are compatible with practices currently in general use for black-and-white photorecording and processing. Both materials can be handled under a Series 2 dark-red safelight; have adequate speed and latitude for most applications; have shelf life comparable to that of black-and-white papers; can be processed in 4-tank stabilization equipment (such as the CEC 23–109 processor) at speeds ranging from 3 to 6 feet per minute; have image stability comparable to that of stabilized black-and-white records; and may be duplicated in monochrome by diazo and other standard reproduction methods presently used for reproducing stabilized black-and-white oscillograms.

KODAK LINAGRAPH 705 PAPER. The photographic and physical properties of this Kodak two-color material are described in the following subsections.

Emulsion. Two-color, "false-sensitized" to blue and green at about 450 mμ and 550 mμ respectively. The blue sensitive layer yields a red dye, while the green-sensitive layer results in a cyan dye. These dyes were selected because of their maximum visual contrast; the overlap of the two yields a neutral (black) color.

Speed. Typical writing speeds for two different oscillographs are given in Table 5.7. Data of this type are necessarily approximate and may be influenced by individual oscillograph and galvanometer adjustments.

Darkroom Handling. Does not require total darkness, but can be handled under a Kodak Safelight Filter, Wratten Series 2 (dark red), in a suitable safelight lamp with a 15-watt frosted bulb at no less than 4 feet from the paper.

TABLE 5.7

TYPICAL WRITING SPEEDS FOR KODAK
LINAGRAPH 705 COLOR PAPER

CEC Oscillograph, Model 5–119 (Tungsten Light Source)—Paper travel speed at 10 inches per second.

Lamp Voltage	Cyan Writing Speed (in./sec.)	Red Writing Speed (in./sec.)
16	1,200	2,700

CEC Oscillograph, Model 5–119 V (Mercury Light Source)—Paper travel speed at 25 inches per second; KODAK WRATTEN Filter, No. 2E, in intensity-control filter position.

Intensity	Cyan Writing Speed (in./sec.) WRATTEN Filter, No. 12	Red Writing Speed (in./sec.) WRATTEN Filter, No. 34
2	18,000	36,000
3	24,000	48,000

The writing speed of the neutral (black) trace will be slightly higher than that of the red trace.

Exposure. Any oscillograph suitable for exposing conventional black-and-white photorecording papers can be used. The only change required involves selective filtration of the light supplied to the several galvanometers. The blue exposure is made through a Kodak Wratten Filter No. 34; the green exposure, through a Kodak Wratten Filter No. 12. The natural (black) color is achieved by supplying both blue and green light to a galvanometer.

A Kodak Wratten No. 6 Filter is used in conjunction with mercury-arc oscillographs. It is necessary to balance the blue and green light so that a suitable neutral (black) trace can be produced.

Surface. Smooth, accepting pen and pencil notations easily.

Base. Paper, ultra-thin. A synthetic base of the same type as is used for Kodak Ektaline Papers. It has excellent dimensional stability, resists wrinkling and creasing during processing, resists scorching on the drying drum, and curl or brittleness after processing. This stock has a basis weight of 10 pounds per 1,000 square feet, which permits maximum footage per given roll diameter.

Storage. Shelf life of Linagraph 705 Paper is comparable to that of black-and-white recording materials, and refrigeration will not normally be required.

Processing. Linagraph 705 Paper can be processed in stabilization equipment, such as the CEC 23–109 Oscillogram Processor, with the chemicals in the Kodak Linagraph 705 Processing Kit. The four solutions contained in this kit and the order in which they are used are:

First Tank	Developer
Second Tank	Stop Bath
Third Tank	Bleach
Fourth Tank	Stabilizer

This system produces adequate densities over the processing speed range of 4 to 6 feet per minute. Although the CEC Processor normally operates at approximately 100 °F, there is also considerable temperature latitude in the 705 process.

Despite its high activity, the developer has a high density-to-fog ratio and low staining propensity. This means that the low background level will not interfere with discrimination of traces.

The stop bath in this process is designed to perform two functions: the acidity of the bath stops the development process while, at the same time, an addition product is formed with oxidized developer which prevents its coupling to form unwanted dye.

At this point in the process, the dye image and the silver image, together with unexposed silver halide grains, are present in the sheet. It would be possible at this stage to stabilize these silver halide grains in a manner similar to that used for stabilizing black-and-white papers. However, if this were done, the silver image would remain and have the effect of desaturating the dye image. To provide maximum brightness of trace color, a bleach is used. This specially formulated bleach converts the silver image to a colorless silver compound that is relatively insensitive to heat, light, and moisture.

The stabilizer, as its name implies, influences the stability of the record. In addition, however, it decolorizes chemicals retained in the sheet from previous solutions and prepares it for efficient drying.

Records processed in these solutions show excellent stability when subjected to a high level of fluorescent illumination and to a warm, humid atmosphere. The stability is equivalent to that of conventionally stabilized black-and-white records.

GAF-ANSCO COLORTRACE COLOR OSCILLOGRAPH PAPER. *General Description.* Ansco color oscillograph recording paper is a thin base, two-layer paper providing four distinct trace colors, and is designed for use in closed-magazine oscillographs having either tungsten or ultraviolet light sources.

Ansco paper employs two color emulsion layers which were specifically selected to give optimum color discrimination through a complete range of densities. A blue-sensitive emulsion creates a cyan (blue-green) color trace when exposed through a magenta (Wratten32) filter. A magenta (blue-red) color trace is produced by exposure of a green-sensitive emulsion through a yellow filter. The dark blue and the purple traces result from different degrees of exposure of both emulsion layers.

The 55 gram, 100 percent rag content base has good dimensional stability with low brittleness and high wet strength.

Important in an oscillograph paper, records have good image stability making them storable under the same conditions as black-and-white papers.

Darkroom Handling. The sensitivity range of both emulsions allows Colortrace paper to be handled under Series 2, dark red, safelamp illumination.

Exposure. Colortrace oscillograph paper is adaptable to simple conversion of the standard oscillograph for using color. The rotating lens turret of the CEC Type 7–300 galvanometer (Consolidated Electrodynamics Corporation) provides a convenient method for holding the required filter. The lens turret, when rotated 90 degrees, makes accessible the depression formed by the back of the lens and the turret frame. A proper filter can be fitted into the depression. Rotating the turret back into operating position locks the filter in place. A set of 20 filter chips each of the required four colors to fit CEC galvanometers, is available from GAF-Ansco offices.

The following color filter adapters are also available from GAF-Ansco offices to fit other makes of oscillographs. The adapters are non-magnetic and are attached to the magnet blocks with screws:

Honeywell 1612. Requires 3 adapters (one adapter per magnet block). Each adapter contains 12 filters, made up of each of the four filter colors arranged alternately in a row.

Midwestern 560D. Contains 14 filters, made up of each of the four filter colors arranged alternately in a row. One adapter per oscillograph is required.

244

Century 409X. Each adapter contains 14 filters, made up of each of the four filter colors arranged alternately in a row. One adapter per oscillograph is required.

Processing

The new FPC-371 Colortrace Processing Kit, adapted to the CEC 23–109 four tank processor, was developed for Colortrace color oscillograph paper. The chemicals may also be used for tray or tank processing. The kit consists of concentrated prebath, developer, and fixer solutions, and three packets of dry developer, Each of the concentrated solutions may be quickly and readily diluted in the processor tanks, and the powdered chemicals contained in the packets dissolve easily in the diluted solutions. The process does not require a bleaching or washing step. The stability of the dyes is such that no degradation is encountered from contact with the 350 °F drying drum during the drying cycle. A total of 800 square feet of Colortrace paper may be processed per kit.

Elimination of the bleaching step in processing Colortrace paper, provides important advantages. It makes possible the retention of the developed silver in the color traces which assures image permanence equal to that of stabilized black-and-white oscillograph records. Eliminating the bleaching process means that there will be no bleach stain to be removed by washing or other means; however, in cases where a high degree of trace-line permanence is important, thorough washing will produce longer-lasting oscillograms. The number and cost of processing solutions, and the time required to prepare them, is reduced.

Optimum developer temperature is 102 °F ± 2 °F. A variation in processing speeds of ± 2 feet per minute from the optimum speed of 6 feet per minute causes no appreciable change in the processed record. There is also no appreciable change when the developer temperature is varied from 100 °F to 105 °F, at a given processing speed. However, processing temperature should not be allowed to drop below 100 °F.

Oscillogram Duplication. The thin base characteristics of Colortrace color paper, coupled with the retention of the opaque developed silver, make possible excellent monochromatic duplicate oscillograms, by diazo or other standard reproduction methods used presently for the reproduction of stabilized B & W oscillograms.

LITERATURE REFERENCES

[5.1] HUNT, H. D., "High-Speed Direct-Recording Papers," *Photographic Science and Engineering*, Vol. 5, No. 2, p104, (March–April, 1961)

[5.2] JACOBS, J. H., "An Investigation of Print-Out Paper," *Photographic Science and Engineering*, Vol. 5, No. 1, p1, (January–February, 1961)

[5.3] JACOBS, J. H. and MCCLURE, R. J., "A Spectrograph for the Evaluation of Print-Out Paper," *Photographic Science and Engineering*, Vol. 9, No. 2, p82, (March–April 1965)

[5.4] CURRENT, I. B., "Sensitometry of Print-Out Papers," *Photographic Science and Engineering*, Vol. 7, No. 2, p104, (March–April 1963)

[5.5] TODD, H. N. and ZAKIA, R. D., *Sensitometry*, Rochester Institute of Photography, (Rochester, New York) 1962

[5.6] HERCOCK, R. J., *The Photographic Recording of Cathode-Ray Tube Traces*, p11, Ilford Limited, (Ilford, England) 1947

[5.7] NITKA, H. F. "Sensitivity and Detail Retention in the Recording of Light Images and Electron Beams," *Photographic Science and Engineering*, Vol. 7, No. 3, p188, (May–June 1963)

[5.8] HORAK, R. J., "Selecting Photographic Paper for an Oscillograph," *CEC Recordings*, Vol. 17, No. 3, p11, (Third Quarter 1963)

[5.9] SMITH, H. I., *Photographic Recording Materials*, Consolidated Electrodynamics Corporation, (Pasadena, California) 1960

[5.10] WYCKOFF, C. W., "An Experimental Extended Exposure Response Film," *SPIE Newsletter*, p16, (June–July 1962)

[5.11] POMEROY, F. A. and WELCH, M. E., "A Two-Color Photographic Paper and Rapid Process," *Photographic Science and Engineering*, Vol. 7, No. 3, p161, (May–June 1963)

[5.12] POMEROY, F. A., WRISLEY, K. L., and FASSBENDER, H. J., "Photorecording in Color," *Photographic Science and Engineering*, Vol. 8, No. 5, p296 (Sept.–Oct. 1964)

VI PROCESSING CHEMISTRY

TYPES OF PROCESSING. Oscillograms that have been recorded on developing-out paper must be chemically processed to make the latent image visible. The processing function generally falls into one of two categories: 1) *conventional;* or 2) *stabilization* processing. Conventional processing requires five steps: development, rinsing, fixation, washing and drying; stabilization eliminates the washing operation. The abbreviated process produces good quality, but the records are nonarchival. (A third category that combines the functions of development and fixation is called *monobath* processing. It is sometimes used in quick-access systems, but has not been widely used for oscillogram processing.)

Records can be processed by hand or by machine in one of several ways. The methods are: 1) tray processing by hand for very short-length records; 2) semi-automatic motorized-rewind tank processing, useful for low-production conditions and long-access requirements; and 3) continuous-machine processing for large-scale operations involving many feet of record and short-access times. All processing methods, whether done by hand or by machine, require solutions made up of water and various chemicals. The user may purchase ready-mixed liquid chemicals, or prepared batch-mixed powders which can be dissolved in water. Alternately, he may mix his own solutions from bulk ingredients. Today, most users find that time can be saved and consistant quality can be maintained with factory-prepared chemicals. The chemicals are convenient to handle and use, and liquid chemicals may be made ready in the processor, thus eliminating the mixing facilities normally required. Modern packaging is important, too: many chemicals come in throw-away containers, which solve storage and container-return problems. Typical containers are plastic, or plastic-in-cardboard, which eliminate the inherent danger from glass-container breakage, from dropping, freezing, and similar causes.

Because of the prepackaging of chemicals in recent years, photographic chemistry has come to be taken for granted. Many users have little or no idea of chemical function, or any appreciation of the care that must be exercised to achieve good results. The following suggestions should be followed when making factory-prepared chemicals ready for the processor:

1) Be sure that the water used for mixing is pure (free of iron, sulphur, etc.). Avoid using "hard" water. Mix chemicals at the specified temperature, and adjust the solution to the correct processing temperature before using.

2) Observe the directions on the chemical carton and the instructions for processing printed by the sensitized-material manufacturer.

3) Be sure that the solution concentration and volume are correct for the processor (and for the size of the material being processed, if this is a factor).

4) Put the proper solution into its respective tank.

5) Be careful about contamination in mixing, spilling, and handling.
6) Replace the chemicals as specified by time or amounts of material processed.

CONSTITUENTS OF PROCESSING SOLUTIONS. Regardless of its form, prepared solution, prepared powder, or user mixed, the constituents of various developers and of various fixers are similar. Developers contain reducing agents, accelerators, preservatives, and restrainers; fixers contain silver-halide solvents, preservatives, acids, and hardeners.

PROCESSING TEMPERATURE. In conventional photography, the normal processing temperature is 68 °F. The rate of development is dependent upon temperature.

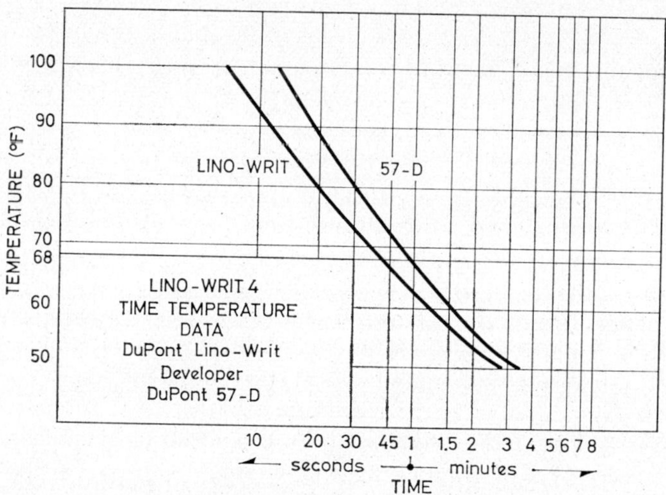

Fig. 6.1. Typical development graph (time-versus-temperature). Curves for two developers are shown

Within certain practical limits, the developing time can be reduced by increasing the temperature of the developer, as illustrated in Fig. 6.1. Most oscillogram processing requires quick access; therefore, paper-oscillogram processors usually are operated at approximately 100 °F. This temperature is satisfactory for paper, but films normally are not processed at high temperatures (except under special conditions, with special developers, or with special processors). To prevent reticulation, frilling, etc., films processed at elevated temperatures may require pre-hardening; alternately hardening agents may be added to the solutions, or they may be used as separate baths in the process. As temperature is lowered, development rate is reduced, in fact the action of some developers may cease altogether. Therefore, it is important to watch the temperature closely when processing without temperature-control equipment. Temperatures, either high or low, must be closely monitored when operating under field conditions, or other circumstances where ambient temperatures may be extremely variable and cause similar changes in the processing solution.

CHEMICAL ACTION OF DEVELOPMENT. The latent image formed during exposure in the oscillograph is made visible by chemical solutions called *developers*. The chemical action that causes this change is the reduction of exposed silver halides ($AgCl$, $AgBr$, and AgI) to metallic silver (Ag). The reduction takes place to those silver-halide grains that have been exposed to light, with practically no effect to unexposed silver grains.

DEVELOPING AGENTS. The chemicals that cause the reduction are called *developing agents*. They are organic compounds, which in ionized form, cause complex chemical reactions to take place, and reduce silver halides by supplying the electrons necessary for the reaction.

Two of the most common developing agents are metol (or Elon, Pictol, etc.) and hydroquinone. Generally, the two chemicals are used in combination. Other developing agents such as Phenidone are sometimes used, especially for high-temperature processing.

ACTIVATORS. The action of a developing agent by itself may be too weak to produce a suitable image. For this reason, an *activator* or *accelerator* is normally added to the developer solution. Alkalies are used for this purpose, and in general, the higher the alkalinity of the solution, the greater the development activity. High-energy developers contain a vigorous alkali, in some cases a caustic such as sodium hydroxide. Most developers used to process oscillograms contain a somewhat less active alkali, usually sodium carbonate. Film developers, especially those used for fine-grain development often use borax or sodium metaborate as an activator.

Control of developer alkalinity is used in the manufacture of liquid chemicals, and for control of developer replenishment in large laboratory installations. Alkalinity is measurable with indicator papers (similar to litmus paper) or with a pH meter. The reading obtained is an indication of the hydrogen-ion content, called the pH vaule, expressed with numbers ranging from 0 to 14. In this scale, it is assumed that pure water is neutral, and that it has a reading of 7. Any solution having a reading of less than 7 is acid, and any solution having a reading from 7 to 14 is alkaline. The scale came into being[6.1] to simplify the exponential form of expressing hydrogen-ion values.

On this scale, a developer for processing paper oscillograms will have a pH of approximately 10 to 11.

PRESERVATIVES. Developing agents by themselves oxidize easily and lose the ability to develop film or paper. When they are combined with an alkali they are oxidized even more rapidly. The developing solution changes in color as it is oxidized; starting as an almost water-clear solution, it becomes darker and darker brown in color as oxidation continues. Old developers are therefore weak and may even cause staining of the record.

Sodium sulfite (and sometimes other chemicals) is used to preserve the developer mixture, and to form colorless rather than colored oxidation products. This increases the storage life, the active life in processing, and prevents staining until the solution reaches the exhaustion point.

The temperature of the developer, both during storage and in use, is a factor that contributes to the amount of sulfite required. Other factors are the types and amounts of developing agents and alkalies used in the formula, and the amount of aeration received in mixing, storing, or processing. Oxidation problems become critical at high temperatures and at high aeration rates.

RESTRAINER. Although the principal action of the developer is to reduce exposed silver grains, limited reduction of unexposed silver grains takes place unless a *restrainer* is used in the formula. The most common restrainer is potassium bromide which retards fog and promotes uniform development. The restraining action of the bromide occurs with both exposed and unexposed silver grains, but the unexposed are affected to a greater degree.

Chemicals other than potassium bromide are sometimes used as restrainers alone or in combination with the bromide. The best known of these is benzotriazole, sometimes sold under various trade names.

OTHER CHEMICALS. Other ingredients are sometimes used in developers. Most proprietary, factory-prepared powders and solutions contain chemicals other than the standard ingredients. Special ingredients can be mixed into standard formulas if the user completely understands their purpose. Sodium hexametaphosphate is often added to prevent sludging caused by calcium or magnesium in the water. Sodium sulphate is used in some formulas for high temperature processing. Its function is to reduce swelling of the gelatin emulsion.

TYPICAL DEVELOPERS. Formulas for typical developer solutions as recommended by several sensitized-material manufacturers are listed below. Chemicals should be dissolved in the order given. Each ingredient should be dissolved completely before the next is added.

It should be noted that these formulas may not be identical with those supplied as factory-prepared powders or solutions, or in kit form for oscillogram processors.

ILFORD 1D-33

	Avoirdupois	Metric
Warm Water (125 °F or 52 °C)	24 ounces	750.0 cc
Metol	75 grains	5.0 grams
Sodium Sulfite (desiccated)	1 oz. 292 grains	50.0 grams
Hydroquinone	120 grains	8.0 grams
Sodium Carbonate (monohydrated)	1 oz. 195 grains	43.3 grams
Potassium Bromde	75 grains	5.0 grams
Add cold water to make	32 ounces	1.0 liter

Use full strength. Develop about 5 minutes.

250

KODAK D-72

	Avoirdupois	*Metric*
Water, about 125F (50C)	16 ounces	500.cc
Kodak Elon Developing Agent	45 grains	3.0 grams
Kodak Sodium Sulfite, desiccated.........	1 1/2 ounces	45.0 grams
Kodak Hydroquinone	175 grains	12.0 grams
Kodak Sodium Carbonate monohydrated	2 oz. 290 ounces	80.0 grams
Kodak Potassium Bromide	30 grains	2.0 grams
Water to make	32 ounces	1.0 liter

Kodak also furnishes a factory-prepared powder with similar characteristics called "Dektol".

DU PONT 57-D

	Avoirdupois	*Metric*
Water (125 °F. or 52 °C.)	24 ounces	750.0 cc.
Metol	22grains	1.5 grams
Sodium Sulfite, anhydrous	285 grains	19.5 grams
Hydroquinone	88 grains	6.0 grams
Sodium Carbonate, monohydrated*	408 grains	28.0 grams
*If anhydrous, use	350 grains	24.0 grams
Potassium Bromide	12 grains	0.8 gram
Add cold water to make	32 ounces	1.0 liter

DEVELOPER RINSE. When development has continued to the correct point, the sensitized material must be removed from the solution and the developing action must be terminated. Usually, further development is stopped by immersing the material in an acid rinse.

For this reason, the rinse is called a *stop bath* or *short stop*. Development is stopped because the acid neutralizes the alkaline activator in the developer. The rinse also prevents yellow-brown stains that could occur if developer on the surface of the processed chart were permitted to carry over and to oxidize.

A mild acid, usually acetic acid, is used in the stop bath. A typical concentration is shown in the following formula:

Acetic Acid, 28 per cent	50.0 ml
Water	1.0 L

NOTE: To prepare 28 per cent acetic acid from glacial acetic acid (U. S. P. XI 99 1/2 per cent), add 3 parts of glacial acetic acid to 8 parts of water (by volume).

FIXATION. After the latent image has been developed, and developer action has been stopped with an acid rinse, the sensitized material must receive further treatment to achieve permanency. At this point in the process, most of the unexposed and undeveloped silver halides still remain in the emulsion. Either a *fixer* or a *stabilizer* is used to convert the remaining silver halides into other silver complexes, so that the record will not be obscured or darkened by exposure to light. A fixer (sometimes called *hypo*) converts the remaining silver halides (those not exposed in the recorder, and consequently not reduced to metallic silver in the developer) to soluble compounds that can be washed from the chart with water. The fixer forms the soluble compounds without destroying the developed-silver trace image.

NOTE: Stabilization processing is described in the section following the discussion on drying fixed records.

The chief ingredient of a fixer is usually sodium thiosulfate (hypo). The solution also contains other chemicals to improve the life of the fixer, reduce staining, and to harden the emulsion.

Sodium sulfite is the most commonly used preservative; it reacts with colloidal sulfur to form sodium thiosulfate.

The life of the fixer is lengthened and staining is reduced by the addition of an acid. A weak acid, such as citric or acetic acid must be used, which in combinations with the sodium sulfite prevents the formation of colloidal sulfur. Alkali products from the developer, which cause many fixing problems, are neutralized by the acid in the fixer.

The fixing bath may also contain a hardener such as potassium alum. Rapid fixing baths are made from ammonium thiosulfate, or have an ammonia compound such as ammonium chloride added to a sodium thiosulfate mixture.

Typical fixing baths are presented in the following formulas.

KODAK FIXING BATH F-24

	Avoirdupois	*Metric*
Water, about 125F (50C)	16 ounces	500 cc
Kodak Sodium Thiosulfate (Hypo)	8.ounces	240.0 grams
Kodak Sodium Sulfite, desiccated	145 grains	10.0 grams
Kodak Sodium Bisulfite	365 grains	25.0 grams
Cold water to make	32 ounces	1.0 liter

Dissolve chemicals in the order given.

This bath may be used for films or papers when no hardening is desired.
For satisfactory use, the temperature of the developer, rinse bath, and wash water should not be higher than 68F (20C).

Kodak Fixing Bath F–5 is recommended for general use. This bath has the advantage over the older type of fixing baths, which did not contain boric acid, that it gives much better hardening and has less tendency to precipitate a sludge of aluminum sulfite.

KODAK FIXING BATH F-5

	Avoirdupois	*Metric*
Water, about 125F (50C)	20 ounces	600.0 cc
Kodak Sodium Thiosulfate (Hypo)	8 ounces	240.0 grams
Kodak Sodium Sulfite, desiccated	1/2 ounce	15.0 grams
Kodak Acetic Acid, 28%*	1 1/2 ounces	48.0 cc
Kodak Boric Acid, crystals**	1/4 ounce	7.5 grams
Kodak Potassium Alum	1/2 ounce	15.0 grams
Cold water to make	32 ounces	1.0 liter

*To make approximately 28 per cent acetic acid from glacial acetic acid, dilute 3 parts of glacial acetic acid with 8 parts of water.

**Crystalline boric acid should be used as specified. Powdered boric acid dissolves only with great difficulty, and its use should be avoided. Films should be fixed properly in 5 to 10 minutes in a freshly prepared bath. The bath need not be discarded until the fixing time becomes excessive, that is, over 10 minutes. Fix prints 5 to 10 minutes.

KODAK RAPID FIXING BATH F-7

	Avoirdupois	*Metric*
Water, about 125F (50C)	20 ounces	600.0 cc
Kodak Sodium Thiosulfate (Hypo)	12 ounces	360.0 grams
Ammonium Chloride	1 oz. 290 grams	50.0 grams
Kodak Sodium Sulfite, desiccated	1/2 ounce	15.0 grams
Kodak Acetic Acid, 28%*	1 1/2 ounces	48.0 cc
Kodak Boric Acid, crystals**	1/4 ounce	7.5 grams
Kodak Potassium Alum	1/2 ounce	15.0 grams
Cold water to make	32 ounces	1.0 liter

*To make approximately 28 per cent acetic acid from glacial acetic acid, dilute 3 parts of glacial acetic acid with 8 parts of water.

**Use crystalline boric acid as specified. Powdered boric acid dissolves only with great difficulty and its use should be avoided.

Dissolve chemicals in the order given.

This bath fixes much more rapidly than Kodak F–5 or F–6, and its useful fixing capacity is considerably greater.

CAUTION: With rapid fixing baths, do not prolong the fixing time for fine-grain film or for *any* paper prints; otherwise the image may have a tendency to bleach, especially at temperatures higher than 68F (20C).

If corrosion is encountered when using Kodak Rapid Fixing Bath F–7 with stainless steel containers, it can be minimized by substituting 8 ounces of ammonium sulfate for the 6 3/4 ounces of Kodak Ammonium Chloride in the 1-gallon formula (60 grams per liter for 50 grams). When changed in this way, the formula is known as Kodak Rapid Fixing Bath F–9.

Records should be agitated vigorously when immersed in the fixer, and then at frequent intervals during the fixing operation. This will reduce the possibility of stains and insure even and complete fixation. Fixing problems will occur if the recommended quantity of material to be fixed per solution volume is exceeded. Additional immersion time will partially compensate for fixing baths that have been used. In addition, the solution should be tested periodically with a test kit, such as the Kodak Testing Outfit, to determine the condition of the bath. The fixer should be discarded before it becomes cloudly or frothy.

WASHING. At completion of fixing, the record is saturated with hypo which can decompose and attack the trace image, causing it to fade or yellow with age. Thorough washing is necessary to prevent such degradation, and to make the image permanent.

The amount of washing required depends upon: 1) the amount of hypo carryover; 2) the efficiency of the water-circulation system; 3) the type of recording material being processed; and 4) the temperature of the wash water. Best results are obtained when hypo is squeegeed or rinsed off the record before it is placed in the final wash tray or tank.

Washing can be speeded up by the use of a hypo clearing bath before final washing. The washing time for paper of films can be reduced to as short as 10 minutes with satisfactory results, as recommended by the manufacturer. The wash water should be 65 to 70 °F for films, and may be somewhat higher for paper. Colder water requires a longer washing period. When maximum permanency is required, films should be washed for 20 to 30 minutes in running water, and paper records for 45 to 60 minutes. Fresh, running water should flow over the emulsion, and the hypo-contaminated water should be continuously removed. In field processing, or in other situations where running water is not available, the records should be

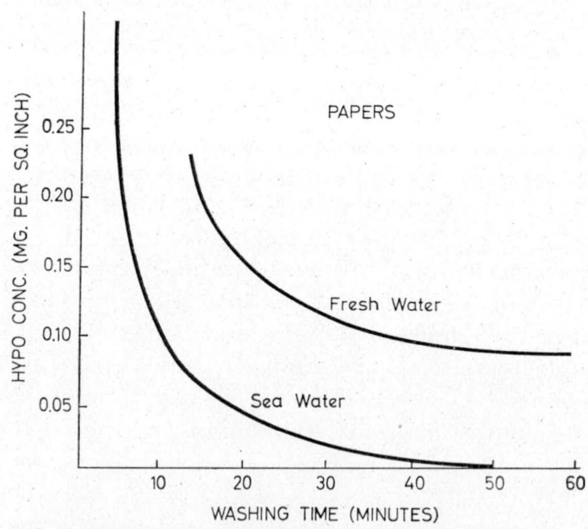

Fig. 6.2. Washing times in sea water and fresh water (E.K. Tech Bits 64-1)

washed in 10 to 12 changes of water, for a period of 5 minutes each. Even sea water may be used, providing the record is given a final wash of about 5 minutes in fresh water to remove the sea salts. (See Fig. 6.2).

DRYING. The processed-paper record can be dried by: 1) running it through a belt, drum, or air-impingement drier; or 2) by exposing the record to room air if a drier is not available. In either case, the record should not be subjected to tension; for air drying, short records may be looped loosely over dowels mounted on a board, which can be attached to a wall. (See Chapter 9, Fig. 9.3).

Water drops left on the surface can cause spots, because of uneven drying. Therefore, surface water should be removed with a rubber-blade squeegee or a damp viscose sponge.

Another way to prevent spots is to add a small amount of wetting agent to the final water rinse before drying. This will reduce or entirely eliminate the need for wiping. However, the record will dry faster if excess surface water is removed prior to drying.

STABILIZATION. Stabilization processing is used as an alternate to fixing and washing, and reduces access time. Although the abbreviated process produces an image that has a shorter life, the record still may last several years under optimum storage conditions. This is many times the life necessary to read the chart and to reduce the data before destroying the original.

Stabilizers for developing-out materials are of two types: one which produces soluble silver complexes, and the other which produces insoluble silver complexes.[6.2, 6.3, 6.4]

Both types produce silver complexes that are relatively stable to light, heat, humidity and atmospheric gases. Regardless of the type used, the oscillogram is not washed, and the converted silver complexes, together with any surplus stabilizer, are left in the emulsion. Films, when stabilized, may be given a short bath or spray rinse with water. This removes surface chemicals that on drying would crystalize into powdery residues on the emulsion and backing.

Generally, the fixing and washing technique (rather than stabilization) is used for tray processing of short-length records. The reason is that in hand processing, stabilization may cause contamination of facilites and equipment. In addition, timing cannot be accurately controlled in a tray, and charts may be less stable. Handling also presents a problem, for the stabilizer will leave white chemical deposits on sinks and cabinets. Stabilizer carried over to the drier will contaminate drier belts so that they cannot be used for drying washed charts or prints. In addition, the belts will deteriorate very rapidly from the chemicals in the stabilizer.

Many of the disadvantages of stabilizers are overcome in machine processors. For example, by continuous transport in high-energy developers, the image is developed so rapidly that the paper is only partially moistened by the developer. This is important in stabilization processing, since any developer in the paper must be subsequently displaced, leaving a surplus of stabilizer to keep the image relatively permanent. The amount of stabilizer left in the paper is important; an excessive amount of an inappropriate formula might cause fading, and too little may cause staining, especially upon exposure to heat, sunlight, or moisture.

The following formula is an example of one type of stabilizer:

Water (120 °F or 48 °C)	850.0 cc
Sodium Sulfite	15.0 gm
Sodium Bisulfite	45.0 gm
Sodium Thiosulfate (Hypo)	240.0 gm
Add cold water to make	1.0 liter

Immerse the cart that has been developed and rinsed in a stop bath in the above solution for approximately 5 to 8 seconds at a temperature of about 100 °F (38 °C). The chart should be squeegeed following treatment, prior to drying.

STABILIZING DIRECT-PRINT MATERIALS. The term *stabilization* also refers to other processes in which chemical stabilizing baths are used to render an image more stable than it would have been without such treatment. For example, print-out paper can be viewed under specified room light for a specific time. However, it can be made more stable by a special chemical treatment, stabilization. A stabilizer for print-out materials is usually different in content and concentration than a stabilizer used for developing-out paper; therefore, the two should not be interchanged. To differentiate between the two processes, at least one manufacturer terms the stabilization of print-out paper *permanizing*. (Copyright, E. K. Co.)

The decision to stabilize print-out records should be made before the chart is degraded by excessive exposure to light, handling, humidity, and heat. If possible, the chemical stabilization treatment should immediately follow recording and latensification. Stabilizing will not save badly fogged records in which the trace has become almost invisible from exposure to excessive light or ambient conditions.

A second reason for stabilizing print-out records is to enable them to be reproduced (without being damaged by the light used in the reproduction machine). After being treated, the record can be run through diazo or other light-operated duplicators as frequently as desired.

Proprietary chemical kits for stabilizing print-out paper are supplied by the paper manufacturers, chemical supply firms, and equipment manufacturers. Typically, the stabilization processor is filled as follows: developer is put into the first tank, stabilizer (usually two different types) is put into the second, third, and fourth tanks. The normal processing speed, on processors having a 2-foot developer path, is 6 to 8 feet at 100 °F for records that have been latensified. Non-latensified records will require a slower speed.

RECORD DEFECTS. Defects of the record can occur prior to processing, during processing, or after processing. Usually the fault will not be seen until processing has been completed, even though the cause was before or during processing.

Table 6.1

PROCESSING FAULTS (AND THEIR CAUSES)

Fault	Cause
Minus density spots	Handling dry paper with wet hands Splashing dry paper with water or chemicals prior to development
Plus density scratches	Dry abrasion Abrasion in developer
Gray appearance (may be mottled)	Light fog (safelight may be too close) Overdevelopment Improper storage of materials Developer too hot
Areas lacking full development	Uneven immersion in developer Lack of sufficient initial agitation Exhausted or overdiluted developer Paper loops not separated in the developer
Peeling of emulsion	Storage in too hot water Exhausted fixer
Yellow-brown stains	Iron in wash water Vegetable matter in water supply Fixer being carried into developer
Image fading or yellowing with age	Insufficient fixing (may be due to exhausted fixer) Insufficient washing
Scum on record	Exhausted fixer Dirty wash water
Purple-brown stain	Insufficient fixing Exhausted fixer Premature exposure to white light

Table 6.1, a summary of processing faults, is reprinted with permission from the Eastman Kodak copyrighted publication, "Kodak Materials for Geophysical Exploration."[6.5] It lists the most usual faults together with the probable cause. The table is useful in determining errors so that mistakes will not be repeated on future charts. Most defects occur from carelessness, and can be prevented by watchfulness and adherence to recommended practices.

In addition to the defects listed in the table, there are accidental occurances that damage charts, sometimes beyond reclamation. In some cases, the chart may contain information so valuable that every effort must be made to salvage it. Frequently, the fault may occur in the recording operation: for example, the recording material may be light fogged due to the recorder magazine being damaged by the test it is recording (such as an explosion, crash of an aircraft, etc.)

It is sometimes possible to specially process fogged records to develop a trace that can be read, even through a considerable amount of fog. The chance of success is greatest on charts that have been given a trace exposure above normal. A high contrast developer, such as Eastman Kodak D-8, modified by the addition of benzotriazole (or other anti-foggant or restrainer) is used in place of the normal developer. Preliminary tests should be made on small strips prior to subjecting the entire record to the special treatment.

Fogged records on developing-out chart paper have been salvaged[6,6] by use of the forebath and the developer from a CEC POP-300 Direct-Print Stabilizer Kit.

Damaged records may be improved by reduction and/or intensification. Such procedures are tricky at best, and call for extreme caution. Test strips should be made before treating an entire roll. Formulas for reducers and intensifiers may be found in *Photo-Lab Index*,[6.7] Kodak Data Book, *Processing Chemicals and Formulas*,[6.8] and in other publications.

Removal of stains is usually just as hazardous as the removal of fog. Removal of common discolorations is described in the Kodak Data Book, *Stains on Negatives and Prints*.[6.9]

Occasionally, a record becomes wet while it is in roll form. This may happen to an exposed but unprocessed record, or to a finished record. The usual mistake is to dry the record in its rolled-up form. This makes the emulsion stick to the back so tightly that the material cannot be unrolled or separated. The correct action is to unroll it while it is still wet and can be separated. Naturally, if the material has not been processed, it is still light-sensitive, and the operation must be carried out in a darkroom; the record may then be processed. In some cases, if the record has been wet by dirty or contaminated water, the record should be washed in running water before processing. It also may be necessary to dry the record, and to re-roll it after washing, depending on the processor to be used.

A similar stuck-roll condition occurs during processing when moist paper is wound on the take-up spool by a careless operator, or by failure of the drier. Such a roll should be unspooled before it becomes stuck together; it should be throughly dried before it is rewound.

The following list of suggestions should be followed if corrective action is required to restore a record:

1) Most records today are stabilized; therefore, be sure to wash the record thoroughly before attempting restorative measures.

2) Remember that the formulas for reduction, intensification, etc., were compounded originally for standard photography—not for oscillography. Therefore, when used to improve oscillograph charts, optimum results may not be obtained without some modification to the formula, to its concentration, or to the procedure for its use.

3) The "ounce of prevention" maxim definitely works for oscillography. Be sure to use fresh and properly stored sensitized materials, and process them in solutions that are correctly mixed and not overworked.

258

LITERATURE REFERENCES

[6.1] EATON, G. T., *Photo Chemistry in Black-and-White and Color Photography*, p49, Eastman Kodak Company, (Rochester, New York) 1957

[6.2] RUSSELL, H. D., "Procedures and Equipment for the Rapid Processing of Photographic Materials." *Photographic Engineering*, Vol. 2, No. 3, p136, (1951)

[6.3] RUSSELL, H. D., YACKEL, E. C., and BRUCE, J. S., "Stabilization Processing of Films and Papers," *PSA Journal*, Section B, (August 1950)

[6.4] BRUENNER, R. S., "Thiourea and its Derivities in Photographic Stabilization Processing," *Photographic Science and Engineering*, Vol. 4, No. 2, p186, (1960)

[6.5] *Kodak Materials for Geophysical Exploration*, Eastman Kodak Company, (Rochester, New York)

[6.6] JACOBS, J. H. and BROWN, R. P., *Private Communication*, Bell and Howell Research Center, 360 Sierra Madre Villa, (Pasadena, California)

[6.7] *Photo-Lab-Index*, CARROLL, J. S., Morgan and Morgan, (New York 17, New York)

[6.8] *Processing Chemicals and Formulas for Black-and-White Processing*, Eastman Kodak Company, (Rochester, New York)

[6.9] *Stains on Negatives and Prints*, Eastman Kodak Company (Rochseter, New York)

VII   MANUAL AND REWIND PROCESSING

TRAY PROCESSING. The simpliest method of processing is tray processing. Obviously, this technique has limited use for oscillogram development because of the long-roll form of sensitized materials used in oscillography. However, for very short-length charts, for calibration purposes, and for occasional field use, it is a speedy and economical method. It is also useful in stabilizing (permanizing) short sections of data cut from direct-print rolls.

The following guide for tray processing is presented for those unfamiliar with photographic procedures. Photographic trays of enameled steel or stainless steel of suitable size are arranged on a work table or in a long photographic sink. At least three trays are required, and a fourth tray is desirable. If the sink has running water, a photographic tray siphon can be fitted to tray number four, and the tray filled with water. The trays are filled with solutions as follows: 1) developer; 2) stop bath; 3) fixer; and 4) water. The chemicals should be prepared in accordance with the instructions of the manufacturer. The temperature of the solutions should be checked before use; it is preferable to tray develop at normal (68° to 75 °F) rather than high temperatures. The material should be processed for specified times in each solution. A darkroom timer should be used. The sensitized material should be handled under the proper safelight. Extreme care should be used to prevent contamination of the developer with stop bath or fixer.

If the chart is longer than the tray, it is advisable to prewet it in *fresh* water (tray 4) before placing it in developer (tray 1).

NOTE: Be sure the water in the tray is *not* contaminated with fixer or other chemicals. The prewetting technique will permit numerous loops of the record to be processed in relatively small trays without irregular development, undeveloped areas, and other adverse effects. A minute amount of photographic wetting agent added to the developer will facilitate handling. The loops should be moved about in the developer to provide adequate agitation. After the specified time has elapsed, the loops are lifted, drained, placed in the stop bath (tray 2), and agitated thoroughly for approximately one minute. They are then lifted and drained of stop bath solution, trasfered to the fixer (tray 3) and agitated again for at least 30 seconds. Periodically, while in the fixer, the chart should be moved about to insure even and complete fixation. When fixation is complete, the chart loops are again lifted, drained, and then transfered to the wash water (tray 4). Under a continuous flow of fresh water, the record will be washed sufficiently in 30 to 60 minutes to have a life of from 5–15 years. Water temperature, and other individual conditions will vary the life and/or amount of wash required.

When washing is completed, the paper chart should be removed from the water, drained, and the surface moisture removed with a viscose sponge or a blade squeegee. If no drier is available, the strip may be festooned over lines stretched across the room, or from pegs mounted in a wall. No tension should be applied to the chart.

PROCESSING NARROW-WIDTH FILMS. Narrow-width films, such as 35 mm or 70 mm oscilloscope recordings, may be processed in conventional spiral-reel tank processors. Tanks of this type for short lengths of film can be obtained from photographic supply firms, in both daylight and darkroom-loading models. Light-tight lids on the darkroom-loaded models permit all processing functions to be carried out in daylight.

Long film lengths (to a maximum of 100 feet, usually), in 16, 35, 70, and 90 mm widths, may be processed in similar spiral-reel tanks. Outfits consist of the reel, a loading device, and three or four nesting tanks. Some systems feature motor-driven loaders, which can also be used to rotate the film for drying. Other processor units have forced-air driers. Manufacturers of spiral-reel processors are Pako, Watson, Nikor, and others.

BATCH-TYPE PROCESSING. The batch-type processor is a manual or motorized semi-automatic tank apparatus, in which an entire roll is treated in one solution and then transfered to the next tank, and so on. The rewind processor, often used for developing aerial films, is the most common batch processor, and is available in different sizes to process various widths of material ranging from 16 mm to 12 inch wide records.[7.1] Most models are adjustable for several widths.

The usual processor consists of a two-reel assembly on which the record is wound back and forth, a motor drive transporting the material from one reel to the other. Automatic reversal of the drive when the end of the roll is reached, rewinds the material back on the first reel. The entire loaded-reel assembly is placed in each processing tank successively to develop, rinse, fix, and wash the roll. Typical rewind processors are illustrated in Fig. 7.1, 7.2, and 7.3.

LOADING. Exposed records are loaded on the reel assembly from an accessory loading plate mounted on the bench top. Working under safelight conditions, the roll of exposed record is removed from the oscillograph takeup reel, positioned on the loading plate, and the free end of paper (or film) attached to one reel of the pro-cessor. The paper should be fastened squarely in the reel slot, to insure correct alignment and smooth tracking. The paper is transferred to the developing reel, and the other end attached to the opposite processor reel.

PROCEDURE. The above operations should be conducted only after the tanks have been filled, and temperatures have been checked. The tanks are usually made to be nested when not in use, thus saving space for storage, and for ease of transport in field operations. The tanks are set up for processing by locating them side by side on a bench or in a sink. They are filled with developer, water, and fixer respectively. The middle tank (water) should have a hose connection to the water supply tap.

262

Fig. 7.1. Fairchild-Smith Roll Film Developing Outfit. Model F-213 processes 70 mm film. Adapters can be used to process 35 and 16 mm film. Model F-214 processes 35 mm film. Adapter reels can be used for processing 16 mm film

The loaded reel assembly is lowered in the tank of water and the sensitized material is allowed to make two passes through the water bath to wet it evenly*. The reel assembly is then lifted, drained, and placed in the developer tank. The motor is turned on and the timer is set for the required development time. While the film is being

---

*At times uneven development or emulsion damage may occur especially with perforated films or with color films that have an inherently soft emulsion. The prewet or prebath solution causes the

263

developed, the required amount of acetic acid should be added to the water in the middle tank to make the stop bath. Be sure to stir it thoroughly.

When the timer bell rings, the motor is turned off, and the reel is lifted from the developer, drained, and placed in the stop bath. The motor is turned on again, and

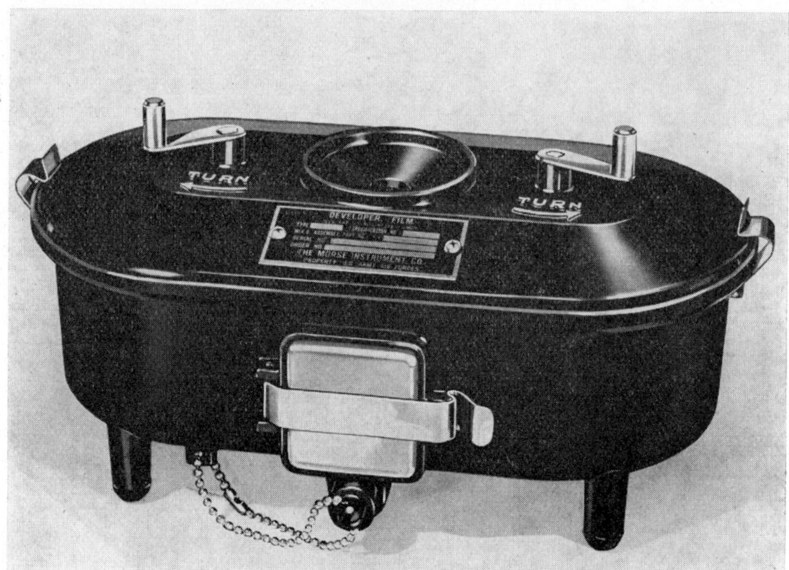

Fig. 7.2. Morse G-3 16/35 mm Developing Tank

the chart allowed to make two passes through the stop bath. The reel is again lifted, drained, and transferred to the fixer tank. The motor is again turned on and allowed to run until fixation is complete. This will take from 5 to 15 minutes, depending upon the type of fixer, its condition, the kind of material being fixed, and its length. White room lights can be turned on after the chart has made a couple of passes through the fixer.

---

damage by seeping between layers of dry film on the feed reel assembly. There it wets and softens the gelatin unevenly causing it to stick to the back of the adjacent film layer. When the film is transported to the other reel, the emulsion is peeled off its support.

If this problem occurs, it can usually be overcome in the following manner. Place the loaded reel assembly of dry film with the axes in a horizontal position, above the prebath tank. Allow the film to come off the loaded reel in a dry, unmoistened state. It should then pass vertically and slowly into the prebath solution below, and wind up on the takeup reel under the solution below, and wind up on the takeup reel under the solution level.

This procedure will insure that the film is evenly and completely wet before it comes into intimate contact with the back of the adjacent layer.

A rectangular tank should be used for this operation, and must be sufficiently large to allow the operator's hand to turn the crank on the takeup reel. After the initial prewet, the balance of the processing may be carried out in the conventional manner.

While the chart is being fixed, the stop bath should be emptied from the middle tank and the tank refilled with fresh water from the tap and hose connection.

When fixing is complete, the reel is again lifted and drained. If processing is being conducted in a sink, a hose spray can be used to rinse the reel and chart of excess

Fig. 7.3. Morse M-10 Developing Outfit

surface fixer. The reel assembly is placed on the sink bottom, and water is sprayed on the chart while it is run from one reel to the other. This will greatly reduce the amount of fixer carried into the wash water, thereby reducing the washing time. The amount of washing required will depend upon the record life desired.

PROCESSING TIMES. The frequency at which the rewind processor reverses direction will affect the processing time, since the number of reversals is dependent upon the length of the roll being processed. The exhausted solution, trapped between the wound-up-film layers on the reels, must be replaced with fresh solution. The rewind processor can only replace spent processing chemicals when the film or paper is being moved between the two reels. At that time only, replenishment and agitation take place. After the material is wound on the reel, processing continues at a much slower rate. When the material is rewound, another activity cycle begins. Because of this function, developer-exhaustion rate is quite important, especially when long rolls are processed. It is important to choose a developer with high activity, generally one which contains a high concentration of developing agents and alkali.

The graph in Fig. 7.4 will guide the user in determining processing times.

265

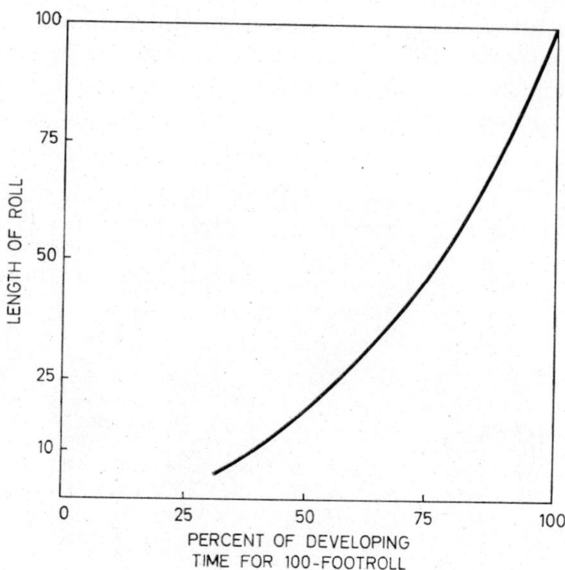

Fig. 7.4. Percent of development time for rewind processors. The curve is used to determine the approximate development time for partial rolls of film. Under average conditions it can be assumed that a 100 foot roll requires 100 per cent of the total time; times for shorter lengths can then be determined from the curve

UNLOADING. Fig. 7.5 shows a reel which may be used to transfer processed material from the developer reel assembly. This will enable the processing-reel assembly to be dried and made ready for the next roll to be processed.

DRYING. The processed and washed roll may be dried in commercial driers used by the photographic profession. Generally, these driers are of two types: 1) film driers; and 2) paper driers. Film driers are air-impingement machines and can also be used to dry paper. Paper driers are drum-belt machines and are not suitable for drying film.

Under field or temporary conditions, film or paper may be dried by exposure to air. Pegs, mounted in the wall, or lines over which the record is looped, are typical air-drying methods. Such improvised techniques are unsatisfactory when a considerable amount of footage is to be handled and good quality is desired.

NARROW-WIDTH FILM DRIERS. Reel driers are inexpensive room-air driers that are satisfactory for narrow perforated or unperforated films. A typical example is shown in Fig. 7.6.

Reel-type driers have a drying-speed limitation determined by ambient temperature and relative humidity. Forced heated-air driers, such as the Fairchild-Smith Model F-304, speed up film drying. The Model F-304 will accommodate 16, 35 and 70 mm films.

266

Fig. 7.5. Transfer Reel. After being washed, the still wet oscillogram is removed from the processor-reel assembly, right, by winding it onto a core having removable flanges. The wet record is then run through a drier

WIDE-FILM DRIERS. Impingement driers, used in wide-roll aerial film processing, are satisfactory for drying either film or paper oscillograms. Machines of this type utilize a centrifugal fan mounted inside a drum that has closely spaced air slots around its circumference. Wooden rollers are mounted at intervals around the drum, and are spaced less than an inch away from its surface. Each roller is driven by means of a sprocket at one end, which is turned by an endless motor-driven chain.

The fan inside the drum causes air to emerge from the slots and to support the film emulsion away from the drum surface. The back of the film, pushed against the rotating rollers, is carried around the drum to a take-up spool. The spool, driven through a slip clutch, winds up the dried film.

The type of drier described above, is often sold in the United States as surplus armed services equipment. Two dryers of this type are designated as the A-5 and the A-10, either of which will dry material in various widths to 9–1/2 inches. The A-10 drier is enclosed in a cabinet, and the heated moist air may be vented outside.[7.2] The A-5 is not enclosed; its shipping case is used as a base when the drier is in operation.

Driers of these types have nominal drying rates ranging from 2 to 8 feet per minute. A discussion of drying aerial films on the A-5 is contained in[7.3]. This report by Michener describes the use of deflector plates, which he added to the drier to improve the drying of gel-backed films. The use of the drier for paper is not discussed. However, it is probable that deflector plates would increase drying efficiency of paper as well as gel-backed films. The sheet metal plates, attached to each side

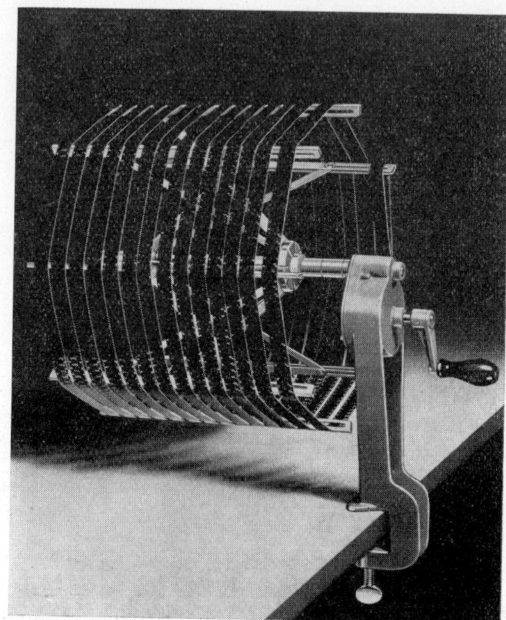

Fig. 7.6. Morse M-30 Reel Drier for narrow-width films

of the drum around its circumference, deflect a portion of the exhaust air to the back of the film to speed drying of that surface.

PAPER DRIERS. Rotary photographic-paper driers used by commercial photographers can be used to dry oscillograms. Commonly called *print driers*, the machines are made in two types: glossy driers and matte driers. Glossy driers have a shiny chromium or stainless steel drum with heaters on the inside. A canvas belt travels around approximately 75 percent of the circumference. In normal commercial photographic practice, prints made on glossy paper are placed on the belt facing the drum. The belt sandwiches the paper to the shiny drum surface, and when the print emerges it has the characteristic glazed, (ferrotyped) finish. Matte prints are dried with the emulsion *away* from the drum surface (i.e., the paper backing toward the drum surface). Since oscillograms are not glossy, they should be dried the same as matte prints.

Matte driers may have one or two belts. If the matte drier has a single belt, it undoubtedly is similar to the glossy drier described above. Therefore, in drying oscillograms, the emulsion should face away from the drum.

If the drier has two belts, the prints (or oscillograms) are fed between the two. Again, it is advisable to insert the paper with the emulsion away from the drum.

Only washed records should be dried in canvas-belt driers. If stabilized prints are run through such driers the belt will soon disintegrate from the chemicals absorbed by the fabric. In addition, the chemicals will attack the metal drum on older machines whose drums lack corrosion resistance. It is a good rule to never put unwashed or even slightly washed records through any drier that is used for archival records or photographic prints. The chemicals will contaminate the belt, and will render impermanent any material run through the drier afterward, even though the later material be completely washed.

Paper driers are sold in widths ranging from approximately one foot to over four feet. It is possible to dry several record strands simultaneously on wide models. Care must be exercised to space the individual strands apart and to see that the tracking is straight so that they do not overlap in the drier.

If a print drier is to be used for drying records at a fairly high production volume, the user should modify the drier by adding a feed-reel support and a driven takeup reel. Both devices must be in alignment with the drum axis to insure straight tracking. The takeup spindle, on which the takeup reel is mounted, can be cantilevered for easy loading and unloading. The spindle may be driven with a small torque motor or with a slip-clutch drive. Takeup of multiple strands on a wide drier becomes complex and is not recommended.

LITERATURE REFERENCES

[7.1] LEVIS, R. E. and FROULA, H. C., "Use of the G–3 Film Processing Tank, "*Journal of the SMPTE*, Vol. 50, No. 5, p 474, (1948)

[7.2] *Handbook for Photo Lab Processing*, AF Manual 95–11, p 5–57, Superintendent of Documents, U. S. Government Printing Office, (Washington 25, D. C.), 1960

[7.3] MICHENER, B. C., *Drying of Processed Aerial Film*, Eastman Kodak Company, (Rochester, New York)

VIII   CONTINUOUS MACHINE PROCESSING

Types of Processors. Continuous processors are made for materials of various widths and generally can be divided into two types: stabilization and archival processing machines. Although some processors will accept both film and paper, the problems differ for handling these two sensitized materials. For this reason, it is common practice to process film on one type of machine, and to process paper on a different type.

Today, the use of photographic paper for oscillographic recording far exceeds the use of film. Most paper oscillograms are processed on continuous stabilization equipment, such as the CEC and Kodak Ektaline processors. The use of these two machines is so extensive that installation and operating instructions are printed in the following sections of this book. This is done to acquaint potential purchasers with the capabilities of the machines; it also provides technicians needing to process an occassional roll of paper with the correct operating procedure.

Other types of continuous equipment, primarily for archival processing of film and paper, are described in successive sections. Specialized processors for microfilms, motion pictures, and other materials are also discussed.

The Consolidated Electrodynamics 23–109B Processor. The information which follows is an edited version of the Consolidated manual, to which the author has added pertinent supplementary material.[8.1]

General Description. The Consolidated stabilization processor, originally marketed in 1954, was designed as a self-contained unit to meet the need for a rapid, reliable machine that would accommodate oscillogram—and strip-photographic-paper records. Requiring only chemical solutions and electric power to function, it may be used under field conditions, with minimum darkroom facilities, and can be operated by personnel having limited photographic experience. However, the processor may also be used in fully equipped photographic laboratories to speed the processing of oscillograms.

Numerous improvements have been made to the original machine, and the current model, the CEC Type 23–109B processor is illustrated in Fig. 8.1. However, aside from certain part numbers, transport speeds, drying speeds, and cleaning methods, all models are similar enough to permit operating manuals to be used interchangeably.

The processor is manufactured in two models by Consolidated Electrodynamics Corporation: Type 23–109B–P4 for use with 115 volt power, and Type 23–109B–P6 for use with 230 volt power. Either machine requires 60-cycle, single-phase power. A modification for 50-cycle operation is available on special order. The processor, measuring 18 inches high, 19 inches wide, and 31 inches long, weighs 160 pounds dry, and is constructed principally of AISI Type 316 stainless steel. Current models have an aluminium drier drum to promote rapid and even heat transfer. Capable of whitelight operation, the processor has a light-tight magazine which will hold up to 250 feet of standard weight paper, or 475 feet of light-weight paper, in widths from 70 mm to 12 inches. The only darkroom operation required is to transfer the exposed record from the oscillograph magazine to the processor magazine, using the correct safelight for illumination. If, under field conditions, transfer is made under subdued white-light conditions, fogging of the first 10 feet of record may be expected.

Fig. 8.1. CEC 23–109B Oscillogram Stabilization Processor

The processor may be operated at ambient temperatures from 70 to 100 °F, at an ambient RH of 80 percent. Under these conditions, the 230 volt model will process records at speeds ranging from 5 to 30 feet per minute, depending on chart width and thickness. The maximum drying speed of the 115-volt model is 10 feet per minute, depending on the relative humidity, paper used and other conditions.

The portable processor uses no circulating water system. Instead, the exposed record is continuously transported through a four-tank stabilization process. The first tank contains the developer, a conventional formula except that it is more highly concentrated and particularly suited for high-temperature operation. At a processing rate of ten feet per minute, any one point of the record is immersed for only eight seconds. Thus the image is developed so rapidly that the paper is withdrawn before it is completely saturated. This fact is of considerable importance, since developer absorbed in the paper must be subsequently replaced by the stabilizer.

After development, the record passes through the stop bath where development is arrested. The remaining two tanks contain the stabilizer solutions. At this point the process ceases to be conventional.

272

The stabilizing agent reacts with silver halides in the emulsion to produce a soluble complex. This complex, which is partly allowed to remain in the record, is relatively stable to light, heat, humidity, and atmospheric gases. After stabilization, the record passes over the drying drum and is wound on the takeup spindle.

The solutions in the processor are maintained at an operator-controlled preset temperature. A thermostatically controlled water bath, contained in a jacket that surrounds all four tanks, provides uniform and stable temperatures, after a short warmup period. This capability exists even when environmental conditions are considerably different from operating temperature.

PROCESSOR INSTALLATION. The processor should be located on a rigid work surface, preferably 26 to 36 inches above the floor. The table may be portable (a rigid cart, for example) but should be level to insure proper operation of the transport system. A power connection is needed at, or near, the working area: the P6 model requires a 15-ampere, 4-conductor, 230-volt, 60-cycle single-phase connection; the model P4 requires a 15-ampere, 3-conductor, 115-volt, 60-cycle, single phase connection.

Ground Connection. The processor case must be adequately grounded. One conductor of the power cable is connected to the processor case. At the supply end of the cable this connection is grounded to the conduit when plugged into an approved "polarized" outlet (4-prong for 230 volts, 3-prong for 115 volts). However, if the outlet is *not* an approved "polarized" type, the furnished adapter may be used to convert the supply end to a pigtail grounding plug. When the adapter is used, the pigtail becomes the grounding connection and *must* be connected permanently to the conduit or an electrical ground before the processor may be operated.

Water is required for mixing chemicals, and for filling and maintaining the level of the constant-temperature water bath of the processor. A sink, with hot and cold running water, is desirable for cleaning the tanks and roller assemblies. A small, short-handled mop, sponge, and bucket are recommended for general daily cleaning.

PROCESSOR OPERATION. Operation of the Type 23–109B Oscillogram Processor should be undertaken only after a through study of the instruction manual, and when familiarization of the controls and their function has been completed.

After the processor has been set up and connected, the operational schematic, Fig. 8.2, together with the following description, will serve as an operating procedure:

1) Release the catches and remove the light tight cover.

2) Release the inner catches and lift out the framework of the lower roller assembly. Note that the proper orientation of this unit is determined by small pins and notches in the supporting flanges.

3) Lift out the upper-roller assembly which is now free. (Figure 8–3)

4) Lift out the four solution tanks.

5) Drain a small amount of water through the bath drain tube to make sure it is free-flowing.

6) Fill the main heater bath tank to the water level mark with 1 1/2 gallons of tap water. Check first to see that the drain valve at the left end of the processor is shut off. Water level dimple is slightly above the solution tank-support rail and the rear side of the tank.

7) Reset the footage counter to zero only when chemicals are replaced.

8) Prepare chemicals and fill the tanks as follows: tank B, developer; tank C, stop bath; tanks D and E, stabilizer.

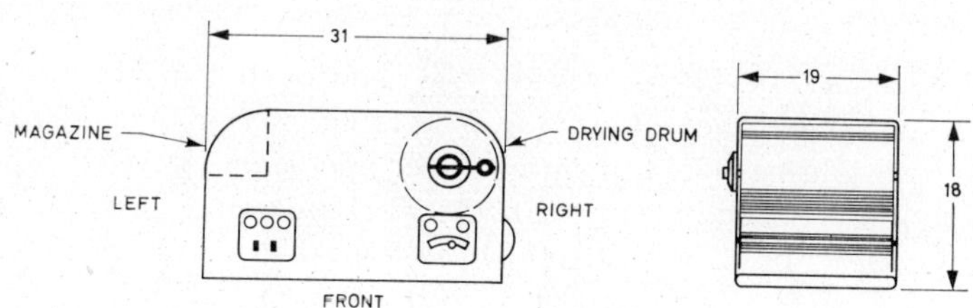

OUTLINE DRAWING

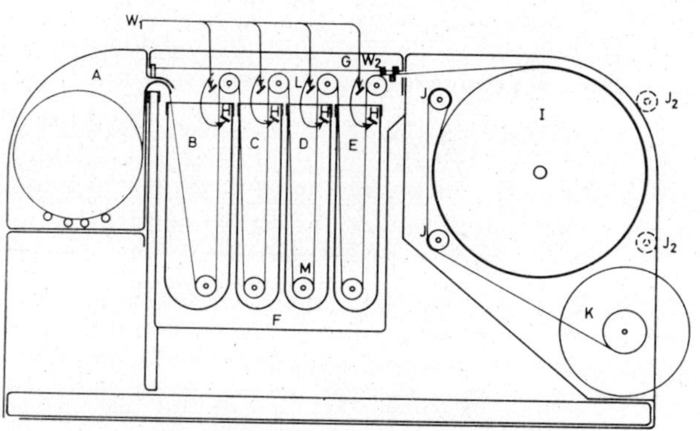

Fig. 8.2. Operating schematic of CEC 23—109B Processor

A PAPER MAGAZINE———CAPACITY { 475 ft. THIN BASE / 250 ft. REGULAR
B TANK NO. 1 — DEVELOPER———CAPACITY 1.5 GALLONS
C TANK NO. 2 — STABILIZER #1——CAPACITY 1 GALLON
D TANK NO. 3 — STABILIZER #2——CAPACITY 1 GALLON
E TANK NO. 4 — STABILIZER #3——CAPACITY 1 GALLON
F BATH HEATER————————CAPACITY 1.5 GALLONS

G LIGHT TIGHT COVER
I DRYING DRUM

J_1 WRAPAROUND ROLLERS IN OPERATING POSITION
J_2 WRAPAROUND ROLLERS IN THREADING POSITION
K TAKEUP SPOOL
L UPPER ROLLER ASSEMBLY
M ROLLER OF LOWER ROLLER ASSEMBLY
W_1 WIPERS
W_2 WIPERS

274

CHEMICAL MIXING. Chemical solutions for the processor may be prepared from liquid concentrates or from powders. Liquid-concentrate chemical kits enable mixing to be conducted in the processor tanks. The kits also have the advantages of convenient storage, and dilution is simple enough so that untrained personnel can prepare the processor for operation.

CEC's No. 156455 Liquid Concentrate Chemicals Kit consists of four plastic throwaway bottles packaged in a compact cardboard box. One bottle contains a

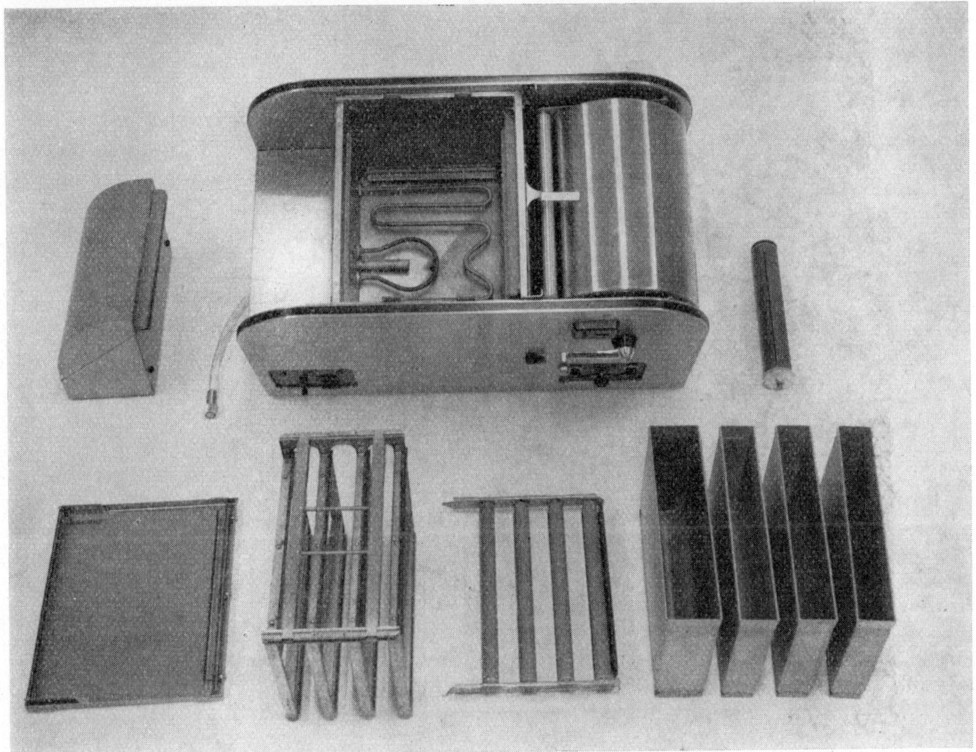

Fig. 8.3. CEC 23—109B Processor disassembled

liquid concentrate developer; the other three contain stabilizer solutions. The liquid concentrates are poured in the four tanks, and require only the addition of warm water which simultaneously dilutes and mixes the solutions.

Consolidated has added a special lubricant to the last two stabilizer solutions which reduces wear on processor wipers, prevents transfer of residue from the drums, and enhances the appearance of the finished record.

The chemicals do not release irritating fumes during the processing cycles. Bottled solutions are ICC approved. Kits are available through all CEC sales and service offices.

NOTE: High temperature processing with some brands of chemicals, particularly of stabilizer solutions, may cause irritating fumes. These can be eliminated with an exhaust fan, the intake of which should be located close to the drier drum. The fan should be turned on prior to processing.

In all cases, adequate ventilation should be provided in processing areas.

The chemical manufacturers' intsructions furnished with their respective products should be followed carefully. Mixing dry chemicals in the solution tanks, especially when they are in the processor, is *not recommended*. This method is time-consuming and haphazard; it increases the chances of contamination and makes it difficult to know when the chemicals are completely dissolved. The following procedure is recommended for preparing solutions from powders: Use a bucket of a material which is unaffected by and does not contaminate the solution to be mixed. Stainless steel, hard rubber, or plastic containers of approximately 3-gallon capacity are recommended. A paddle of similar material is required for stirring. A motor-driven stirring device will greatly speed up the mixing operation and for this reason is recommended.

Dispose of used chemicals only in a well-ventilated area, particularly if they are at the operating temperature. Avoid flooding a large shallow area, such as a dark-room sink with chemical solutions. Wash down sinks and wipe up spilled chemicals so that there is no residue, streaks, or deposits to cause contamination, staining, or corrosion.

9)   Turn the selector switch to PREHEAT.

10) Throw the power switch to ON. Allow approximately 15 to 20 minutes for the bath and chemicals to heat to the operating temperature, 100 °F, before starting to process. (Or allow the BATH indicator light to cycle twice (on-off-on).

Paper may be loaded and threaded during the warm-up period. The sequence of steps is as follows:

Loading the Magazine

1) Remove the magazine by lifting up and back. This can be done only when the main cover is removed. The magazine cannot be opened while it is on the processor, as its cover-release buttons are not accessible.

2) Press the catch buttons on each end of the magazine to release its cover, and open. Observe the curved strip that forms the lower lip of the light seal, and, when the cover is in place, permits the record to pass from the magazine to the tanks without danger of light leak.

3) Unsnap the guide or spacer plates from the graduated bar and adjust them to the width of the record to be processed. Simultaneously, center them longitudin-ally in the magazine.

4) The four rollers in the bottom of the magazine are used to support records eight inches or less in width. The mgazine spindle is used to support records which are more than eight inches in width. This spindle passes through the core of the roll and is installed, or removed, by deflecting the end supports. When it is not in use, it may be stored in the magazine.

5) Records wound on oscillograph takeup spools which do not have removable

end flanges, should be supported on the single magazine spindle in the same manner as wide records. To accommodate the additional space required by the end flanges, one magazine guide plate support has a small slot on the end adjacent to the calibrated bar. This slotted spacer plate support should be positioned to permit the slot to fall over the barrier on the calibrated bar, thus giving an additional space adjustment of 1/8 inch.

> *CAUTION.* When records eight inches or less in width are supported by four rollers in the bottom of the magazine, be certain that caps, flanges, or plugs have been removed from the ends of the core of the roll, as they may interfere with the guide plates.

6) In a darkroom, open the oscillograph magazine, take out the exposed record and remove the oscillograph takeup spindle from the exposed roll.

The tail end of the record (the last portion to be processed) must be capable of detaching itself from the core upon which it is wound and be free to pass through the light seal of the magazine. Records that are properly taped to the cardboard core, when threaded onto the takeup spool of the recording oscillograph magazine, will meet the requirements above. Records that are mechanically clamped to an oscillograph takeup spool must be rewound and released.

7) Place the record in the processor magazine so it will unwind from the bottom up on the side adjacent to the light seal.

8) Pull the end of the record from the roll and lay it over the light-seal lip so that the paper extends three or four inches outside of the magazine. The sensitized side of the record *must be up.*

9) Close the processor magazine cover securely. The magazine may now be carried out into the light and placed in position on the processor.

Use of Records Leaders

1) If the last significant data recorded is less than 20 feet from the end of the record, it will be necessary to attach a leader to the record. This will allow the processor to be threaded without the danger of loss of record due to exposure.

2) Leader splicing and spooling should be done only in a darkroom under a Wratten Series 2 Safelight.

3) Leaders may be made from unprocessed record material, or the following standard leader material stocked by CEC. These widths are available:

	CEC Part No.
3 inches by 500 feet	33680–203
5 inches by 500 feet	33680–205
7 inches by 500 feet	33680–207
8 inches by 500 feet	33680–208
12 inches by 500 feet	33680–212

Required individual leader length, 20 feet.

Leader material should be the same width as the record to which it will be attached. For ease of handling, it is recommended that a quantity of leaders of the same width and 20 feet long be spooled in such a manner that they may be pulled off the roll as needed.

Leaders may be attached to the exposed record with Scotch Brand 3M polyester film tape #853, 2″×72 yards (CEC Part No. 217040). The dispenser for the above tape may be obtained under CEC Part No. 217041. This tape must be used because the splice is subjected to considerable tension, warm chemical solutions, and high drying temperatures and ordinary tape is unsuitable.

CAUTION. The oscillogram processor has a silicone-treated surface; after contact with the surface, and before attempting to make any paper splices, wash hands with soap and water.

Contact of the tape on the silicone surface will cause the tape to lose its adhesiveness.

Cut tape according to the following requirements:

Paper Width	Number of Strips
12	4
7	3
5	2
3–1/2	1

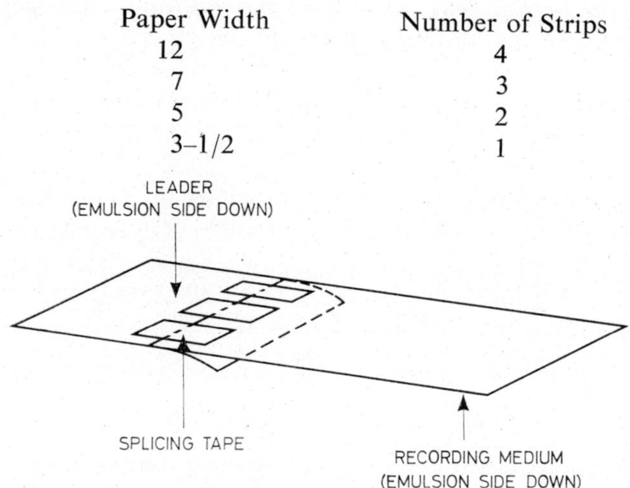

Fig. 8.3a. Leader splicing

In a darkroom, illuminated only by the light from a Wratten Series No. 2 Safelight, unroll about 10 inches of the record roll. Place leader material on flat surface with emulsion down. Overlap the leader material with recording paper, emulsion side down, by about 2 inches.

Make certain that the edges are carefully aligned. Apply the designated adhesive tape in strips approximately 6 inches long, with no more than 1 inch between adjacent strips or between the paper edge and a strip of tape. Apply firm pressure over the entire tape backing to insure a positive bond.

It is not necessary to apply tape to the emulsion side of the overlap joint since the trailing edge will not catch on the wipers.

278

When the splice is complete, wind the leader on the roll, checking carefully to make sure that the edges of the leader are even with the record. The roll may now be loaded into the magazine in the normal manner. The emulsion side of the recording medium should be face up when processing the record.

1) A 20-foot length of record, or of leader material, is required for threading the processor. If leader is used, splicing is required, as described above, and must be done prior to attaching the magazine to the processor.

2) Place the magazine in position on the processor by hooking the light-seal lip over the wall of the water bath tank. Make certain that the two locating screw heads on the top edge of the tank wall enter the mating holes in the flat underside of the light-seal lip of the magazine.

3) On the upper roller assembly, note the arrows marked DRUM. Install the assembly with these arrows pointing toward the drying drum. This means that the two rectangular supports projecting diagonally outward from the side rails will be positioned adjacent to the magazine.

4) Place the wrap-around rollers in the threading position by depressing the latch lever and rotating the crank 180° in a counter-clockwise direction. Wear on the latch will be reduced if the roller crank is lifted slightly before the lever is depressed.

5) Pull the end of the record protruding from the magazine (sensitized side up) across the top of the upper roller assembly and over the drying drum. Drop the end of the record down between the drying drum and the wrap-around rollers.

Pull the end of the record approximately 16 to 18 inches beyond the lower wrap-around roller. Remove the takeup spool by lifting upward and slightly away from the processor. Remove the *takeup spool* cap (the end opposite the gear) by pulling and turning in either direction.

Hold the takeup spool in the right hand by the gear end, and slip the split end over the edge of the record from right to left. Continue moving the spool across the record until it is roughly centered. Replace the spool cap. Turn slightly while pressing firmly to insure that it is all the way on.

With approximately six inches of record extending beyond the takeup spool, rotate the spool so that the end of the record passes over the top and lies against itself with the emulsion sides together. Center the record by the index lines on the spool. The takeup spool and record may be aligned by accurately matching the edges of the tail of the record to the edges of the adjacent portion of record. While maintaining this edge alignment, simultaneously pull and rotate the spool to "crease" the record at the spool slots. Wind up sufficient record to permit replacement of the spool in the processor. Engagement of the gear drive will prevent unwinding.

An alternate method of squaring and starting the winding of the record on the takeup spool is as follows:

After the spool has been slipped across the record and the spool cap put on (see above), replace it in the processor with the record passing vertically through the spool. With the right hand hold the center of the width of the record in against the end of the processor case. While pulling the record straight down firmly, turn the motor switch to ON and allow the transport system to wind up about one-and-a-half or two turns.

6) Swing the wrap-around rollers to their operating position by again depressing the latch and turning the crank 180° clockwise until it catches. The additional paper necessary will pull from the magazine.

7) The alignment between the upper and lower roller assemblies is determined by pins and notches in the mating side members of the units. Note the arrows marked DRUM on the cross bars of the lower roller assembly. These should point toward the drying drum when the assembly is installed. This means that the wiper blades attached to the frame will lean toward the drying drum. Press the lower roller assembly down between the upper rollers to carry loops of record into the solution tanks. The additional record necessary to do this will again pull from the magazine. When this roller unit is inserted, it should be kept as level as possible. If the rollers are allowed to tip down at one end, the record will "run" over to the opposite side. Attempt to have the entire "threaded" portion of the record centered in the processor as evenly as possible.

8) Secure the roller units in place by means of the "Release" handle.

9) Replace the top cover and secure with the two latches on the partition adjacent to the drying drum. The two short light-baffles on the underside of the top cover go adjacent to the magazine.

10) The processor is now completely threaded.

The processor will have reached its correct operating temperature by the time the magazine has been loaded and the processor is threaded.

1) When the BATH indicator light goes out, throw the selector switch to OPERATE.

2) Select PROCESSING SPEED and TEMPERATURE setting (See 5 below).

3) When the DRUM indicator light goes out, processing may be started.

4) Place the MOTOR switch to ON to start processing.

5) Two factors affect the optimum processing rate: 1) the development of the trace image, and 2) drying characteristics of the paper. The developer is formulated for best performance at speeds of 8 to 10 feet per minute, although it is possible to operate considerably below and above those speeds without seriously impairing development.

Choice of best speed for drying hinges on the width and thickness of the paper, drum heat, and residual water in the paper as it leaves the tanks. The amount of water remaining is largely dependent on the condition of the wipers; if they are worn or improperly adjusted, moisture will be unevenly removed.

Because many different types of recording papers are used, and drying characteristics vary greatly, it is not possible to state here optimum processing rates for each type and width of material. In any case, make sure the record is completely dry as it goes from the drum to the takeup roll. If moist, the chemicals will tend to fuse the roll together as it dries.

Until the drying behavior of the paper is known, the processing speed should be experimentally set to 8 feet per minute, and the drum temperature control set to 8. At the beginning, the drying should be checked by feeling the emulsion side of the record, particularly the edges, between the takeup spindle and the drying drum. The processing speed and drum temperature can be quickly adjusted as required.

6) Diagonal contraction wrinkles appearing on the record just after it comes in contact with the drum are caused either by too slow a processing speed or too high a drum temperature.

7) The MOTOR switch should be turned to OFF when the trailing end of the record comes onto the drying drum. As the last of the processed record pulls through the wiper blade in the top cover, spooling tension ceases and the loose end of the record will not dry because it will not be pulled tight on the drum. Allow this loose end to rest on the drum for a short period to insure that the emulsion side will not stick as the record passes over the wrap-around rollers. The MOTOR switch may be thrown to ON to complete the spooling.

> *CAUTION.* The wrap-around rollers must be in their operating position before the lower roller assembly is inserted.

8) Remove the takeup spool and record from the processor. Remove the spool cap and *pull the record off the spool while still hot.* Records that are allowed to cool on the takeup spool are more difficult to remove.

9) Change chemicals before exceeding the maximum record length recommended by the manufacturer.

10) An excessive accumulation of solidified processing chemicals on the drying drum will act as a thermal insulator and poor drying will result. The greater part of this residue should be removed from the drum after every 500 linear feet of record processed when the instrument is in continuous operation. It is not necessary to allow the drum to cool before cleaning.

11) Sometimes the lever that follows the takeup roll diameter will not return to the UP position when the record drive is not running. The lever should move up to contact the coil as soon as the paper transport drive is started. If it does not position itself automatically, all accessible pivots and bearings in the linkages should be cleaned and oiled. The lever should not be positioned by hand.

CLEANING THE PROCESSOR

During Operation.

1) Wipe off all spilled chemicals as soon as possible.

After Each Record is Finished.

1) Rinse cover, top and bottom rollers in hot water after every roll processed.

2) With a soft, damp cloth wipe off the drying drum, the wrap-around rollers, and the panels around the drum. The drum will be hot but, if the operator is careful, he can reach all these surfaces by use of a mop or sponge attached to a wooden

handle. Do not compound cleaning problems by melting a cellulose sponge on the drum.

3) Each time the tank cover is removed, wipe off the lip of the magazine, the under side of the cover, and the accessible areas around the top of the tank.

After Each Day's Run.

1) Wash out the chemical tanks with water.

2) Thoroughly rinse the upper and lower roller assemblies and top cover.

3) Clean the drying drum with sponge and water. Do *not* use a cellulose sponge. A cellulose sponge can melt on the drum surface if the drum is excessively hot.

4) Thoroughly clean the wrap-around rollers, the top surface of the water-bath tank adjacent to the underside of the top cover, and the projecting light-seal lip of the magazine. Clean the walls of the cabinet adjacent to the drum.

At the End of Each Week.

1) Scrub all surfaces with a non-abrasive cleaning powder and flush with clear water. Clean around pivots and bearings. The inaccessible parts are just as important as those that are easily reached.

2) To clean processor tanks, empty tanks of processing chemicals, rinse with water and fill with Kodak Developer Systems Cleaner. When cleaning action is complete, empty and rinse tanks thoroughly before filling with processing solutions. Residue on the roller assemblies should be removed by soaking in a tray of this cleaner.

3) Keep machine and thermostats clean; use small brush to clean thermostats. *Make sure all power is off.* To keep stainless steel clean, use a damp rag and wipe dry with soft paper towel. Periodic maintenance of thermostats may be required. When the tolerance of the minimum-to-maximum setting of the thermostat is too great, the temperature of the drum will cycle excessively, and the resulting roll will vary in width throughout its length. This condition can be held to a minimum by frequently checking (with a surface thermometer) the surface temperature of the drum under off and on conditions. If the tolerance is excessive, the contacts should be cleaned and adjusted, or if they are in bad condition, they should be replaced. New models of the CEC processor have drums made of aluminum instead of stainless steel. This change has improved heat transfer and evenness of temperature. In addition, the larger mass of aluminum with its greater specific heat, reduces the temperature variation of the drum with respect to its on and off cycles. The new aluminum drums are interchangeable with the old, and are available as replacement parts for older models of the CEC processor.

4) The paper will cause wear of the squeegees, therefore they should be checked periodically, and reset or replaced when necessary. Blade pressure should be adjusted as evenly as possible. Excessive or uneven drag on the wet paper may cause the record to wrinkle.

5) The drying speed is materially increased when the final set of wiper blades is

adjusted to remove as much liquid from the record as possible before it goes to the drying drum. Efficient wiping above each tank reduces loss of solution, thus conserving the chemical. The wipers also prevent contamination which reduces efficiency of the process in the next tank.

6) Proper adjustment of the squeegees can be checked by running paper through the processor with the top cover removed. Either water or processing solutions may be used while checking the blade adjustment. Do not use water if a check on the processing rate is desired. When records are run through four tanks of water, instead of solutions, the drying rate will be considerably reduced.

CAUTION. The 23–109B Processor is constructed of AISI Type 316 stainless steel. Abrasive cleaners and steel wool should *never* be used on the machine. See instructions in Chapter 9 for maintaining stainless steel.

APPLICATOR PROCESSORS. The CEC 23–109B stabilization processor just described is a compact processor using conventional tanks. The exposed record is passed through the tanks to achieve contact with the processing chemicals. However, there are other ways of achieving the necessary contact; the alternates are being used more and more in the design of processors where high speed and compactness are important considerations.

To produce a compact processor, the tanks, together with their large solution volumes, must be eliminated. Tanks are replaced by photo-solution applicators, devices that deposit developer, fixer, or other chemicals on the photosensitive materials for the process function. Thus, the applicators conserve weight and space, achieve a higher surface transfer-rate, and effect more speed, efficiency, and uniformity of operation.[Refs. 8.2, 8.3, 8.4]

The desirable characteristics of an applicator are: evenness of application, fast transfer rate coupled with high turbulation, no leakage, and high reliability. The applicators must transfer the solutions at specific rates and temperatures to produce uniform density. Further, by applying solutions to the emulsion side only, the drying problem is minimized. For rapid processing, chemical solutions may have to be capable of being heated to elevated processing temperatures without loss of energy, decomposition, or adverse effects to the sensitized material. Various types of applicators have been used to achieve these goals, some of which may be classified as follows: 1) impingement; 2) roller; 3) belt; 4) capillary; 5) slot; and 6) blade.

The CEC Datarite integral recorder-magazine processor was described in a previous chapter. It uses a slit applicator to achieve processing in a small amount of space, and to effect immediate access.

As photographic technology advances, deep-tank processors are giving way more and more to applicator-type processors. Examples of equipment of this type are the Rapidata equipment manufactured by Photomechanisms, Inc., 15 Stepar Place, Huntington Station, Long Island, New York; various camera-processor-viewers manufactured by OPTO mechanisms, Inc., Engineers Hill, Plainview, Long Island, New York; and integrated camera-processor-projectors developed by Kelvin

Hughes Limited for the Royal Air Force (available in the United States from Kelvin Hughes America Corporation, P. O. Box 1951, Annapolis, Maryland). Although these devices are most frequently used in conjunction with radar and display functions, they also have some application for CRT recording in which immediate access (within a few seconds) is required.

Rapid access for wide oscillograph paper records is also being accomplished with applicators instead of deep tanks. An example of such a machine is the Kodak Ektaline 200 Processor, manufactured by Eastman Kodak Company.

THE KODAK EKTALINE 200 PROCESSOR. The following information is condensed from the publication, "Kodak Ektaline 200 Processor, Model 1M, Installation and Operation", furnished through the courtesy of Eastman Kodak Company.

The machine is a stabilization processor first marketed in 1962. It is equipped with automatic-threading and automatic-takeup features.

GENERAL DESCRIPTION. The Kodak Ektaline 200 Processor, Model 1M, shown in Fig. 8.4, is designed for high-speed processing of photo-recording papers; specifically Kodak Ektaline 12, 16, or 18 papers. It will process and dry these papers in

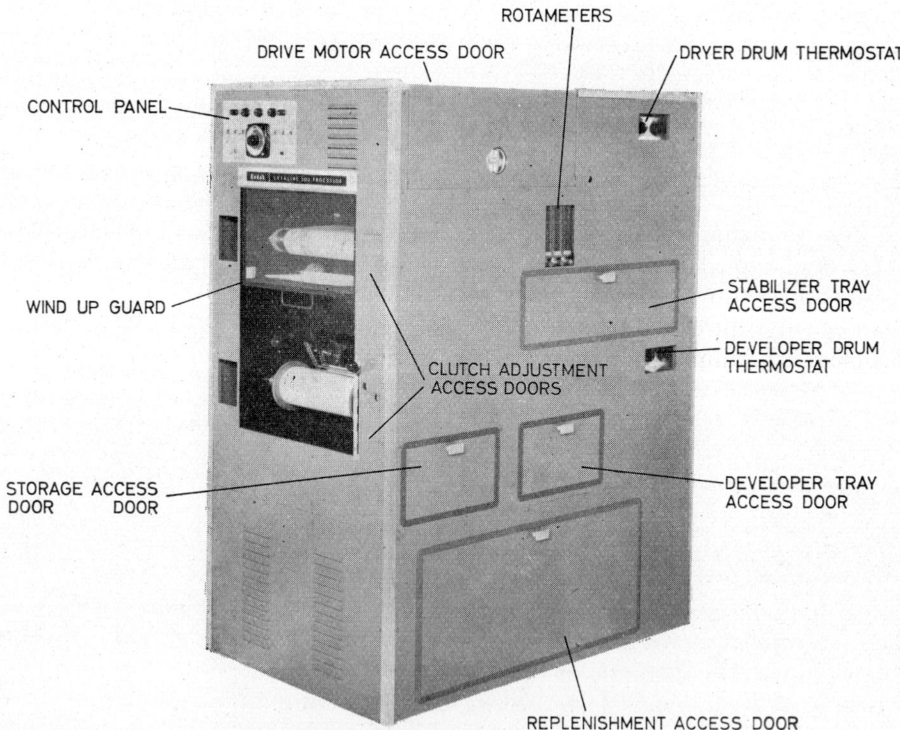

Fig. 8.4. Kodak Ektaline 200 Stabilization Processor

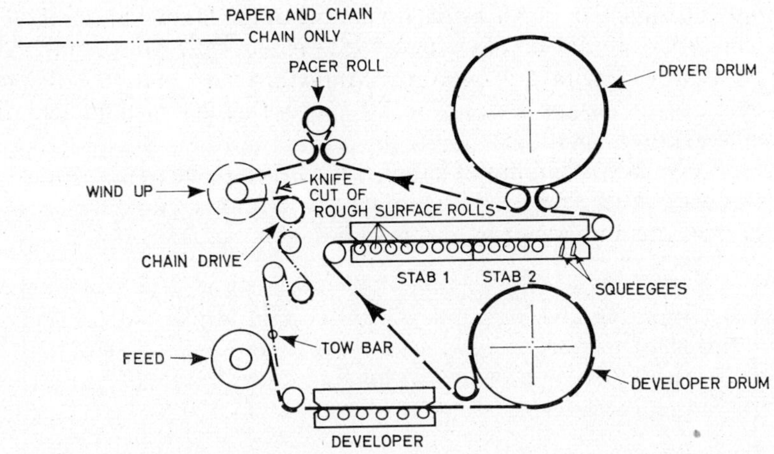

Fig. 8.5. Kodak Ektaline 200 Stabilization Processor threading diagram

rolls up to 13 inches wide and 6 inches in diameter at speeds of 25, 50, 100, or 200 feet per minute. For best results, processing at 200 feet per minute is recommended. The processing path is shown in Fig. 8.5.

The installation procedure is not difficult, but it requires attention to the detail and procedures described in the manual to obtain optimum trouble-free performance.[8.5]

Space Requirements. The processor is designed for *darkroom operation only*. The darkroom area in which the processor is to be used will vary from installation to installation, depending upon the needs of the particular department. A recommended layout and information regarding processor dimensions, drain locations, etc., is shown in Fig. 8.6.

The processor weighs approximately 1000 pounds and occupies approximately 10 square feet of floor space (31 inches wide, 49 inches long, and 66 inches high). A service clearance of not less than three feet is recommended on the feed and wind-up end, and the access door sides of the processor. Allow minimum of two feet for the electrical service side of the processor. The darkroom ceiling should be a minimum of seven feet.

Service Requirements

Electrical. The processor is normally supplied for 120/208-volt, 3-phase Wye (4-wire), 60-cycle service. Power requirements are: 208-volt, 3-phase (4-wire), 90 amps per leg, 30 kw maximum.

Darkroom Illumination. The darkroom should be equipped with separate, switch-controlled, white and safelight circuits. A Kodak Utility Safelight Lamp, Model C, with 10×12 Kodak Safelight Filters, Series 2 and a 10-watt bulb should be installed at the ceiling at each end of the processor. These safelight lamps should be at a height of 6 1/2 to 7 feet.

Water. No water is needed during processing. However, a hot and cold water source with a mixing faucet will be needed for processor maintenance and chemical mixing. This water source should be located toward the back end or the operating side of the processor.

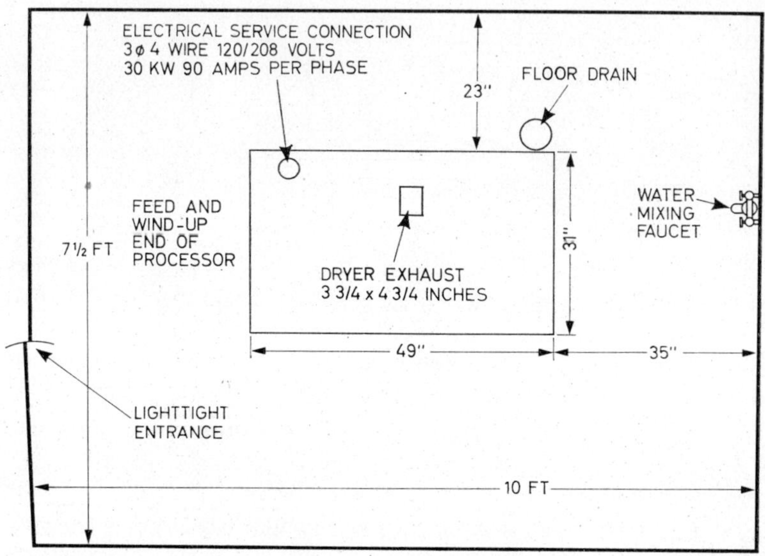

Fig. 8.6. Suggested darkroom space and service connections for Kodak Ektaline 200 Processor

Drain. A floor drain is not absolutely necessary since the liquid chemical overflow is small. The waste chemicals and clean-up water can be collected in a portable waste tank. However, it is preferable to provide a 2 to 4-inch cast iron floor drain. This drain should be flushed daily with fresh water to minimize corrosion. The processor drain connection consists of a length of flexible tubing located at the left rear of the processor. Extend this tubing to the floor drain (if provided), or to a portable waste tank.

Air Exhaust. A 3 3/4 by 4 3/4-inch rectangular exhaust opening is provided on top of the processor. This is to be directed into the building exhaust system (open connection) or through duct work through the wall or ceiling. A 6-inch duct is recommended to handle the drier's exhaust of approximately 200 cfm at 100 °F.

REPLENISHER SYSTEM. Developer, stabilizer 1, and stabilizer 2 replenisher tanks, are located in the bottom section of the processor. The capacity of each tank is 10 gallons. Replenishment is automatic. During the process, the replenishment solutions are pumped from the tanks through rotameters and into processing trays. The pumps are wired to the electrical drive circuit so that each solution tray is replenished while the paper drive is operating. Manual control switches on the electrical control panel provide independent control of each pump for initial filling of each tray.

OPERATION. 1) Mix the developer and the stabilizer solutions in accordance with the instructions packaged with the chemicals. The developer may be mixed in the developer replenisher storage tank. Since the stabilizers are more difficult to dissolve, they must be agitated for several minutes in separate containers to insure proper mixing. Transfer the mixed stabilizer solutions to their respective replenisher storage tanks.

2) Close the three solution-tray drain valves.

3) Turn on the main power wafety disconnect switch.

4) Open the three rotameter valves wide; then start the replenisher pumps by placing each of the three pump switches in the "Fill" position to fill the solution trays. The trays will fill in approximately 1 1/2 minutes with the rotameters opened for maximum flow. If necessary, jog the switches to bleed any air from the piping system. If the piping system was flushed with water, allow at least half a gallon of liquid to go to the drain before closing the three drain valves.

5) When the trays are full, judged by observing the solution overflow, set the rotameters to the proper replenishment readings. Place the three pump switches in the "Replenish" position.

6) Replace all access doors and panels to insure proper drying.

7) Turn on the fan and heater switch. Check the setting of the two drum thermostats to be sure they are set at the recommended operating temperatures. Allow 20 to 30 minutes for the drum temperatures to reach equilibrium. The temperatures may require adjustment as there may be a variation in electrical service and room conditions at each installation.

8) Place a paper core of the proper diameter and width on a wind-up spool and install in the wind-up section. Wrap the paper core with double-coated tape in a helix with approximately a 2 to 3-inch pitch. Use Scotch Brand Pressure Sensitive Tape No. 666 with liner, 1 inch by 36 yards, or equivalent.

9) Place the roll of paper to be processed on a feed spool. Center it on the spool and attach the flanges; then install in the supply section. Be sure the paper will unwind from the spool emulsion side down.

10) Thread the end of the paper around the bottom of the tow bar over and under the top tow bar rod.

11) Pull out the wind-up guard and lower to its closed position against the two magnetic catches. The tow bar cannot be actuated unless the wind-up guard is closed.

12) Set the delay timer according to the speed the processor will operate and to the type paper to be processed. Position the drive switch to either "Full Heat" or "Controlled Heat" as recommended. This will start the delay timer.

13) The processor drive motor will start after the delay timer reaches "0". To

start the process, push the auto-thread button. Observe the rotameters to be sure they are adjusted properly and that the replenishment pumps are operating.

14) Immediately after the roll of paper has been processed, push the main drive switch to its center or "Off" position.

15) Follow steps No. 5 through 14 to process additional rolls of exposed paper.

Maintenance

Every Day. 1) Drying Drums—Each day before the drums are heated, wipe the surface of each drum with a clean, soft cloth moistened with water. *Do not* use any abrasive material to clean the drums. After continued use, some discoloration may appear on the protective finish. However, no attempt should be made to remove it.

2) Processing Solution Trays—At the end of each day of operation, drain the trays and flush them with water.

3) Applicator Rolls—At the end of each day, thoroughly rinse the developer and stabilizer applicator rolls. This is done when flushing the trays.

4) Idler and Pacer Rolls—Each day, before starting processing, wipe all the rubber-covered rolls (8 idler and 1 pacer) with a damp sponge to prevent a build-up of residue on the rolls. Do not use any scouring materials.

Every Two or Three Days. Soluton Replenisher Tanks—Drain and flush the developer tank and lines with water. Drain and flush the stabilizer tanks and lines as needed for good housekeeping

Every Two or Three Months. Applicator Rolls—Remove all rollers and soak them in a 10 percent solution of citric acid; then scrub them gently with a brush and rinse with water thoroughly. This will remove any residue buildup after extensive processing. There will be no need to soak either the upper or lower hold-down rack in a citric acid solution. Rinsing the racks daily with water should be sufficient.

Continuous Archival Processing

Continuous Fix-wash Processors. The continuous processors discussed in the previous sections utilize the stabilization process. For those who desire a conventional fix-wash process, there are a number of tank machines available which are described in the following sections. An evaluation of batch and continuous processing is contained at the end of this chapter.

The Hi-speed Oscillograph Paper Processor. The Hi-Speed Oscillograph Paper Processor, Model P-10M, is shown in Fig. 8.7. It is a self-contained machine, made in two sections: a darkroom loading section, 32 inches wide, 28 inches long, and 78 inches high; and a daylight processing section, 30 inches wide, 80 inches

288

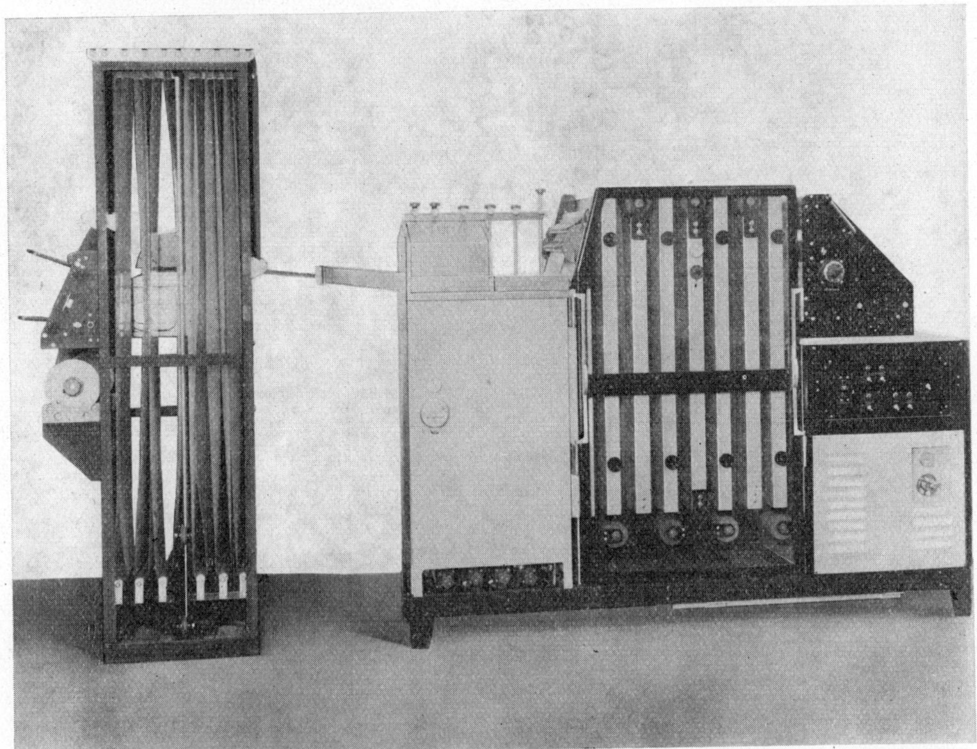

Fig. 8.7. Hi-Speed Oscillograph Paper Processor, Model P-10M

long, and 60 inches high. A typical installation floor plan is shown in Fig. 8.8. The machine will process both regular and thin-base papers up to 12 inches wide, at 30 feet per minute (within 5 minutes, dry-to-dry). Other black and white photographic papers, and black and white film can be processed on this equipment.

Loading is normally conducted at the darkroom end of the machine, but a magazine and light-tight loading elevator can be provided for full daylight operation.

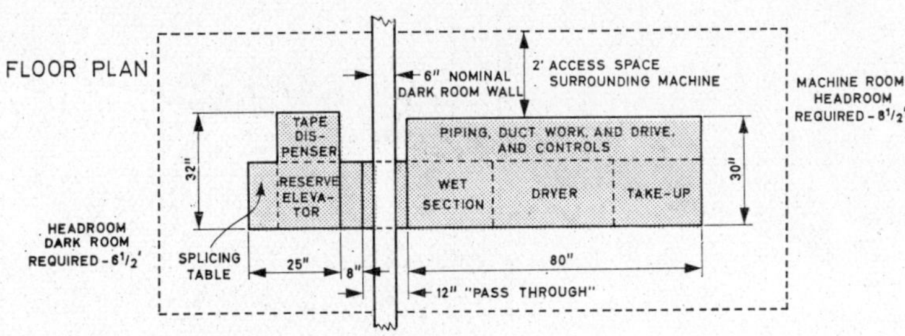

Fig. 8.8. Typical installation floor plan for Hi-Speed Model P-10M Processor

The wet section, drier, and takeup station are located at the daylight end of the machine. The wet section contains five solution tanks: one light-tight developer tank, two fixer tanks, and two tanks used for immersion and spray wash. The paper is dried in its passage through a heated-air impingement drier. The dry record is taken up on a standard paper core, which is rim driven and supported by two phenolic rods.

All materials in contact with solution are Type 316 stainless steel or plastic. The integrated machine requires only connection of power, water, and drain for operation. The power requirement is 40 amperes, 208 volts, 3 phase, 60 cycle. A total of 3 gallons of water per minute is blended in the built-in thermostatic valve from warm and cold water. The water, at the preset temperature is fed to the spray wash, which in turn feeds the immersion wash. Developer and fixer are recirculated through separate systems. Developer temperature is controlled to within $\pm 0.5\,°F$ by a combination of immersion heater and cooling coils. A squeegee is used to reduce developer carryover, and excess water is removed prior to the drier entrance with a squeegee.

HOUSTON FEARLESS EQUIPMENT. The processors described in the following subsections are manufactured by the Houston Fearless Corporation, 11801 West Olympic Boulevard, Los Angeles, California 90064. Laboratory equipment, portable laboratories, and film viewers manufactured by the firm, are described elsewhere in this book.

Fig. 8.9. Houston Fearless HTA-3CM Processor

HOUSTON FEARLESS HTA PROCESSORS. The current HTA series of Houston Fearless processors will continuously process to archival quality, black and white, negative or positive film and photographic paper from 70 mm to 9–1/2″ wide. Film transport speed is variable from 5 to 60 feet per minute.

Developer and fixing solutions can be constantly recirculated and filtered, and temperatures of solutions are automatically controlled. The cascade wash system provides archival quality with economical use of water. Film is dried by impingement of heated air from plenums in the drying cabinet.

The HTA-3C processor measures 21 feet 10 inches long and maximum width is 4 feet 2 inches. The HTA-3CM model, intended for installation in mobile and other space-limited laboratories, measures 17 feet 10 inches long, and 4 feet 1 inch wide. It has a maximum height of 5 feet 10 inches over raised turbulation rods. The HTA-3CM processing machine is illustrated in Fig. 8.9.

HOUSTON FEARLESS LABMASTER PROCESSORS. The Labmaster series of continuous processors are economically priced and modularly designed to process negative, positive, or reversal materials, including microfilms in 16, 35, and 16/35 mm sizes. The standard model, shown in Fig. 8.10, features in-line Type 316 stainless steel tanks, refrigeration temperature control, turbulation of developer, 1200 ft. magazines for daylight operation, spray wash, air squeegees, forced filtered air and infrared heating for drying, and clutch driven takeup.

The size of the processor is 16 inches wide and 55–3/8 high (approximately 100 inches with elevators raised). The length varies from 101–1/2 inches on the negative positive machine to 140 inches on the reversal model. The machine operates from 115/230 V, 60 cycle ac power, and requires approximately 10 gpm of water.

Other models include spray and color Labmasters. Optional equipment includes replenishment stations, chemical mixing and transfer stations, gum rubber or air squeegees and air unit, and aerosol dip tank.

HOUSTON FEARLESS SEPRATRON PROCESSORS. The Sepratron continuous, table-top immersion processor shown in Fig. 8.11a, produces high-quality processing with short access time. The unusual design enables the film to be processed, washed, and dried as it travels in a straight horizontal path through the solution tanks and drier. Because nothing but fluids touches the film and because it is not flexed as it would be in standard processors, it can be developed at elevated temperatures without harm to the emulsion.

The unique Sepratron principle is illustrated in Fig. 8.11b. The processing liquid is confined to its own tank and prevented from mixing with solutions in other tanks by air gates that separate solutions at the entrance and exit of each tank. The air gates also serve as air squeegees.

A short immersion time in each tank is sufficient to complete processing and washing of the film at the high temperature and high turbulation used. The processing time from dry-to-dry in the 16/35 mm Sepratron is 12.5 seconds at 10 feet per minute. Solution capacity is 2 liters per tank, maintained from one-gallon capacity

Fig. 8.10. Houston Fearless Labmaster Processor

feeder bottles located at the back of the machine. The Houston Fearless Water Conservation System (WCS), described in a later section, may be integrated with the Sepratron. Although regular tap water may be used, the WCS feature enables film to be processed to archival quality without connection to water supply and drain lines.

The 16/35 mm Sepratron is 55 inches long, 15 inches deep, and 25 inches high. The processor weighs approximately 185 pounds (dry), and operates from a 30 ampere, 115 volt, 60-cycle power source.

Wide-film, dual-strand, reversal, and color Sepratron processor models are now being developed by Houston Fearless.

HOUSTON FEARLESS WATER CONSERVATION SYSTEM. The Houston Fearless Water Conservation System (WCS) is used in archival-processing wash systems. It effectively reduces the quantity of fresh water consumed by making 1 gallon of water do

292

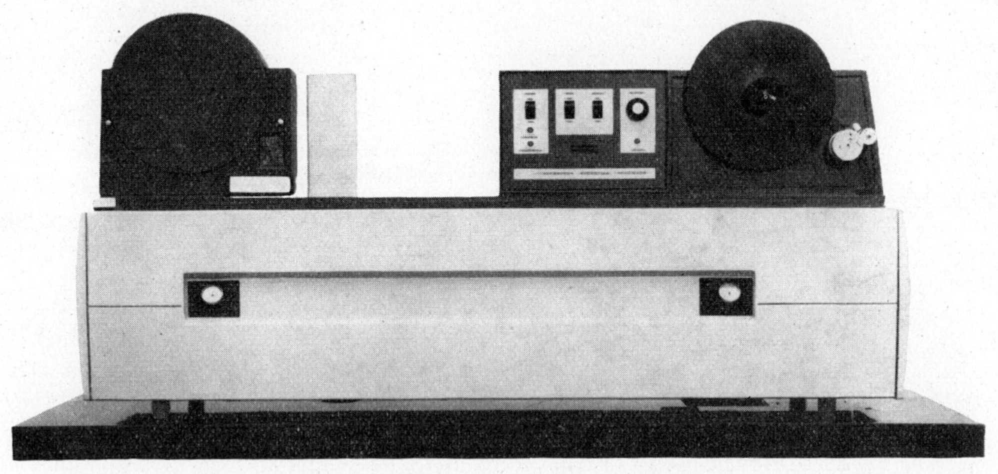

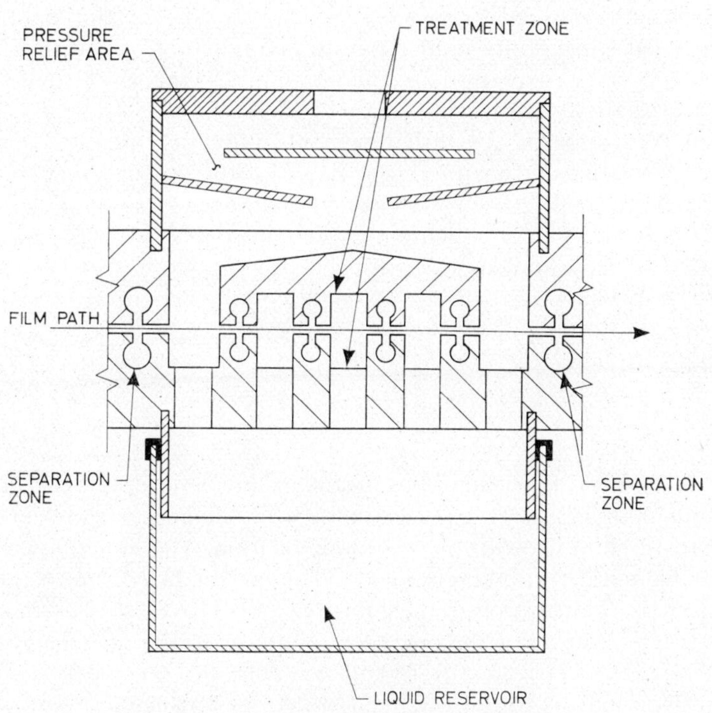

Fig. 8.11. Houston Fearless Sepratron Processor.

Top: Front view of table-top model; *Bottom*: Cross section schematic of Sepratron Processing Module illustrating the principle of solution separation

the work or 1253 gallons in film processing operations. Its advantages are especially significant for those conditions in which water is available only in limited quantities, is too impure or inconvenient for processing use, or where weight of the water is a factor (as in situations in which water cannot be piped but must be transported).

The WCS is based on a proprietary concept of accurate control of water flow and contaminant level throughout the wash cycle.

The WCS is available as an accessory unit to any of the Houston Fearless film processors.

OSCAR FISHER PROCESSORS. The Fisher Processall comes in two models; Model G-6 and Model G-12. The Model G-6 will handle all types of film and paper in roll form from 16 mm to 6 inches in width and in lengths up to 1000 feet. The Model G-12 will handle film and paper up to 12 inches in width, as well as 16 mm film and paper.

The Model G-6 measures 43 inches long, 20 inches wide, and 68 inches high and weighs 390 pounds crated. It operates from 110 volt, 60 cycle electrical power.

The Model G-12 is 7 inches wider than the G-6, weighs 475 pounds crated, and operates from a 220 volt source.

Washing is accomplished by means of a spray wash; development and fixation is effected by means of spray immersion. Spraying is performed through the solutions to reduce aerial oxidation to a minimum. Drying is effected by means of turbulent heated air.

The speed of the unit may be varied from 3 to 12 feet per minute and is controlled by means of a positive drive with an overdrive for the takeup through a slip clutch. The time in the developer and fixer may be adjusted by raising or lowering the rollers. Temperature control is maintained by means of thermometers, thermostats, and a heat exchanger, and may be varied to suit the desires of the user. The unit is self-threading. The weight of the cover is counter-balanced for ease of operation.

Portable Processor

SIEMENS & HALSKE PORTABLE PROCESSOR. A portable developing machine that measures only 30 by 32 by 41 cm is available from Siemens & Halske for processing records 60 to 127 mm wide at the recording location. The unit, shown in Fig. 8.12, will operate from a 220 v, 50 cycle electrical supply; a 24 vdc unit is also available. The electricity is used to power the film-transport drive, and to thermostatically maintain the temperature of the developer at any preset temperature within the 20 to 30 °C range. If electric power is not available, the sensitized material may be transported manually by use of the hand crank. Under these conditions, processing is conducted at room temperature.

The processor contains developer and fixer tanks which have a capacity of one liter each. There is no provision to water wash or force dry the record. Instead, squeegees remove the surface moisture so that the record can be air-dried by exposure to the atmosphere. The developing time may be varied from 1 to 12 minutes. Chemicals can be replenished without disassembly of the unit. Accessories for replenishment come in a separate accessory case, which measures 21 by 41 by 35 cm.

Fig. 8.12. Siemens & Halske portable processor

Comparison of Batch and Continuous Processing

EVALUATION OF THE PROCESSING METHOD. The history of the continuous processor with relation to oscillogram processing was considered in Chapter 1. Chapters 7 and 8 have describes both hand and machine processors. Before proceeding to Chapter 9, which describes laboratory installation and function, the reader should determine what method he will use. His choice is dependent upon numerous factors which should be critically analyzed in making the decision.[8.6, 8.7, 8.8, 8.9] Table 8.1 has been prepared as an aid to the reader in this evaluation.

The evaluation can be summed up for most cases in the following manner. A continuous processor is usually a justifiable investment if the volume is sufficient to operate the machine (at least 20 per cent of the time). If utilization is less than 20 per cent, the capital expenditure may be justified by the need for quick access to the recorded data, or by the fact that less space is necessary, or that more preparation cost may be involved in setting up the laboratory for batch processing than for a compact stabilization processor.

Hand-processing is an advantage when volume is low, chart widths are narrow (and/or short), access time is not critical, and archival quality is needed. The decision to use this method may also be influenced when an existing photographic laboratory is already set up and may have some or all of the equipment necessary. The capital expenditure for equipment is low, and often surplus or used equipment can be purchased. In addition, the need for stabilizing (permanizing) short lengths of print-out material may be the only use for processing at all.

Several factors may influence the choice of continuous stabilization or conventional fix-wash processors. Among these reasons are the use of the equipment to process materials other than oscillograph paper, and the need for archival quality.

TABLE 8.1

EVALUATION OF BATCH AND CONTINUOUS PROCESSING TECHNIQUES

Consideration	Batch	Continuous
1. Volume	Low	High
2. Future Volume	Little Growth Potential	Large Growth Potential
3. Capital Investment	Low for Equipment but more for laboratory	3 to 4 times and up of rewind equipment
4. Operating Cost	High	Low
5. Maintenance	Low	Moderate
6. Archival Capability	Yes	Yes ; certain types
7. Stabilization Capability	Not Recommended	Yes
8. Access Time	Slow	Quick
9. Installation	Fixed or Field use	1. Fixed, but may be cart mounted for moving to several locations. 2. Portable. Van installations
10. Space Required	Twice that of small stabilization processors	Compact
11. Multiple Use	Yes, for other photographic processing	No (or limited)

Folding Finished Records

OSCILLOGRAM FOLDING MACHINES. Some of the advantages which result from folding long rolls of paper into an accordion fold are as follows:

1) They are more easily handled for analysis. Unrolling and rerolling are eliminated. Any part of the record is readily available for examination—as easy as opening a book.

2) They may be stored in conventional folders and file cabinets. Thus they may be kept with other information concerning the test.

296

3) They may be bound into reports and transferred or mailed in conventional envelopes.

An operator can fold charts five to ten times faster with automatic folders than by hand. Neatness and accuracy of fold spacing are far superior to hand-folded records. Operation is simple and requires no special training.

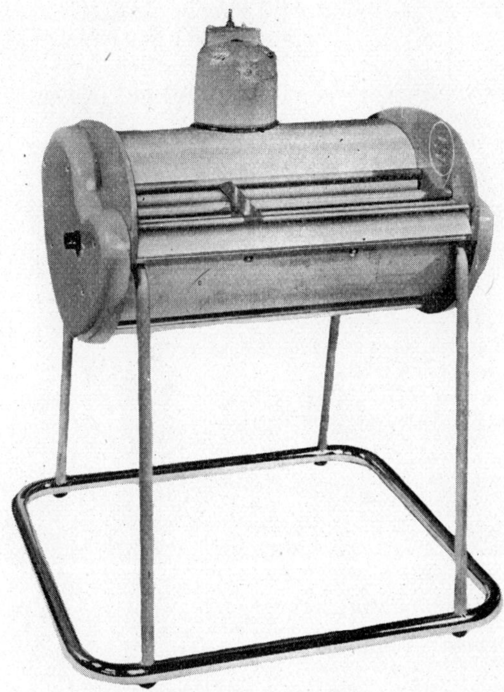

Fig. 8.13. Oscillogram folding machine (Benson Lehner Rollafold)

BENSON–LEHNER ROLLAFOLD. This automatic folding machine, illustrated in Fig. 8.13, is a portable folding machine for producing neat, accordion folds in a long strip or roll of paper.

The body of the machine is supported by four legs and a tubular frame in a manner which provides space for the folded chart to accumulate. Rubber bumpers contact the surface on which the machine rests to prevent marring.

The overall size of the model B is 22 by 20 by 28 inches high. It weighs 55 pounds (approximately). Fold spacing is adjustable from 7.5 to 13.5 inches, at a fold rate of one fold per second (approximately). The power requirement is 120 vac, 60 cps, 1.7 amp (maximum). The Rollafold will accommodate rolls up to 6 inches in diameter, up to 16 inches wide. Other models capable of handling larger chart sizes are available on special order.

LITERATURE REFERENCES

[8.1] *Type 23–109B Oscillogram Processor, Operation and Maintenance Manual*, 2062, Consolidated Electrodynamics Corporation (Pasadena, California) 1961

[8.2] HERSH, S. L., and SMITH, F., "Rapid Processing: Present State of the Art," *Photographic Science and Engineering*, Vol. 5, No .1, p48, Jan.–Feb. 1961

[8.3] BARNES, J. C., and FORTMILLER, L. J., "Factors Involved in Design of Rapid-Reversal Processing Systems," *Photographic Science and Engineering*, Vol. 7, No. 5, p269, Sept.–Oct. 1963

[8.4] HOADLEY, H. W., "A Compact, Rapid-Access, Wide Film and Paper Processor," paper presented at the annual conference of the Society of Photographic Scientists and Engineers, Cleveland, Ohio, May 20, 1965

[8.5] *Kodak Ektaline 200 Processor, Model 1M, Installation and Operation*, 334–9–62, GLP–B. (Also Parts List 769044, July 1963). Eastman Kodak Company (Rochester, New York) 1962

[8.6] *A Short Annotated Bibliography on Rapid Processing*, Research Laboratories, Eastman Kodak Company (Rochester, New York) October 1960. Approximately 85 references to conventional, modified conventional, rapid, and ultra-rapid processing.

[8.7] FORMAN, A. J., Editor, "New Rapid Film Processing Enhances Scope-Camera Use," *Special Report on Electronics*, Chariot Publishing Company, 65 Broad Street, (Stamford, Connecticut) June 1963

[8.8] MASON, R. P., "An Advanced Technique for Short Delay Processing of Photographic Emulsions," *Photographic Science and Engineering*, Vol. 5, No. 2, p 79 March–April 1961

[8.9] JACOBS, J. H., Rapid Access Methods in the United States, *Photographic Journal*, Vol. 105, No. 10, p 271–292 (1965); 47 references.

IX   THE PROCESSING LABORATORY

GENERAL CONSIDERATIONS. While it is possible to process oscillograph records with modern equipment under adverse conditions, high-quality records, if produced economically and on a consistently high-production basis, require the environment of a well-designed laboratory. In planning this function, the following factors must be carefully considered before a layout can be made: location, volume, number of personnel, size of equipment, and flow of work.

LOCATION. The location, and size of the area in which the processing is to be carried out should be determined primarily on the basis of volume, the type of process used, and the organization of personnel operating the oscillograph and of those doing the processing. If volume is low, and a tabletop processor is used, it is conceivable that the most efficient arrangement might be to locate a small darkroom and laboratory near the recording equipment. In this case, the oscillograph operator would be trained to process the records which he exposed.

If the volume is high, space near the recorder is at a premium, and personnel on a full-time basis are required to run the processor(s), the function can best be carried out in a laboratory facility combined with other photographic operations. This is because it is more economical to construct the facility initially to provide its specialized water, power, and other services, and then to staff, operate, and produce optimum records on a day-in, day-out basis.

Location is also dependent upon other factors. For example, the proximity of utilities and supplies, the ease with which the area may be made ready, the relation of the processing operation to the acquisition of the data, and the utilization of the record after processing. The priority for space, classified projects and/or areas, and trained personnel also may be factors that need to be sifted in the analyzing phase. Obviously, the laboratory must not be in an area where radioactive emission will fog sensitized materials. Areas that are too damp, too hot, or too cold should not be used, or if they are the only spaces available, they must be corrected to proper photographic-environmental conditions.

Provision should be made for possible future growth by allocating additional adjacent space for expansion of the laboratory. The requirement for future utilities should also be anticipated by installing electrical, plumbing, and drain lines having sufficient capacity to handle additional equipment and facilities.

VOLUME OF PRODUCTION. The volume of processing is dependent on the types and quantity of recorders, the speeds at which they operate, the extent to which they are used, and the kind of records made; e.g., permanent, quick-access, or quick-access with after treatment.

Volume estimates should be based on expected load, keeping in mind that there will be peak loads that must be handled in some manner. If delivery times permit, peak-load processing may be handled by putting on an extra shift. However, additional space, processors, and personnel will be needed if delivery times are critical.

Future increases in work load and changes in the type of work should also be taken into consideration. The final plan and layout should allow for expansion and flexibility to meet such changes. Should the work load increase without corresponding expansion of facilities, services will be disrupted. This will result in customer dissatisfaction; annoyance with poor schedules, or the degraded quality resulting from the overloaded equipment and the overworked personnel.

After a survey has been made of the quantity of processing to be handled and a delivery schedule has been agreed upon, the personnel and equipment requirements can be determined.

EQUIPMENT. The size of the processor and its support equipment, multiplied by the number of such units, must be taken into consideration in determining the square footage and the layout of the laboratory. The equipment descriptions throughout this book list approximate sizes of equipment. Sufficient space must be allowed beside the processor to permit proper operation and access for cleaning and maintenance. The operation manual and installation drawings should be studied by the architect engaged to design the laboratory, so that service connections to the equipment can be direct, functional, and still be unobstrusive.

The following items should be checked and provided for if they are found to be necessary:

1) Electrical requirements: voltage, amperage, phase, number of wires.

2) Water supply: consider volume, filtering and temperature control, water hardness and chemical content.

3) Drains: floor, wall, or sink; type of traps, flushing.

4) Vents: consider fumes, heat, humidity.

5) Attachment of equipment or processors to other systems such as chemical mixing, silver recovery, etc.

6) Can the equipment be operated in a light room, or is a darkroom required?

7) What support materials or equipment should be immediately adjacent to the processor?

8) Are the doors of the room large enough to permit passage of the equipment, or must an access panel be provided?

9) Does the equipment require or create any problems concerning the heating (or cooling) of the equipment or laboratory, and is the relative humidity unduly affected?

Storage cabinets, work tables, sinks, and other cabinetry will be required, in addition to the major pieces of equipment. These can be prefabricated or custom built, and in general can approximate the cabinet sizes, configuration and construction used in kitchens. The counter height should be a nominal 36 inches, but if a

piece of equipment is located on the counter, the working height should be convenient for the operator. Space between the base cabinet and the wall cabinet above it
should be sufficient for any equipment used on the base. A typical cross-section
of a counter and wall cabinet installation is shown in Fig. 9.1.

FLOW OF WORK. It is quite important to develop a work-flow pattern that is
orderly, progressive, and provides for all of the functions that must be carried out
by the laboratory. This can best be visualized from a layout. However, even before
the layout is made, it is well to list all of the functions. These should then be organized so that nothing is left out and so that there is no conflict between functions.
The following check list will aid the reader in planning a work flow arrangement:

1) Storage
   a Chemicals
   b Sensitized materials
   c Equipment not in immediate use
   d Departmental work records
   e Processed oscillograms

2) Lightrooms
   a Office functions
   b Chemical mixing
   c Unpacking supplies
   d Equipment maintenance
   e Washing, drying
   f Processing in "daylight" processors

3) Darkroom
   a Unloading and loading magazines
   b Processing
   c Splicing
   d Loading portable processing tanks with exposed, sensitized material

4) Light traps
   a Passboxes
   b Light-trap entrances

To achieve maximum efficiency, all operations, that do not have to be done in a
darkroom should be scheduled for the white-light area. These operations should
then be arranged to progress in the most direct manner, so there is no lost motion,
and no interruption of other work or personnel.

Both the lightroom and the darkroom should be separated into "dry" and "wet"
zones.[9.1] The white-lighted dry area will contain counters and cabinets, desks, storage
for packaged sensitized materials, viewers, etc. The wet area will contain washing
and drying facilities. If powdered chemicals are mixed in large quantities, a separate
room should be provided.

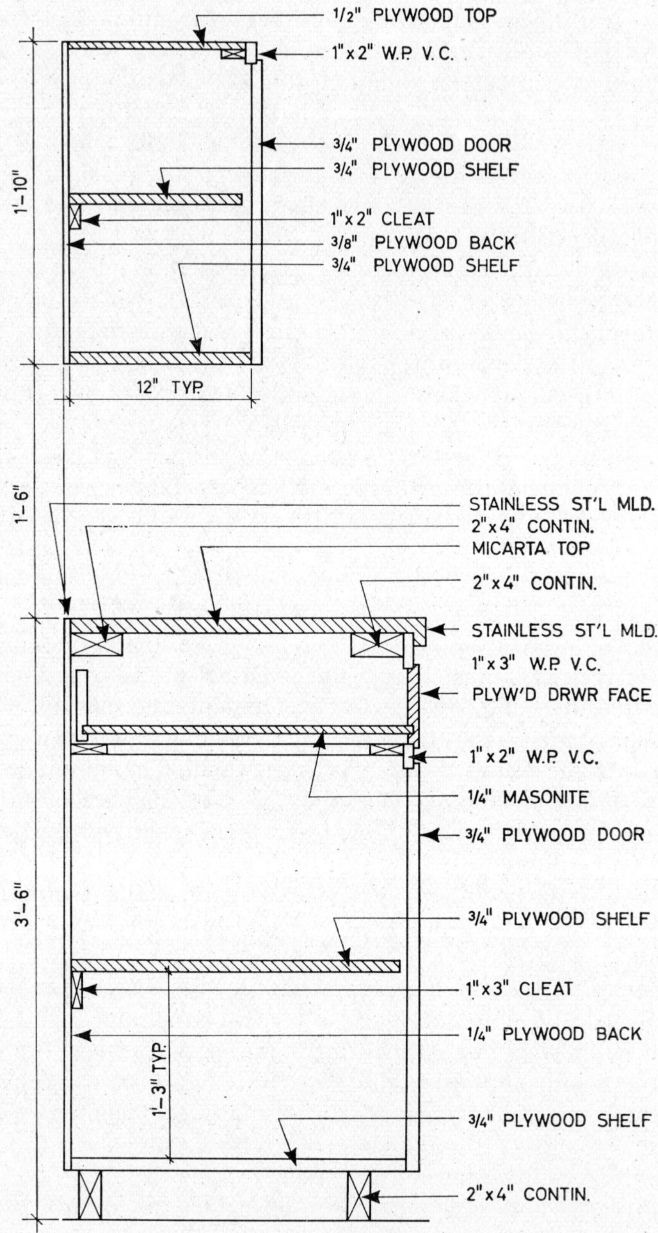

Fig. 9.1. Typical cross section of laboratory base and wall cabinets

In the darkroom, the sink, processor, or other "wet" operation should be isolated from magazine loading and unloading, so that sensitized materials will be protected from solutions that might be splashed. An aisle, three or four feet wide, will form a convenient separation between the two functions. Wet and dry areas can be on opposite sides of the room; or a center "island" wet area (with aisles on either side, three sides, or perhaps all the way around) can be in the center with dry areas on either side of the room. Cabinet areas that meet in corners should be avoided; the resulting "dead" corner is lost as an efficient table-top work area, and the storage space is almost inaccessible.

Entrances to the darkroom should be placed at the ends of aisles, if possible. This will prevent cutting up wall space and cabinets, will be a more convenient working arrangement, and will minimize interference of operating personnel.

The entrance to the darkroom may be one of three types: a single door, a light lock, or a labyrinth. A single light-tight door is adequate for one or two technicians; and generally in a small laboratory there will not be sufficient space to install a more elaborate entrance. The door should be capable of being latched from the inside, but should be provided with a safety-type knob so that it can be opened quickly from the outside in an emergency. A light lock should be installed, especially if there is more than one darkroom. This type of entrance is a small hall having a door into the white-lighted room, and doors at the sides, or at the opposite end, to the darkrooms. Door closers should be installed, and interlocks, or a light or buzzer system, added as a precaution against opening both doors at one time. If interlocks are used, safety releases must be provided. Doors or walls should be fitted with light-tight ventilating panels to permit easy opening and closing of the doors and to allow passage of air, so that the ventilation system will not be unbalanced. Doors should swing into the rooms if the lightlock is small. The minimum size should be about three feet square. Sliding doors may be used in place of swinging doors if space is at a premium. However, they present light sealing problems and generally are awkward and troublesome.

The light lock should be illuminated with a small safelight with a suitable filter. Another plan is to use a flush-mounted receptacle with a ruby lamp or other approved light source, installed about 16 inches from the floor. Even a small lamp at this distance will enable a person entering the light lock to watch his footing. The walls should be finished in a light color, and a vertical panel above and below the door knobs should be finished in a contrasting color.

A labyrinth should be considered only if there is to be a large staff that must continually enter and leave the darkroom. The labyrinth eliminates doors by trapping light within its maze. It requires considerable floor space, and for oscillograph processing offers few advantages over a light lock.

Several labyrinth and light lock arrangements are shown in Fig. 9.2. Any white-light from the lightroom, falling on the walls of the labyrinth (painted a dull black) is eliminated by repeated reflection and absorption before it can reach the darkroom. Since there are several light-absorbing surfaces in the maze, the small amount of remaining white light can be ignored unless it falls directly on sensitized-material handling areas. Such areas should receive additional screening treatment and should be thoroughly tested before operations are begun.[9.2]

LABORATORY LAYOUT. By the time that consideration has been given to all the aspects of oscillograph processing discussed so far in this chapter, some basic ideas will have been formulated on the size and the approximate shape of the finished lab. With this preliminary planning carefully in mind, scale templates can be cut of the equipment, cabinets, and work benches. These can be moved into various positions

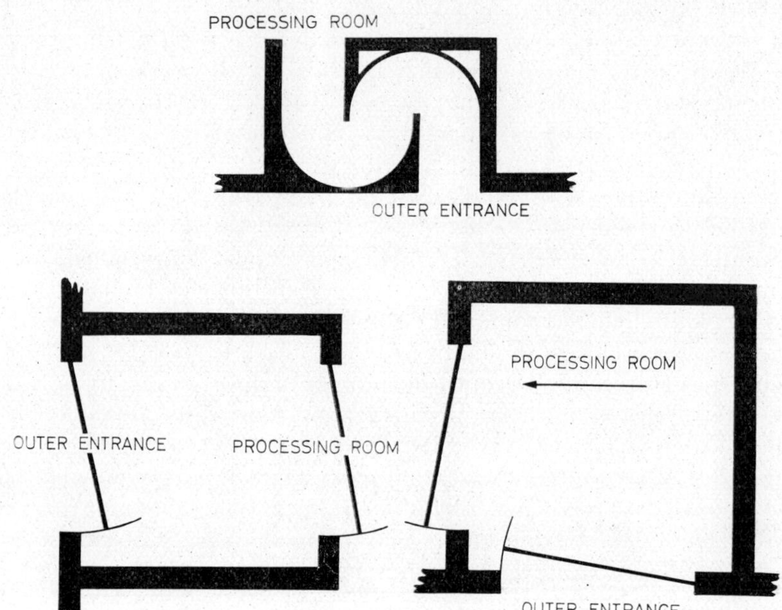

Fig. 9.2. Various labyrinth and light lock arrangements

on cross-hatched paper, until an appropriate floor plan is worked out. Chart-pak materials, available at drafting supply stores, will aid in making the layout. A good scale to use is 1/4 inch to the foot.

Several layouts are shown in Figs. 9.3, 9.4, and elsewhere in this book. These will aid in working out the best arrangement. In addition, the following tips will be helpful in certain instances:

1) Arrange the equipment in the order in which it is used in the work.

2) Allow plenty of working space for each operation; be sure the work counters are at the proper height.

3) Be sure to provide adequate material and accessory storage adjacent to the piece of equipment with which these will be used.

4) Remember the basic rules of dry and wet areas; light and dark rooms.

5) Don't scatter equipment and operations haphazardly around a large room; rather, equipment and work tables should be grouped efficiently, and extra space allocated for future growth or a larger storage room.

6) If the preliminary layout indicates a lack of space, it will be necessary to cut down aisle widths, shorten work counters, and provide short-term storage only.

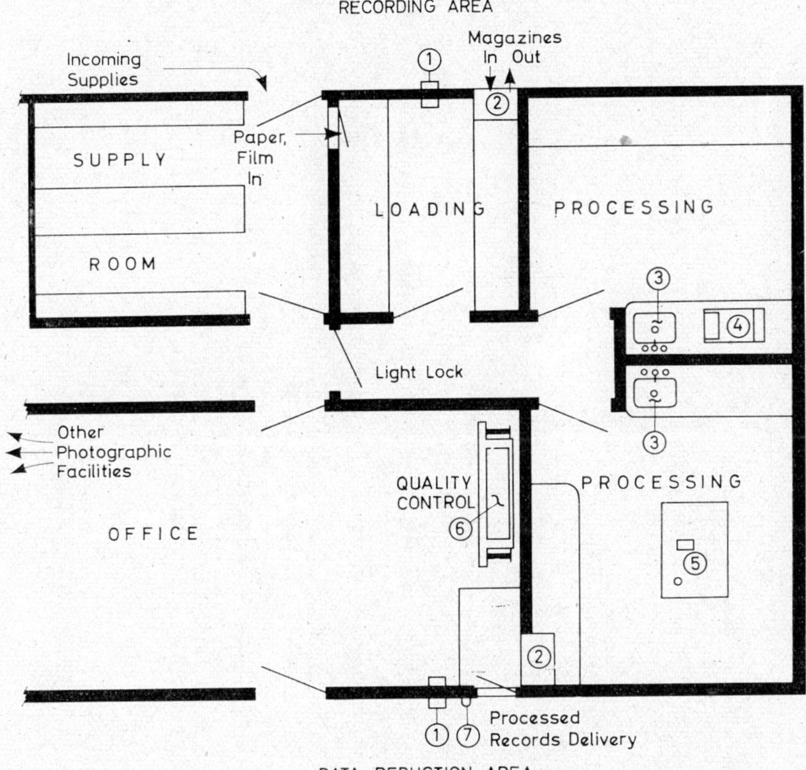

Fig. 9.3. Processing laboratory floor plan. Efficient work flow and provisions for administration, supply, processing, and quality control are conveniently grouped in this layout.

1. Intercom; 2. Pass box; 3. Sink; 4. Processor, CEC 23-109B; 5. Processor, EK Ektaline 200; 6. Inspection light table; 7. Signal light

However, do not jeopardize the efficiency of the lab by trying to utilize an untenable space allocation. The false economy should be resolved immediately.

7) Keep in mind the number of persons who will be working, and provide them with enough elbow room and an efficient work plan.

8) It is most economical to consolidate utilities. For example, if water lines and drains are in one wall, avoid placing another sink on the opposite wall. But if two darkrooms are installed, try to have plumbing and electrical facilities back-to-back.

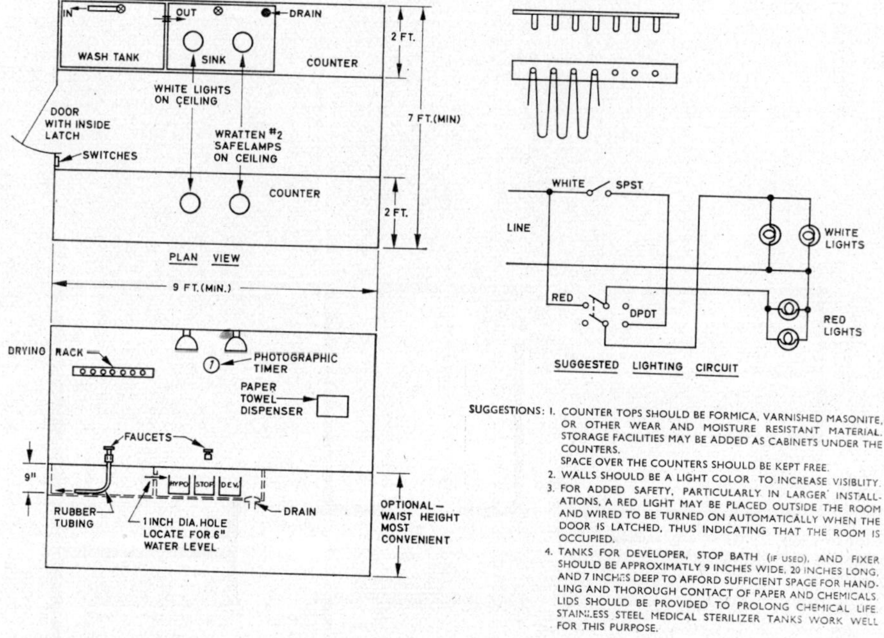

Fig. 9.4. Suggested darkroom floor plan. The peg rack for drying records *(top right)* has pegs about 6 inches apart to reduce the risk of emulsion sticking between the loops.

UTILITIES. The following utilities are the minimum requirements for any laboratory:

1) Electricity
2) Water, hot and cold
3) Drains
4) Ventilation

In addition, individual requirements, size, location, building codes, or other factors may make it advisable to provide the services listed below:

1) Vents and exhaust fans
2) Air conditioning, heating, cooling, and humidity control
3) Gas
4) Water—temperature control, filtering, conditioning, etc.
5) Telephone
6) Intercom
7) Vacuum
8) Compressed air
9) Bells, lights, or other signals
10) Automatic sprinkler system

306

SAFETY AND FIRE PROTECTION. If the construction drawings are not prepared by an architect, local building codes should be checked before progressing too far with the design. The codes will prevent oversight of most hazards. However, it is well to remember that a darkroom creates its own special hazard—darkness.

Fire protection is mandatory and will require compliance with underwriter's regulations. This will mean not only proper wiring, heating, plumbing, ventilation, etc., but the use of specified fire extinguishers, automatic sprinkler systems, and similar devices.

Individual items pertaining to safety may be further classified into the following categories:

1) Water Supply, Plumbing and Drains
   a Overflow drains
   b Floor drains
   c Vacuum breakers
   d Gas
   e Hot water tanks

2) Insulation
   a In walls and ceilings as a fire-protection measure.
   b Pipe insulation over steam and hot-water pipes for personnel protection.

3) Electrical Service and Precautions
   a Circuit breakers
   b Instrument-ground connections

4) Fireproof vaults.

WATER SUPPLY, PLUMBING AND DRAINS. Poorly designed water supply systems and inadequate drains are the two most common failures in darkroom installations. If the water supply is rusty or full of sediment, filters will have to be installed ahead of the photographic supply system. If the supply is "sulphur water", or if it is contaminated with other chemicals, trouble may be expected unless the contaminants are removed with a softener or by other chemical and filtration methods. Filtering or softening equipment should be installed at the entrance, ahead of water heaters, refrigerated supply systems, and all other equipment and outlets. NOTE: If the supply system will be used for other photographic processes in addition to photo-oscillography, check the water-hardness requirements. Some color materials are adversely affected by water that is too soft.

Usual recommendations call for plumbing lines to be run along or within the outside walls of the laboratory. The "outside" may mean an exterior wall, or a wall bounding the laboratory and separating it from other departments or rooms within a building. The advantage of running plumbing along the circumference of the laboratory is that partitions within the area can be installed easily, and at some future time may be moved without major remodeling. However, more economy will result if plumbing is installed in internal walls when a back-to-back darkroom installation is planned.

Insulation should be installed on pipes and this becomes especially important if pipes are in exterior walls. and are subjected to hot or cold weather conditions.

INSULATION. Room-temperature problems may arise from outdoor temperature fluctuations and/or the sun shining on the wall. This can cause temperature changes in sinks or other equipment located along the inside of the wall. The best safeguard against this hazard is to insulate the wall thoroughly. This should also be done on interior walls that divide the laboratory from another area where temperatures may be extreme or at variance with the controlled temperatures within the laboratory.

Ceiling insulation is equally as important as sidewall insulation. Starting with the ceiling itself, the contemporary construction is acoustic tiles on an existing ceiling or stringers, or acoustic plaster. Darkroom ceilings made of joined tiles must be inspected carefully and *periodically* for light leaks. Tile ceilings may separate at the joints and allow light to pass through, thus fogging sensitized materials. This problem can be solved in those cases where insulation is needed in the plenum or attic area, by installing a blanket-type insulation above the tile ceiling. The blanket will keep out light, as well as heat and cold, reduce noise transfer and offer additional fire protection. If the attic still becomes excessively hot from sunshine, the roof should be painted white or aluminum, and an attic fan should be installed.

Where existing walls need treatment on the interior, a layer of cork provides an attractive appearance and furnishes excellent acoustic characteristics. The cork may be combined with a foil underlayer to combat radiant heat, Interior (and/or exterior) insulation can also be used on solid masonry walls, or other walls that cannot be insulated with "blown in" materials.

Cold floors are usually associated with concrete poured directly on the ground. However, it is possible to insulate floors made in this manner. One method is to mix Vermiculite in the concrete before pouring the floor. Another method is to install cork tile over the concrete.

TEMPERATURE CONTROL OF WATER. There is no necessity for temperature-controlled water or sinks if all processing is done in automatic processors which have built-in temperature control units. However, if processing is done in rewind processor tank systems, or if the oscillograph processing facility is part of an entire photographic laboratory, it may be advisable to install water temperature-control equipment. This will consist of one or more thermostatic mixing valves fed by hot and cold water. If the temperature of the incoming cold water exceeds the required final temperature it will be necessary to install a refrigeration system. The system will cool a tank of water to a temperature lower than the temperature in the sink, and this water will be fed to the cold side of the thermostatic mixing valve. Refrigerated water may be needed only at certain times; for example, during the summer months when the cold water temperature is too high. Fig. 9.5 shows a schematic of a variable temperature thermostatic mixing valve, together with its associated plumbing. This system provides water of the proper temperature regardless of whether or not the incoming water temperature is above or below the specified working temperature. The operation of the valve is described in the caption that accompanies the diagram. Additional information is given in reference[9.3].

Regulators that supply water at a fixed temperature are also available. These units are lower in price than variable-temperature controls and are adequate where working temperatures do not have to be changed. However, such a valve can only be used if it meets the required temperature. Such valves are useful in wash-water supply lines, where it is desirable to prevent the excessively low temperatures that occur during winter months.

The following list is offered as a guide for satisfactory operation of thermostatic mixing valves:

1) Be sure there are check valves in feed lines to each side of the valve

2) Follow the manufacturer's recommendations for the valve model required to obtain satisfactory output.

3) Install recommended pipe sizes to the valve, and supply hot, cold, and refrigerated water at recommended pressures

4) Keep strainers ahead of the valve clean

5) Remember that sufficient water must flow through the device to cause it to regulate the temperature properly.

In addition to temperature-controlled water outlets in each sink, there should be a hot and cold mixer spout with hot and cold valves. A swing spout with a threaded outlet for a thin-walled flexible plastic hose is preferable. Its location should be high enough above the bottom of the sink to accommodate all containers used in the sinks. Long sinks may require outlets at two or more places. If the sink is wide, the valve can be located in the wall, under the sink, with a valve stem extension to the front of the sink. In this way, the valve can be controlled without reaching across the sink.

VACUUM BREAKERS. Vacuum breakers should be installed at each sink ahead of the outlet. This safety feature is required by most local codes, but should be installed even if it is not a mandatory requirement. The function of the vacuum breaker is to prevent solutions or contaminated water from being syphoned back into the supply lines. This can occur under certain conditions if a hose is left in a tank of solution, and the main water supply is turned off.

ASPIRATORS. An aspirator (or eductor, as it is sometimes called) is useful for emptying processor tanks. It is a device that works on the Venturi principle to create a low pressure or vacuum for syphoning liquids. Wall-mounted aspirator units are available from laboratory supply houses, and portable types may be obtained at most hardware stores.

DRAINS. It is good practice to install a floor drain in each darkroom, and in each room where chemicals are likely to be spilled. A floor drain should also be installed in any room where there is a possibility of water overflowing from a processor or sink. Of course, sinks and other equipment should have integral safety overflows, but these sometimes plug up, and then flooding may occur.

The drainage lines should be resistant to photographic solutions and to rapid changes in the temperature of solutions discharged into them. Stainless steel, Type 316, is satisfactory, but is costly. Brass and copper are easily corroded by the acids from fixer solutions. These acids also attack black steel and galvanized iron drainpipes. If cast iron is used, it should be made of Grade Double Extra Strong and should comply with Federal Specification WWP-401. The cast iron pipe will eventually rust in the presence of corrosive solutions, but the life can be greatly extended by diluting the solution with running water while the solution is being discarded, and then flushing with hot running water immediately afterward.

The diameter of the drain and its slope are the two factors that determine the amount of waste solution the pipe can carry. The drains should be pitched a minimum of 1/4-inch per foot.

In general it is poor practice to discharge photographic chemicals and wash water into septic tanks. Usually, the activity of the septic tank is retarded only temporarily by the action of photographic chemicals, and it will return to normal after the chemicals have been diluted. The volume of water discharged from processing operations is a more serious consideration, since the septic tank normally cannot handle its regular load and the large gallonage of photographic waste water, too.

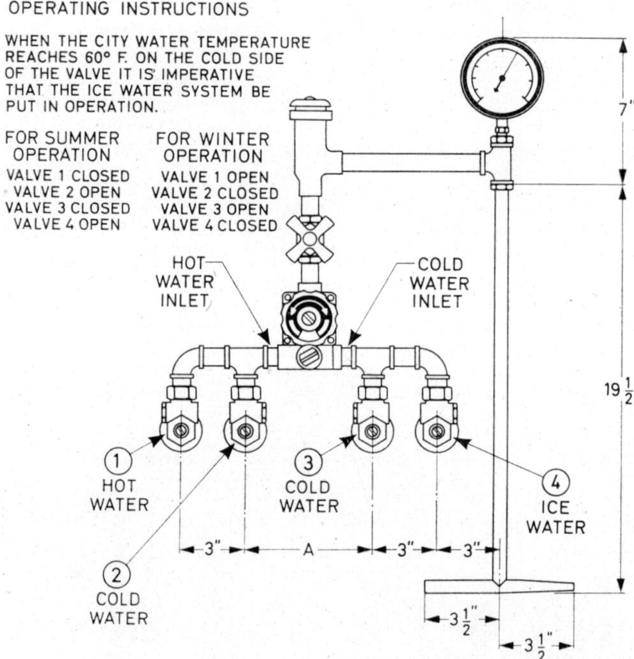

Fig. 9.5. Lawler thermostatic mixing valve for use in areas where cold water supply temperatures may exceed the processing temperature

ELECTRICAL SERVICE. Electricity is needed in the laboratory for white lights, safelights, equipment, and sometimes for other services, such as water heaters. The electrical planning should start with a list of equipment and lights that will be installed, and the voltage and amperage requirements should be itemized. When determining the incoming line sizes, a substantial factor should be allowed for future expansion of the load. For example, if one processor requires 30 amperes of current, and there is a good possibility that the future volume of work will warrant the addition of a second machine, sufficient power should be available to handle two machines.

The incoming supply should feed a power-control and distribution panel that is preferably within the laboratory, or at least closely adjacent to it. Having it in the laboratory will greatly reduce the possibility of the power inadvertently being turned off while processing is being conducted.

The control panel should have a main switch, a separate circuit breaker for each major piece of equipment, and breakers for the lighting circuits that are separate from equipment breakers. Reset-type circuit breakers rather than replaceable fuses should be used, so that service can be quickly restored when it is interrupted. Except for any equipment requiring higher amperage, each circuit should be designed for 30 ampere loads (or the maximum amperage, if the code specifies less than 30 amperes).

All wiring and equipment must conform to the regulations of the National Board of Fire Underwriters and to local, state or municipal codes. To meet these conditions, it will be necessary to design and install the wiring, lighting fixtures and equipment to meet the approved procedures not only in new laboratories, but whenever new equipment is installed in existing laboratories, or when the laboratory is remodeled.[9.4]

ELECTRICAL PRECAUTIONS. The greatest hazard in the darkroom is that of electricity. Darkness, wet hands, spilled solutions on the floor, and wires dangling near plumbing, are the chief reasons for the ever-present danger. However, there is little to fear if the proper precautions are taken.

All processing equipment, safelights, timers, viewers, intercoms, and any electrically-conductive housing or frame of equipment that is operated by electricity must be grounded. Employee-furnished tools or other items which do not meet these standards should not be permitted in the laboratory. Frequently, employees will bring to work radios, coffee makers, and other electrically operated devices that are not properly grounded. These present a threat to the safety record of a laboratory that has otherwise been very carefully planned to guard against electrical accidents.

The following precautions should be taken:

1) All surface wiring should be in conduit to prevent any spilled solutions from contacting the current-carrying wires.

2) Switches, outlets, and connectors should be made of a non-metallic insulating material.

3) All fixtures, equipment and other items should have an "instrument" ground. If endorsed by the code, this may consist of a ground wire, integral with the current-carrying cable, one end of which is connected to the housing of the device, and the other ending in a grounding plug. In this way, the instrument cannot be used ungrounded because it automatically becomes grounded when it is plugged into the socket.

In lieu of the integral method, a separate grounding-wire may be attached to the device and the other end suitably grounded with a mechanical fastener (not a solder connection). Check local code.

4) Outlets should be high enough from the floor to prevent them from getting wet, and to keep attached cords off the floor.

Preferably, outlets should not be placed within arm's reach of sinks and plumbing fixtures but if they are, they will require a switch. In any case, the laboratory technician should learn to use one hand for handling cords, connectors, and switches, keeping the other hand and body away from anything that might complete an electrical circuit (or ground) through his body.

5) Any cables on the floor constitute a special hazard because they are subject to physical damage and to moisture or water. They should be covered with a durable protective covering to resist both conditions, and should be suitably grounded.

6) All fixtures, equipment, and wires should be securely fastened mechanically, so that they cannot be accidentally dislodged into solutions, tanks, or sinks.

LIGHTING. The lighting of the laboratory should be planned to provide even and adequate illumination in both lightrooms and darkrooms. The design will depend upon the characteristics of the room, its size, height, and reflective surfaces, and the seeing task. Special conditions called for by the photographic techniques to be used will also dictate the type of equipment that should be installed. In general, the rooms may be planned with the following basic considerations in mind:

1) Fluorescent lamps in ceiling-mounted fixtures, spaced according to the manufacturer's recommendation are satisfactory for lightrooms. In general, a level of 50 to 200 foot candles of illumination will be required for light rooms.

2) Incandescent lamps should be used in darkrooms (rather than fluorescent lamps which sometimes continue to glow after they are turned off; the glow can fog sensitized materials unless the lamp is suitably filtered or covered).

3) Safelights should be fitted with the filter that is recommended for the material being used.

4) The correct wattage lamp should be used in the safelight.

WHITE LIGHTS. In keeping with the necessity of extreme cleanliness in the laboratory, it is desirable to eliminate as many dust-catching fixtures, chains, and wires as possible. Recessed ceiling fixtures, having glass covers flush with the ceiling, offer one way of reducing dust and will facilitate cleaning. Recessed fixtures can be obtained for both fluorescent and incandescent lamps. Aid in determining types and spacing of lamps may be obtained from bulletins published by electrical-fixture manufacturers.[9.5]

Safelights. Safelight fixtures are available from several manufacturers. A safelight fixture consists of a housing that is completely light-tight except for one area that can be fitted with interchangeable filters. The filters should be made of stable colored glass or plastic, and should pass only those wavelengths of light that will least effect the emulsion with which it is to be used. Usually, a seal is used around the edge of the filter to prevent any unfiltered light from leaking around the flange. Typical safelights are shown in Fig. 9.6. Table 9.1 lists the appropriate filter-sensitized material combinations for one manufacturer.

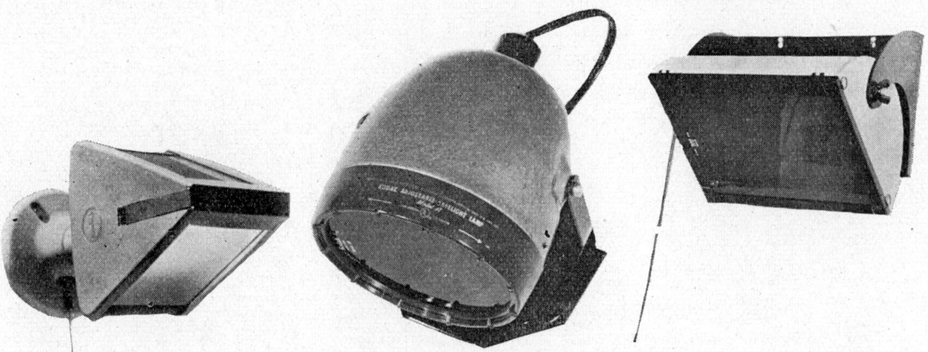

Fig. 9.6. Various types of safelight manufactured by Eastman Kodak Company

Darkroom operations should be carried out under the highest level of illumination that is safe with the material being used, for the time that it is normally exposed to the "safelight" illumination. Since no light will be 100 percent safe when sensitized material is left exposed to it for long periods, or the lamp is too bright, a compromise must be made. In most cases, an indirect low-level light can be used to provide general illumination. Small direct lamps which can be turned on and off as desired, can be installed for those operations requiring a quick look under a higher intensity of light. In some cases, the direct lights can be operated with a foot switch; in this manner the lights will be on for the minimum time, and both hands can be free to work. Before the laboratory is put into operation, a test should be made to test for fogging of sensitized materials. Two tests should be made, a stray-light test and a safelight test.

1) Stray-Light Test: Inspect the room, doors, and ceiling for any visible light leaks from sources outside the darkroom. This test should be made only after the eyes have become fully accustomed to total darkness. Correct any visible light leaks.

Next, in total darkness, place a test sample of high-speed oscillograph paper on the working counter, and cover a portion of it with a heavy, completely opaque material, such as a piece of metal (coins are sometimes used). Be sure it is flat and that no light can leak around its edges. After 15–20 minutes, develop the sheet in fresh developer for the normal time, rinse, fix, rinse, and dry. Examine the sheet in white light to see if the covered portion and the uncovered portion are alike. If the uncovered portion is gray, unsafe light is present, and will have to be eliminated. When there is no longer any stray light present, proceed with the

2) Safelight Test: In total darkness, place a piece of high speed oscillograph paper, approximately 8×10 inches, on the work counter. Cover it with a heavy flat opaque cardboard or sheet of metal along one edge so that approximately 2 inches will receive no exposure. Turn on the safelight and expose for 1 minute. Cover an additional 2 inches of the paper and expose for another minute. Continue the step-and-repeat exposure until the final exposure added to the previous exposures gives a total about twice the time the sensitized material is normally handled under the safelight.

TABLE 9.1.

KODAK SAFELIGHT FILTER-SENSITIZED MATERIAL

Wratten Series	Color	Materials used with
0C	Light Amber	Electrocardiograph 553
1	Red	Electrocardiograph 797 Linagraph Recording
1A	Light Red	Linagraph Survey; Fine Grain Positive
2	Dark Red	Linagraph Ortho, Linagraph Drift Survey, Ektaline 12, 16, 18; Linagraph 44, 77, 480, 483, 809, 697; Kind 1622
3	Dark Green	Verichrome Pan, Linagraph Shellburst, Linagraph Pan (NOTE: Turn on safelight only after development is 80 percent complete)
6B	Amber	Electrocardiograph 553
7	Green	Infrared

Be sure to leave one strip with no exposure. Develop the test sheet in fresh developer, rinse, fix, rinse, and dry. If any portion of the sheet is gray, the safelight can be moved further away from the counter, or its lamp can be replaced with a smaller wattage bulb. The test should be repeated after these changes are made.

Safelight filters sometimes fade from heat or excessive moisture. They may crack from heat because of too large a lamp or from being dropped. Occasionally, the safelight housing may crack or become distorted from these same causes. Therefore, periodic safelight tests should be made.

VENTILATION AND TEMPERATURE CONTROL. Proper ventilation of the laboratory is necessary for the health, comfort, and efficiency of the personnel. Proper ventilation of a laboratory is more difficult to achieve than ventilation of an office, for example. The type of processing that is conducted may also dictate special requirements that have to be considered when the laboratory is planned.

314

Associated with ventilation are the controls for temperature, dust, and humidity. This is because the simplest solution to these problems is to sense and control the temperature and humidity conditions through the ventilation system. The dust can be effectively reduced by installing filters in the ducting, and by maintaining a slight positive air pressure inside the laboratory. The pressure will reduce the amount of dust carried into the area through doors, and through crevices around windows.

The most important considerations in planning ventilation, heating, cooling, and humidity systems are as follows:

1) There should be 7 or 8 changes of air per hour. The range can be from 5 to 10 changes depending upon the number of personnel and other factors.

2) To control moisture and chemical-odor diffusion, air should come into the room in "dry" areas and be exhausted at "wet" areas.

3) Exhaust ducts, preferably with fans, must be located above driers, and over areas where there are fumes. There has been tremendous improvement in stabilization chemicals, but stabilization processors can sometimes contaminate the air with irritating and noxious sulphurous odors. Heat and moisture emitters should not overload air-conditioning equipment; outside air can be drawn in, utilized to absorb heat and moisture at its source, and discharged to the outside atmosphere.

4) For darkroom operation, air may be drawn from or discharged into another area through a light-trap ventilator mounted in the wall or in the connecting door. Light-stopping labyrinth louvres can be purchased in various sizes that are made up of black, nested, W-shaped metal strips.

5) Building codes will be the final authority for the ventilation system. The design should be planned by a competent heating and ventilating engineer, who will see that code requirements are fulfilled.

6) Precise humidity control is not generally necessary for oscillograph processing. However, ambient conditions should not be below about 40 per cent RH, nor should they reach excessively high amounts. In general, if the heating unit has provision for humidifying the air, and a refrigeration type air conditioner is installed, the RH limits will fall within an acceptable tolerance.

7) Evaporative-cooling systems are not compatible with photographic processing requirements.

8) Extreme humidity can be corrected with refrigeration-type dehumidifiers. Silica gel (or other chemicals) can be used for desiccation of material stored in a container, or in a small, closed area, or for temporarily dehumidifying a larger area. (See Storage of Materials).

Processing of oscillograph records on paper is usually carried out at elevated temperatures; therefore, the requirement for maintaining room temperatures at film-processing tolerances is unnecessary. Although the comfort-zone temperature will vary from person to person, a room temperature of 76 °F and an RH of 40 to 60 percent should be satisfactory.

COMMUNICATIONS. The laboratory should have adequate facilities for communicating with other laboratory employees, as well as with personnel in other departments.

Among the devices that can be used are public address systems, telephones, intercoms, buzzers, bells, lights, and speaking tubes. If possible, the system should be planned and the wiring for these devices should be installed when the laboratory is constructed. A telephone extension in a darkroom can be a handicap to a busy technician if there is no one to answer calls outside the darkroom. However, he should have a way of communicating with people outside the darkroom when he cannot be called away during processing operations. Sometimes a plug-in phone, which can be moved from one location to another is satisfactory. A wall phone, centrally located in a common light lock for several darkrooms, will provide easy access for all personnel.

If the laboratory is spacious, or has numerous rooms with a central office, a public address system will be very useful. This can be supplemented with intercoms so that the person paged can answer the call. The more costly intercom systems incorporate circuits for general paging at all stations, and provide for private answering from the remote to master station (s). If many stations are installed, wiring can become quite involved, and more expense is incurred. However, no wiring is necessary on some of the new wireless units.

Between two rooms or a hall and a room, a simple speaking tube, or even a light trap louvre may be adequate for communication.

Other simple and inexpensive forms of communication devices are buzzers, bells, and lights. Such a signal can be used to notify another department when a roll of oscillograph paper has been processed and is ready to be picked up. A light signal can be used to indicate when a magazine has been unloaded and refilled with fresh paper. There are numerous other applications for such signals in the laboratory.

VACUUM AND AIR SUPPLIES. The practice of using pressurized or "shop air" for cleaning magazines and for other uses in the laboratory should be discouraged. The air may contain dirt, rust, and liquid condensates which will serve only to complicate the cleaning job. Even if a filter or trap is located in the air line, the escaping air starts a circulation cycle of dust and lint that eventually settles on the surfaces that should be clean.

A vacuum cleaner, or a central vacuum source, can be used for cleaning magazines, and will serve many other useful purposes in the laboratory. Some adaptation of conventional systems and special attachments may be necessary to obtain satisfactory operation.

CONSTRUCTION MATERIALS[9.6, 9.7, 9.8] *Floors.* The laboratory floor should be resistant to deterioration from water and chemicals, and to the staining action of chemicals. It should be structurally capable of withstanding all loads imposed on it by equipment, materials, storage, and personnel. The floor covering should be comfortable to walk and stand on for long periods, should not be slippery, should be quiet, and should be easy to maintain in a dust-proof manner. It may also require insulating properties.

Wooden floors without a suitable covering over them, are unsatisfactory for laboratory use. This is because such a floor is not waterproof, and its cracks and grain will collect spilled chemicals.

Concrete floors are unsatisfactory unless they are either treated or covered. If left untreated, the surface will discharge particles of dust into the atmosphere from foot traffic. Treatment consists of sealing or painting the surface with special materials for this purpose. Resurfacing must be repeated when the sealer wears off, usually at frequent intervals. A far more satisfactory solution to the problem is to cover the floor with asphalt or other composition tiles. Linoleum cannot be recommended as a suitable floor covering, because it is easily stained and eventually will disintegrate from chemicals and water spilled on its surface. Vinyl, rubber and synthetic materials will resist damage, but some types of these materials will stain easily.

Ceramic tile floors with suitable joint mortar are satisfactory, but are tiring, may be slippery, particularly when they are wet, and may stain. Terrazo, a blend of colored marble chips imbedded in mortar and then polished, has the same disadvantages. Rubber mats may be located at those places where personnel have to stand for long periods.

Asphalt tile, laid over the existing floor, will probably be the most desirable. The floor should be level, except for a slight pitch around the floor drain. All equipment should be fitted with flat furniture slides to minimize damage to the flooring. Slides, attached to threaded studs, will enable equipment requiring precise leveling, to be easily installed. Heavy equipment should rest on pads that are integral with the concrete slab.

If the laboratory is located over other rooms, the floor must be watertight so that the ceiling below will not be damaged from water or chemicals that could otherwise leak through. The floor must have no cracks and must be structurally sound before being waterproofed. The waterproofing procedure consists of laying an open mesh cotton fabric over the floor and saturating it with pitch. Saturated felt may be used over concrete slab.

Wall Finishes. Walls should be light in color to reflect the maximum amount of illumination from white lights and safelights. If the ceiling is white, walls can be pleasing pastel shade of beige, ivory, or even very light sage green and will give adequate reflection. Moreover, the slight tint will be esthetically more pleasing (in white light) than a completely stark white room, and will not show soil as quickly. A semigloss finish is more satisfactory for walls than a gloss finish. Cabinets and wood trim, particularly on the "wet" side, can be finished with a suitable stain-and-corrosion-resistant gloss coat. A resin finish can be used for this purpose. Synthetic-resin materials offer a high degree of resistance to chemical action and the clear solution may be brushed or sprayed on.

Cabinets, benches, and woodwork that have an attractive grain, such as maple or birch, may be given a glaze finish. This consists of an application of a pigment of the chosen color, usually tan, gray or green, which is wiped off after a short setting period. When dry, the wood is given one or more coats of clear resin or varnish.

This treatment enhances the grain, and the variation in color reduces the house-keeping chore, and makes wear less conspicuous against the glazed background.

Solvents in some finishes may fog sensitized materials. Therefore, be sure that unprocessed paper or film are not stored in the area being finished (or refinished) during the painting, lacquering, or varnishing operation.

Cabinets, Sink Bases, and Sinks. Cabinets and sink bases can be constructed of solid wood, waterproof plywood, or stainless steel. Modular units of each type are available as cupboard units or cabinets for kitchens, restaurants, dairies, and laboratories. Custom units, constructed to specifications and built into the laboratory, can be contracted to cabinet shops. (Typical wooden cabinet construction is shown in Figure 9.1). Base cabinets should feature toe space at the bottom, phenolic tops and backs on the "dry" side, and adjustable shelves. Sink-cabinet tops and splash-backs should be stainless steel or fiberglas.

Cabinet tops should have rounded corners and streamlined pulls should be used on drawers and doors. This will reduce the possibility of bruises that usually occur from bumping into sharp corners and knobs in the dark. The appearance, sturdiness, and life of wood cabinets will be increased if hard woods are used. For example, waterproof maple, birch, or ash plywood will be much better than fir. It will also be easier to obtain a good finish, and the finish will stand up better. Suggestions on finishng have been given in the preceeding section.

If stainless steel is selected, AISI (American Iron and Steel Institute) Type 316 should be used. This is a low-carbon, chrome-nickel stainless grade steel modified with two to three per cent molybdenum to provide extra corrosion resistance to the chemical action of developers, stabilizers, fixers, and other chemicals.

If a custom installation is planned, proper stainless steel fabrication methods should be used.[9.9] The material can be sheared, punched, drilled, formed, welded, ground, buffed, and polished, but flame cutting and burning techniques should not be used. Metallic arc welding is preferable, and copper chill plates should be used to draw off excessive heat. Jigs, clamps, or tacking will be needed to prevent buckling or warping, because stainless steels have a high coefficient of expansion. Welded material may require pickling, or grinding and polishing. A number 4 finish is desirable. It is a bright, satin-like surface, is very reflective, easily cleaned, and has high corrosion resistance. The resistance is attained through naturally formed oxide films, and is highest when the surface is chemically clean and free of imbeded foreign matter. To achieve this condition, the final operation is to "passivate" the fabricated part for 1/2 hour in a 20 percent solution of nitric acid in water, at 140 to 160 °F. The acid should be thoroughly rinsed off with water after the passivation treatment is complete.

Prefabricated stainless steel sinks come in a variety of shapes and sizes, and may be adapted or combined with cabinet work in a customized installation.

FIBERGLAS INSTALLATION. Recent development in fiberglas bonding techniques for marine purposes lend themselves to the fabrication of sinks, splashboards, and drain tops for laboratory use. Prefabricated sinks may be purchased and installed in the cabinets, or a custom installation can be made. If the latter choice is made, the basic sink structure can be fabricated of marine plywood joined with waterproof glue and brass or stainless steel screws. It is desirable to send this structure to an experienced marine firm who will apply the fiberglas cloth and epoxy or polyester resin. This will prevent the noxious odors of solvents used in the process from permeating the laboratory. The principal difference between coating boats and sinks is that boats have a convex surface and outside corners, and sinks have inside corners. Therefore, this difference may make the covering operation more difficult for the technician who normally works on boats.

A typical sink-covering operation will be conducted somewhat as follows: First the plywood is sealed with epoxy or polyester resin. When dry, the fiberglas cloth is cut to fit, and cemented to the sealed surface with more resin. A second fiberglas layer be applied in the same manner. Final coating of the cloth is applied, and may contain pigments giving the desired color. Tan is advised because it will show less stain from developer solutions. A spattered finish can be attained with a Zolotone gun, using other color pigments, usually lighter and darker shades of the background color. This effect is desirable to reduce cleaning, and will give a marble-like appearance to the sink. A clear resin coat should be applied for a final finish and added protection.

MAINTENANCE OF FIBERGLAS SINKS. Fiberglas sinks are susceptible to damage from two sources: high temperature hot water, and physical damage to the covering. Hot water should not be allowed to run continuously on one area, or the fiberglas will separate from the wood and form a buckle. Sharp objects or pointed corners of tanks can easily damage the coating, allowing water and chemicals to work beneath the surface and destroying the sublayers and bond.

In time, developer will stain the surface of fiberglas. The stain may be bleached out, but prolonged action of dichromate-sulphuric bleaches should be avoided.

Local repairs to physical damage can be made with materials obtained from a marine supply firm. General recoating, to restore the glossy, chemically-resistant surface, can be made by thoroughly cleaning the dull surface of any stains, and then applying one or two coats of a clear resin.

MAINTENANCE OF STAINLESS STEEL. Stainless steels in sinks, tanks, and processors, that become contaminated from atmospheric exposure and from chemicals used in the photographic process, should *never* be cleaned with steel wool, coarse fabrics, or abrasive cleaning powders or other compounds. In addition, even though Type 316 stainless steel is resistant to most chemicals used in photo-processing, it may be attacked by certain bleaches used in reversal processes, by sodium hypochlorite bleaching agents used in liquid bleaches and by chlorine powders used in cleansing powders. Therefore, these materials should not be used.

Cleaning of minor accumulations of dirt, water spots and residues of developer and fixer can be accomplished with a soft cloth wet in a solution of warm sudsy water containing a liquid detergent. This should be followed by rising with clean water and a soft cloth.

The dairy and food industries use a great deal of stainless steel, which always must be kept hygienically spotless and sanitary. One cleaning agent used for this purpose is Delchem 790.

Another cleaner made especially for stainless steel is a passivating polish manufactured by Oscar Fisher. In addition to its cleansing properties, the compound contains a passivating material to help restore the oxide film on the surface of the stainless steel.

Oily-base polishes are available in spray cans to restore a shine to the surface of stainless steel. If such sprays are used, they should be applied to exterior surfaces only — never on the inside of tanks or on anything that will come in contact with sensitized materials or processing solutions.

Excessive accumulations of dirt and chemicals may create light rust marks or stains. An effective cleanser for such marks is made from a 10 per cent solution of sodium citrate. A soft cloth is moistened in the solution and rubbed vigorously on the stained surface, on which a light layer of whiting (calcium carbonate powder) has been sprinkled. Following this treatment, the surface should be rinsed thoroughly and driedwith a soft cloth.

After continued use, the inside of the developer tank may acquire a buildup of silver and developer by-products. Preventive maintainance, in the form of daily cleansing with Kodak Systems Cleaner, can keep such a buildup from forming. In stubborn cases, it may be necessary to take the tanks to the plating shop for more rigorous treatment. This may consist of a descaling of pickling treatment, or a chemical electropolishing operation.

STORAGE OF CHEMICALS. Photographic chemicals are supplied in two forms: powder and liquid. The powder form is more stable; it can withstand both higher and lower temperatures, and still have a long life. However, it must be protected from high humidity and moisture. Liquid chemicals, on the other hand, are unaffected by moisture, but are very much affected by excessively high or low temperatures. Therefore, if both types of chemicals are used, the chemical storage area should be planned with the following considerations in mind:

1) Provide an area that is easily accessible so that heavy, bulky containers can be moved into it, and easily dispensed. If possible, the storage area should be adjacent to the chemical-mixing area. A door should be provided, wide enough to permit a cart to pass through.

2) The size of cabinets, spacing of shelves, and general arrangement to be followed in locating the chemicals should be determined from the sizes of the chemical containers. Additional space should be added for ease in moving containers onto and off the shelves.

3) The area should be cool and dry. Ambient temperatures much below normal room temperature can cause precipitation or crystallization of agents contained in highly concentrated stock solutions. In addition, the solidifying point of glacial acetic acid, for example, is approximately 60 °F. Therefore, while a cool temperature is desirable, excessively low temperatures cannot be tolerated.

CHEMICAL MIXING. Sink-top processors can usually be serviced by weighing and mixing powdered chemicals in the adjacent sink. If liquid stock solution chemicals are used, mixing can be done in the processor tanks because the highly concentrated solutions only need to be diluted.

If prepared powder mixes are used, these too can be mixed in the sink, in a stainless steel or plastic bucket or similar container. Weighing facilities should be available and use of the metric system will enable quantity and volume changes to be made with ease. Hot (but not steaming) water (approximately 110 to 125 °F) will facilitate mixing. In general, water should not be any hotter than necessary to dissolve the chemicals easily. Use about 75 per cent of the total water volume called for, after dissolving the chemicals add sufficient cold water to achieve the total volume. The water added can be cold or refrigerated water so that the temperature of the final solution will be close to the required working temperature, thus reducing time-wasting temperature adjustment.

Extreme care should be exercised to prevent spillage of chemicals, since the chemical dust can cause spots on paper or film. Mixing should be carried out so that any spillage can be flushed into the sink; or compounding can be done on a sheet of paper, which can be discarded after the chemicals are mixed.

If large volumes of chemicals are to be handled, it is wise to install chemical mixing facilities. This can consist of a simple mixing-transfer tank, or a sophisticated chemical weighing, mixing, and storage room, with chemical-carrying pipes routed to each darkroom. A chemical mixing-transfer tank is shown in Fig. 9.7. If a complex system is planned, it should be engineered by a photographic firm experienced in the design of such systems.

LOADING AND UNLOADING SENSITIZED MATERIALS. The loading and unloading room preferably should be separate from the processing darkroom. It should have cool, dry storage space for sensitized materials; storage for magazines, both loaded and unloaded; a transfer or pass box to the outside; and a means of communicating with the personnel who deliver or pickup magazines. A vacuum cleaning system, previously discussed, will be of use in keeping magazines clean.

A pass box of the type used in X-ray laboratories can be obtained as a finished unit from X-ray equipment supply houses. The unit requires only to be mounted in the wall. It enables a magazine or cassette to be put into the box from either side. However, the opposite door cannot be opened when the first door is open. Therefore, sensitized material can be handled freely in the loading room without danger of being fogged, while someone leaves a magazine (or picks one up) from the pass box.

STORAGE OF SENSITIZED MATERIALS. Proper storage of sensitized materials before exposure, after exposure, and after processing, is an important contribution to good record quality. Storage areas in the laboratory should be located well above the floor level. The height should be sufficient to preclude the possibility of sensitized materials ever getting wet from drains or sinks that could overflow due to accidental flooding from storms, bursting water pipes, and other causes.

Condensation on walls, or condensate dripping from cold, uninsulated water pipes is another potential moisture hazard.

The sensitized materials should not be stored in warm areas such as attics, top floors of uninsulated buildings, metal sheds; in automobile glove compartments, trunks or near the sides of metal van bodies; or near radiators, or hot water pipes, and should never be in direct sunlight.

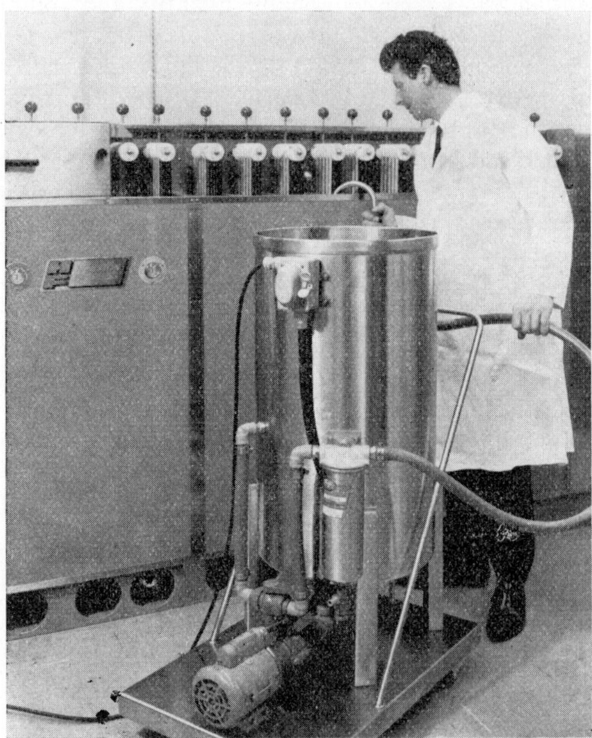

Fig. 9.8. Chemical solution is pumped from Houston Fearless Chemical Mixing and Transfer Tank into Houston Fearless Labmaster Processor

In general, the temperature conditions of basements are satisfactory for storage of sensitized material. However, the relative humidity may be too high, and walls may have condensation, or may leak. Therefore, if materials are to be stored in a basement, it is wise to put them into a sealed, waterproof container, or to use an electric dehumidifying unit. Such units are available from Carrier Corporation, 300 South Geddes Street, Syracuse, New York; Frigidaire Division, General Motors Corporation, Dayton, Ohio, and other firms engaged in the manufacture of air conditioning and refrigerating equipment. A dehumidifying unit should not be confused with a window air conditioner; the air conditioner removes some moisture from the air, but its main function is to cool the air for comfort. Whether or not an air conditioner can be used successfully for sensitized-material storage will depend upon the cubic feet of air in the area and its relative humidity. This should be checked into thoroughly before making such an installation. The writer has successfully dehumidified (and simultaneously air conditioned) a basement that had a relative humidity that was excessive in the summer months. An air conditioner, installed in the basement window was adequate to reduce the relative humidity to approximately 50 per cent, when the unit was operated at regular intervals each day.

In addition to being protected from heat and moisture, the material must not be stored where it will be physically damaged; or where harmful gases, X-rays, or radioactive materials can fog it or otherwise degrade its quality.

The writer has also converted a walk-in cafeteria refrigerator to store sensitized materials. If the conversion of such a unit is undertaken, a check should be made of relative humidity conditions, and controls added (if they are needed) to obtain the optimum storage conditions. Film stored in sealed metal cans presents no problems until the seal is broken. However, sensitized paper, in non-moisture-proof containers may become dehydrated if the RH is too low.

Moisture condenses on an object if its temperature is below the dew-point of the surrounding air. In summer, the dew point may reach 70–90 °F, and materials that have been stored at 50 to 60 °F may have moisture condense on them when they are removed from refrigerated storage. This condition will not show on paper or cardboard cartons, because the porosity of the material will absorb any moisture before it reaches droplet size. To prevent damage to the sensitized material, the sealed package should not be opened until the contents are warmed to above the dew point temperature of the outside air. This requires time, but may be speeded up by separating the packages. A good rule of thumb is to work a day or two ahead of schedule; that is, take an anticipated day's supply out of the refrigerated storage 24 to 48 hours in advance of the time it will be used.

The life of sensitized materials may be extended even further by reducing the temperature to below freezing. In general, the lower the temperature the longer the sensitized material may be kept. However, temperatures below 50 °F are not necessary for oscillograph materials that will be used in a few months.

In general, the following points should be remembered:

1) Store material in a cool, dry place;

2) Store at a relative humidity of 40 to 60 per cent.

3) If the material is to be stored for more than a few months, it should be refrigerated. An ordinary household refrigerator can be used if quantities are small.

4) Storage is even more important after exposure than before exposure. The exposed record should be processed as soon as possible after it has been exposed to obtain the optimum quality. If it cannot be processed immediately, then it should be stored under cool, dry conditions as previously mentioned.

5) Recommended storage conditions for raw (unexposed) oscillograph paper, for periods up to 6 months are 55 °F, at a relative humidity of 40 to 60 per cent.

MOBILE LABORATORIES. Portable laboratories are available in the form of rail, sea, or road transportable vans and trailers, and even as air-transportable single-

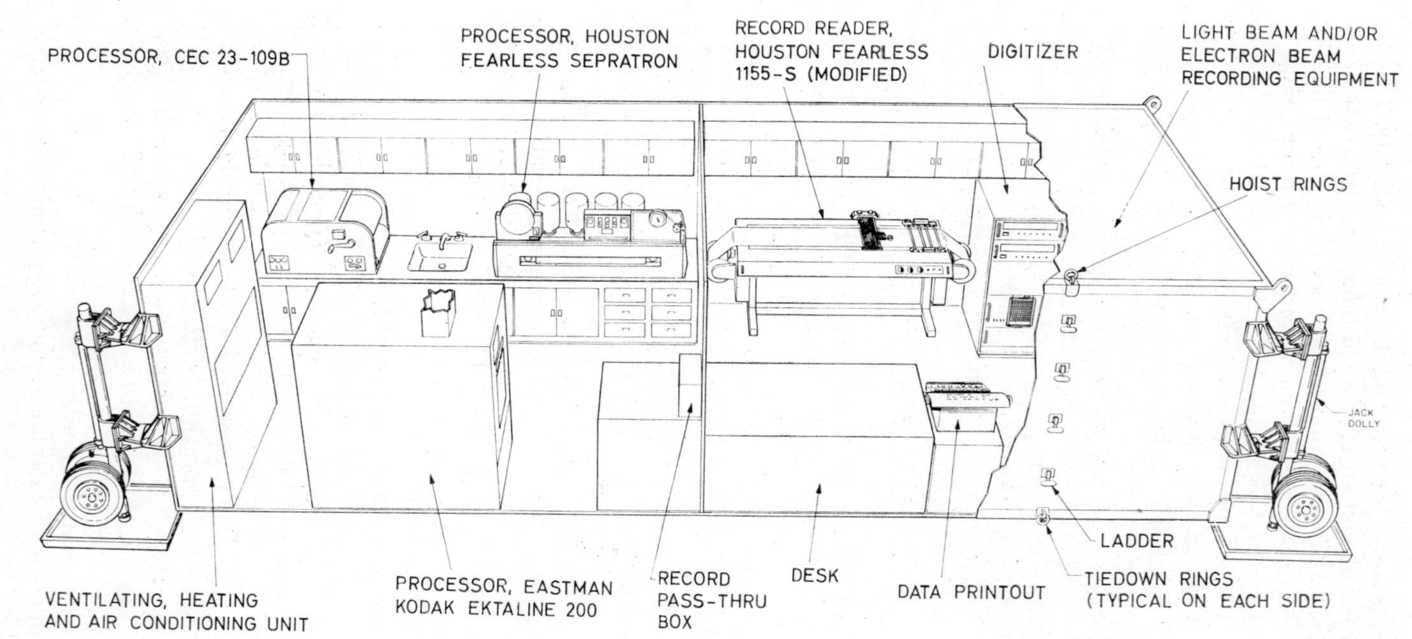

Fig 9.8. Mobile recording, processing, and data reduction laboratory. This proposed air-transportable laboratory is similar in construction and configuration to other portable laboratories produced by Houston Fearless for processing and evaluating all types of films. Entrance to the shelter is at the right end. The darkroom area, left, is entered through a door in the center partition

unit shelters or multiple-unit laboratory complexes. Heating, ventilating, air-conditioning, and other functions previously discussed, must be installed to enable the shelter to be capable of operating under the widest temperature, humidity, and other environmental extremes. The properly controlled interior enables high-quality photographic results to be obtained regardless of external weather conditions. Design of such equipment is a speciality; equipment must be lightweight, compact, and bolted or shock-mounted in position. In addition, operating supplies and perhaps utilities and water, may have to be provided.

An example of an air-transportable recording, processing, mensuration, and evaluation shelter is illustrated in Fig. 9.8. Portable laboratories are available from Houston Fearless Corporation which specializes in their design and manufacture on a custom basis to meet individual user requirements. The proposed configuration provides both white light and darkroom areas complete with the following facilities: recording on equipment selected by the user; stabilization processing of develop-out paper; stabilization of print-out papers; processing of color-recording papers; archival processing of 16 and 35 mm films; and record reading, digitizing and evaluation. The illustrated shelter will provide an interior temperature of 65° to 75 °F under an exterior temperature range of $-40°$ to $+110$ °F. Facilities are provided for chemical mixing and for storage of equipment and supplies.

LITERATURE REFERENCES

[9.1] *New Light on Darkrooms, for More Efficient Operation*, Calumet Manufacturing Company, (6550 North Clark Street, Chicago, Illinois) 1956

[9.2] *Darkroom Design and Construction*, K–13, Eastman Kodak Company, (Rochester, New York) 1958

[9.3] RICH, L. A., "Controlling Water Temperature," *Industrial Photography*, Vol. 14, No. 2, p 24, (1965)

[9.4] *National Electrical Code*, NBFU 70, National Board of Fire Underwriters, (85 John Street, New York 38, New York) 1962

[9.5] STALEY, K. A., *Fundamentals of Light and Lighting*, Bulletin LD–3, General Electric Company, (Cleveland 12, Ohio) 1960

[9.6] *Construction Materials for Photographic Processing Equipment*, No. K–12, Eastman Kodak Company, (Rochester, New York) 1962

[9.7] RYAN, R. T., "Plastics for Motion Picture Laboratories," *Journal of the SMPTE*, Vol. 68, No. 8, p 542, (1959)

[9.8] *Specification for Testing the Photographic Inertness of Construction Materials Used in Photographic Processing*, ASA Standard PH 4.31, American Standard Association, (10 East 40th Street, New York 16, New York) 1962

[9.9] *Eastern Stainless Steel Sheets, A Condensed Handbook for the Engineer and Layman*, Eastern Stainless Steel Corporation, (Baltimore 3, Maryland) 1945

X SYSTEMS AND PROCEDURES

Managing the Department. The purpose of this chapter is to present information, procedures, and techniques having application to photographic processing in general, with specific details for oscillogram processing. Among the instructions are discussions pertaining to personnel and management. It is not the author's intent to present comprehensive information on supervision and management, but rather to point out those aspects that differ from standard office or business routines. Standard supervisory and management techniques should be followed, and adjusted to fit the special criteria invoked by the photographic medium.[10.1]

Job Descriptions. Job descriptions should be prepared for each position in the laboratory. The written description should list details important in finding the proper person to fill the job and should also include all the duties that are a part of the assigned tasks. Minimum and maximum age, sex, minimum schooling, special training, experience in running certain equipment, ability to use specific tools and instruments, and past related experience, are among the major significant items to be included in the job description.

It may be possible to use one person for several different types of work. For example, a darkroom technician might be required to make contact prints and enlargements in addition to his usual responsibility of processing oscillograph records. It is important to list any such part-time duties, and to make sure that each new employee clearly understands what will be expected of him. The following sample job descriptions may be used as a guide:

Supervisor, Oscillogram Processing

Position Summary. Supervise the oscillogram processing laboratory and the functions necessary to support the making, developing, reproducing, and filing of oscillograms.

Work Performed. Supervise and coordinate laboratory activites; outline details, methods, and assign work to laboratory personnel; check finished work, maintain schedules, and maintain quality standards.

Interpret and enforce security and safety regulations as they apply to laboratory activities. Set up and supervise methods for securing, classifying, and filing detailed information on all oscillograms. Prepare weekly reports; budget, progress, and special reports; procedures and recommendations. Develop forms as necessary.

Design, as required, new oscillogram processing facilities and equipment. Determine requirements, and requisition equipment and supplies. Conduct liaison with vendors and other companies on oscillogram-processing matters.

Develop cost budgets, and determine manpower requirements and schedules.

Qualifications. Well-rounded knowledge, and skill in all phases of photography, oscillography, and processing. Thorough knowledge of oscillograph and processing equipment uses, limitations, accessories, quality, cost, etc. Thorough knowledge of sensitized and chemical supplies and supply sources. Knowledge of photographic chemistry, physics, and mechanics as related to oscillography. Ability to plan, organize, and direct laboratory work.

Training and Experience. Two years of college, plus a minimum of five to six years of industrial experience including supervision.

Technician

Position Summary. Performs all functions necessary to process exposed oscillograph materials, and such supporting activities as minor equipment maintenance, chemical mixing, supply requisitioning, and minor record keeping.

Work Performed. Processes exposed oscillograms in accordance with established procedures and recommended practice using manual, semi-automatic, or continuous processing machines.

Requisitions sensitized materials and chemicals as required, maintaining the supplies in accordance with recommended minimum-maximum requirements.

Mixes chemicals such as developer, short stop, stabilizer, fixer, permanizing solution, tray cleaner, etc.

Cleans, lubricates, and conducts minor maintenance of laboratory equipment. Initiates requests for preventive or corrective maintenance to keep equipment in good working order.

Loads and unloads magazines; checks magazines for damage; maintains records showing magazine numbers and their respective recorders, periodic inspection, location, etc.

Training and Experience. High school graduate, with a minimum of one year of previous experience in photographic processing, such as might be obtained in commercial or industrial photography, photofinishing, motion picture processing, etc.

Wage and Salary Ranges. Most large firms today have a wage and salary department which establishes rate ranges and periodic-review systems. If there is a union, the union scale will prevail; if not, the wage and salary administrator will survey the "going rates", in similar industries in the locality, for each job classification. From this information, a rate range can be developed that reflects company policy and individual variations of skills and duties.

The problems attendant upon processing oscillograph records require alertness and skill from operating personnel. The engineer seeking speed and quality in his record is very dependent upon the operator. If the equipment fails or jams, or a record tears, it is the operator who does not panic who can quickly and systematically remedy the trouble.

In automatic processing of motion picture film, the processing-solution condition is under the constant check of a chemist. In oscillograph processing, the engineer is usually dependent upon the operator to change or replenish solutions on a footage basis. Often this is neglected for various reasons — sometimes the volume of work forces an operator to replenish when he should change chemicals. Often a customer will say: "I don't care what it looks like — just as long as I can read it." Under adverse conditions he may be able to process a satisfactory record exposed at slow writing speeds, but he may lose a record exposed at high speeds and processed under these same conditions. He may also grow careless if quality standards are not set and speed and pressure are constantly given more consideration than quality.

Naturally, all these problems are compounded when higher speed processors are installed, when there are more operating personnel and when there is more than one shift operation.

In most companies, oscillogram processing operators are at the bottom of the pay scale, are classified as a helper or are in some other lower classification, have the least seniority, and the least experience and technical ability. These people are then entrusted with irreplaceable records on which the outcome of a test or the future of a project may depend. Many men become operators only because this offers the only means of getting into photography. Soon wanting to get behind a camera, they become bored and careless in handling oscillograph records. The photographic supervisor must juggle the desire for quality against the lack of experience, the low pay against the desire of the man to get into a different branch of photography where he can move ahead.

The motion picture industry has recognized a similar problem and established classifications and pay scales to correct it. A man who processes original negatives is classified as an "Operator A" and is paid a high rate. The man who processes prints is classified as an "Operator C" and receives a lower rate. The margin (more than 10 percent) is the premium paid to the man who handles the original material.

EMPLOYEE TRAINING. The new employee will need to be trained so that he can efficiently perform the tasks assigned to him. Training should consist of orientation (company and department), familiarization with equipment and methods, and assignment of routine duties. Instruction should come from the man's supervisor — not by a "rubbing-off" process from fellow workers.

There will probably be sufficient printed material concerning company policy, safety, security, and departmental procedures for employee orientation. Equipment manuals, handbooks, and standard photographic technical references can also be used for training. However, the new employee should be made aware of where his responsibilities begin and end, and what his exact duties are.

NUMBER OF PERSONNEL. Just as the laboratory space requirement is dependent upon volume, so is volume dependent upon personnel and the equipment they operate. Therefore, it is possible to make a determination of the number of persons who will occupy the laboratory, once the volume and the processing methods have been established.

If automatic processors are used, and one is immediately adjacent to another, it may be possible for a single operator to run both machines. However, such a practice can only be endorsed with this reservation: standby help should be made available. If such tightly-scheduled production is planned, it is well to keep in mind that the operator will require immediate assistance in case of equipment failure. Therefore, provisions for handling such emergencies must be made in advance.

Sufficient foresight should be used in planning small laboratories to accommodate two or more persons, even if the original requirement is for only one man. This concept will result in a laboratory having sufficient aisle space, counter space, and light and darkroom areas to avoid conflict when the staff is increased. This does not mean that the laboratory cannot be compact; if it is well organized, it can be small and still be efficient.

Incidental to manhour output are personnel incentives, individual skill, training, and experience.

SAFETY. The special safety problems in photographic laboratories were considered in previous discussions on stabilization processing and darkroom design. The chief hazards described were concerned with fumes, darkness, and the use of electricity in close proximity to chemicals and water.

There is another potential danger in photographic processing: skin irritation caused by repeated contact with processing chemicals. Generally, this reaction occurs only to some individuals, and may be likened to an allergy. Usually the reaction manifests itself as a skin rash on the fingers and hands. Water blisters form and break, and the skin peels away leaving the hands very raw, painful, and unprotected. Once started, the condition will progress from one finger to another, and then on to the palm of the hand.

The condition is usually caused by certain properties in the developing agent P-methyl-aminophenol sulphate, and is called "metol poisoning." Other developing agents such as paraphenylenediamine and color processing chemicals are also toxic. Another probable cause of dermatitis is the acid used in stop baths and fixers.

This condition can be checked by preventing either dry or liquid processing chemicals from contacting the skin. Hands should be kept dry and out of solutions. If contact or immersion of hands in chemicals is necessary, clean, dry, plastic or rubber gloves should be worn. A prime cause of chemical transmission (and possibly of infection) is from the typical contaminated, soggy, darkroom towel. The use of disposable paper towels will prevent such transfer.

For some uses, a "liquid glove" can be used. This is applied to the hands, and allowed to dry before contact with water-type solutions. Care should be used in handling sensitized materials to prevent finger marks and smudges from the lanolin or similar oily base material in the protective cream.

Many (if not most) photographic processing solutions, especially those used in

330

color processing, are harmful or poisonous if swallowed. Vomiting should be induced and *a physician should be summoned at once.* Strict attention should be given to the warning notices printed on chemical containers. All working surfaces should be kept clean and free from spilled solutions or dry chemicals. Employees should be discourged form using the laboratory as an eating area, and from storing food or snacks near harmful chemicals.

With watchfulness and proper care, safety can be assured to personnel using even the most hazardous photographic chemicals. The list in Table 10-1 summarizes some operations and associated chemicals that require care.

In addition, color processing utilizes harmful color developers, and bleaches. Bleaches usually contain potassium ferricyanide and sodium thiocyanate. Other potentially harmful chemicals used in some color-processing solutions are formaldehyde, ethylene glycol, benzyl alcohol, and methanol.

TABLE 10.1

IRRITATING AND/OR HARMFUL PHOTOGRAPHIC CHEMICALS

Operation	Possible Reaction	Probable Harmful Chemical
Develop	Skin irritation	1. Sodium hydroxide 2. Trisodium phosphate 3. *p*-methyl-amino-phenol sulphate 4. Hydroquinone 5. *p*-phenylene-diamine
Stop bath	Skin irritation	1. Acid, usually acetic or citric
Fix	Irritation of skin, nose and throat from sulphurous fumes	1. Acid, usually acetic or citric 2. Fumes, such as sulphur dioxide
Stabilize	Irritation of skin, eyes, nose and throat due to release of fumes during high-temperature drying	1. Acid, usually acetic or citric 2. May contain poisonous chemicals such as thiocyanates

DEPARTMENTAL RECORD KEEPING. Procedures should be established to keep a record of the flow of sensitized material through the laboratory. The example of one such system, described in this section, combines procedures of several companies and may be modified to suit individual requirements.

The system should fulfill the following needs:

1) Accountability for the material used.

2) A material-control system and a charge system for the material and for the processing labor.

3) A step-by-step record of the personnel who handled the record.

4) A record of the magazines, tests, and incidental items; this is helpful when a failure occurs, a magazine is misplaced, or a record cannot be found.

5) A record of the footage and also the width if various widths of material are handled.

6) Incidental information including dates, times, classifications, emulsion numbers and other data as necessary or helpful.

Magazine and Roll Numbering. Even through precautions are taken in the laboratory and in other storage areas to insure that the material is properly protected, there is no guarantee concerning its handling once it has left the laboratory control. For example, a magazine may be left for long periods in direct sunlight, or in a moist atmosphere, or in fumes that could degrade the record. Therefore, a loading record should be kept.

When there are numerous oscillograph recorders in use, it is advisable to number the feed and takeup magazines with matching numbers. The equipment may already have an inventory tag or number that can be used. However, such numbers are usually long, cumbersome numbers to write and to remember. One or two digit numbers, stenciled onto the magazines will suffice for accurate and easy identification.

A positive check on loading, material, and magazine malfunction can then be obtained if the rolls are numbered, and the magazine and roll number are entered on a log sheet.

Register Sheet. To implement the above requirements, log sheets, similar to the one shown in Fig. 10.1 are needed. Starting at the top of the left column of the Loading Register, each line is numbered consecutively. A Bates (or similar) numbering machine can be used to number each roll of paper as it is loaded, giving it a

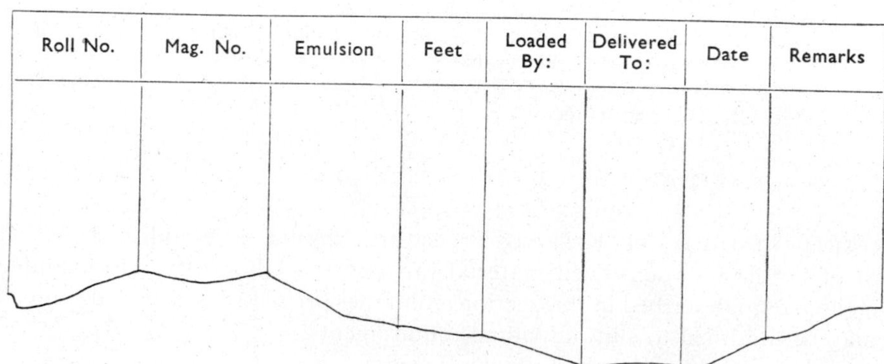

Roll No.	Mag. No.	Emulsion	Feet	Loaded By:	Delivered To:	Date	Remarks

Fig. 10.1. Loading Register

number corresponding to the number on the log. The technician then enters on the log the magazine number, the type (and if desired, the emulsion number) of this material, the footage, and his initials. When the magazine is picked up, the technician who receives the magazine, initials and dates the log.

In addition to the register of magazines loaded, a magazine card should accompany each magazine. The card, shown in Fig. 10.2, will show the material, footage, date loaded, and other details.

OSCILLOGRAPH RECORD

PACK NO.

KIND OF FILM

EMULSION NO.

CELL	TEST	RUN	

O. ENGINEER

PHONE NO.	CHARGE NO.
RECEIVED BY	DATE

(Heading on reverse)

*This side of card for
Photographic Services use only.*

LOADED BY	DATE
PROCESSED BY	DATE

DATE-TIME RECEIVED BY Photographic Services

Fig. 10.2. Magazine card

To obtain processing service, the magazine, and a properly filled-out processing request, shown in Fig. 10.3, are delivered to the laboratory. The data from the request, together with the date, and other information, are entered by the technician on his log for processing service and material.

For accounting purposes, charge numbers can be applied against the roll. A requisition for the total number of rolls used on the project can then be made out.

PRIORITY

OSCILLOGRAM PROCESSING
ORDER

PHOTO JOB
ORDER NO. ____________

CONTRACT
NUMBER ____________

FACILITY ____________ RUN TIME ______ AM PM

RUN NO. ____________ MAG NO. ______
A B C D E

TYPE
PAPER ____________ OPERATOR ______

PRIORITY
AUTHORIZED BY ____________ EXT ______

TIME RECD
PHOTO DEPT. ______ AM PM RECD BY ______

TIME
COMPLETED ______ AM PM PROCD BY ______

CALLED BY ____________ TIME ______ AM PM

COMPLETED
JOB RECD BY ____________ EXT ______

Form 602 -B Rev. 8-57

Fig. 10.3. Requisition for processing service

This will show the quantity used during a day (or a week, or other convenient period), and will be used to maintain a working level of material in the laboratory.

If only a part of a roll is to be processed, the processing request should indicate (as an example) "part of roll 5287", and should show the approximate footage. In this way, the darkroom technician will know that there is still a partially filled feed spool in the magazine, and can conduct follow-up if necessary.

If the roll has been processed using stabilization chemicals, the finished roll should be stamped with a notice that it contains such chemicals. A typical rubber-stamp notice of this nature is shown in Fig. 10.4.

WARNING

This oscillogram contains residue from processing chemicals. Do not –
1) Store or comingle it with photographic prints, negatives, microfilms, or motion pictures, either developed or undeveloped.
2) Wash hands thoroughly after handling this record, before handling any of the above materials.

Fig. 10.4. Rubber stamp warning for stabilized records

When the roll is finished, the laboratory technician notifies the appropriate department that the work is ready for delivery. The person who picks up the finished roll signs the log sheet to evidence receipt of the roll and acceptance of accountability for it.

WEEKLY REPORT. The volume of oscillograph material processed should be reported on a weekly or monthly basis. A tally sheet should be distributed to each laboratory technician, who will enter the amount of lineal footage processed each day. At the end of the week, he will add the daily footages together to get the total for the week. After the balance of the report has been completed, it is turned in to the department secretary, who supplies a new blank for the following week.

The secretary totals reports from all technicians and posts the totals to a master record. It is also wise to make up a graph so that trends in volume can be quickly spotted. The graph is important to compare work loads over long periods and is a forceful tool to convey to management the need for space, manpower, and equipment. It is also indicative of supply inventories; if there is a steady rise or fall in volume of work handled, there must be a corresponding adjustment in the supplies that are carried.

SUPPLY REQUISITIONS. The department manager can determine the amount of supplies that are needed from the weekly report and from the types of work being done. From this information he can establish a perpetual-inventory system. The system will have a stock card for each item, and the card will show maximum and minimum amounts, vendors, lead time, price, acceptable substitutes, and other information. It is possible that the stock and inventory functions will be the responsibility of the material control department, rather than of the laboratory. In that case, the laboratory manager should assist in organizing the records so that there is always sufficient stock on hand, and to see that the stock is properly stored.

Depending upon company procedure, the laboratory will probably furnish either a stores requisition or a purchase requisition for materials that are needed. Assuming that a stores requisition is used, the chemicals and sensitized materials needed for the week can be drawn from stock using the charge numbers from the previous week's work. The information pertaining to charge numbers and footage processed, taken from the processing requests, is the most reliable information on the allocation of materials. Therefore, it is the basis of accurate charge information.

To start the system, a week's supply of materials will have to be drawn out against an overhead charge or a loan ticket. At the end of the first week of operation, requisitions can be made out to balance the accounts. From then on, the supply will be maintained from the work actually processed, and requisitions will be written against those specific numbers.

PERPETUAL INVENTORY. To maintain a perpetual inventory, an adequate "float" must be on hand to provide material during the ordering period. Sufficient "lead" time must be allowed to process the order internally, and to permit the vendor to obtain the material if it is not on hand and then to make delivery. The material must be received, inspected, and delivered to stores, before it can be requisitioned. Therefore, minimum inventories must be set high enough to allow for the system time, and must take into account occasional potentially-high usages. The department manager should periodically issue a form memo requesting advance notice of any abnormally large requirement for film, paper, or processing supplies.

OSCILLOGRAM STORAGE The method used for the storage of oscillograms is determined by the type of photographic material and the way it is processed, the length of storage time, the frequency of referral, the need to protect the record from fire and other hazards, and the classification of the document.

Material and Processing. Records will be on either film or paper, and will be print-out, completely processed, or stabilized types. The storage of film is usually better handled if film and paper records are segregated. In addition, those records that have been processed for archival life should be stored only under approved archival conditions.[10.2] Records processed by stabilization should not be comingled with those that have received archival processing, with photographic negatives or prints, or with motion picture footage. Stabilized records should never be stored with raw stock (unexposed or exposed sensitized materials).

Personnel who have handled stabilized records should not then handle archival records. This is because some of the stabilization residue will adhere to the fingers and will be deposited on the archival records. The chemical action that follows is likely to leave finger marks.

Film-handling gloves, used to edit motion pictures, can be worn for analyzing, handling, and reading film records.

Storage term. If the initial study determined that stabilization processing would be used, that decision was based in part upon the fact that long life of the record was not required. However, the life of a stabilized image can vary greatly depending upon the method of storage. A cool, dry storage condition is required for maximum life of the record. These same storage conditions also apply to archival quality records.

Unless there will be nitrate-base film in the storage room, no special fire protection measures are necessary. Acetate-base film and paper-base records require compliance with local building and fire codes only, and the code may require that automatic water sprinklers have to be installed, whether acetate film is stored there or not.

The main floor of the building is best for storage if the building is not air conditioned. Do not store records in damp basements, top floors that are too hot, or in rooms that may be hot from the sun shining on the wall. Storage shelves should be high enough off the floor to prevent accidental water damage from rain, flood, water pipes, or drains. The temperature of the room should be below 80 °F, and a relative humidity of 25 to 60 per cent is satisfactory. The room does not have to be heated in winter except for the comfort of personnel.

Chemical fumes such as hydrogen sulphide and sulfur dioxide (usually present in coal-burning areas, from gas, and other contaminates of urban areas) cause degradation and fading of the records. In sufficient concentration, sulfur dioxide also causes a deterioration of the folding endurance of paper.

Need for referral. Once the data has been read and analyzed, there may be very little need to keep the record. However, in most cases, it is not wise to destroy it immediately. In light of further testing, a need sometimes develops to review the record of a previous test. Therefore, a decision on when the record should be destroyed will be based on the individual circumstances, and these may fix the destruction date anywhere from the end of the test to the completion of the project.

File space may be insufficient to keep all the records. If this is the case, then just a portion of the records might be kept, or the records might be microfilmed. Microfilming of records is discussed in Chapter 11.

Frequency of referral. The frequency of reference to the records will determine the place of storage, the method of handling the records and the type of file system that should be used. If records can be placed in a dead file after they have been assessed, the storage place can be at a remote location. More elaborate planning of the file space will be necessary if the records are to be kept active.

Protection of the record. The record must be protected from physical damage, and if it is a document containing confidential or secret information, it must be stored in accordance with the requirements of its security classification.

Physical damage includes accelerated deterioration from heat and high humidity
and destruction by fire, flood, chemical fumes, and other causes. In addition, if the
record is on unstabilized print-out paper, light will cause eventual obliteration of
the image by fogging the paper. Unstabilized print-out records should be stored
in light-tight containers. The original container that was used for the unexposed
paper can be used for this purpose. Short-length rolls can be stored in cardboard
tubes with capped ends.

Classifying. As in any filing system, the material to be filed must be indexed
according to pre-determined classifications. For example, an electrocardiogram
would be filed with the case history, by the patient's name; an engine test oscillogram, by the test and run number, etc. Other categories are: by model number, by
location, by date, by customer, by sequence, or by any system that will fit the
requirements of the operation.

Finder Card. A large volume of records will require a finder card. The card will
be filed by the classification system, and will bear the filing address of the document
in its cabinet or vault (for example, Rack C, Shelf 2).

Oscillogram Format. The types and volume of recordings will dictate how the
oscillograms are to be filed. They may consist of any one or several of the following
types: ink-paper, pressure-paper, heat-paper, photographic print-out paper, photographic developing-out paper, photographic film, photographic microfilm. The
photographic papers may be archival quality or stabilized quality, and the way
such records must be handled will have a bearing on how they are filed. The paper
records may be clipped and in sheet form, or in rolls, or folded, and will have to be
filed in accordance with their physical characteristics. Rolled and folded records,
and the equipment for folding rolled records are discussed in Chapter 8.
The volume will determine if a file cabinet or a vault is necessary to house the
oscillograms, and storage activity will be a deciding factor in location of the storage
facilities.

Library Functions. The two most important functions of the oscillogram library
are storage of records and provision for finding records.

Numbering System. Oscillograms can be numbered by assigning a number to
each document when it is received by the librarian. An alternate system is to use
the number assigned to the roll by the laboratory. The latter system with its lifetime number, enables follow-up to be conducted easily from one department to
another without cross-indexing of two or more numbering systems.
Partial rolls bearing the same number can be filed by assigning dash or point
numbers to each roll. For example, if the principal number assigned by the laboratory was 5287, and the roll was partially used and processed, the first part received

by the librarian would be numbered 5287.1, the second part 5287.2, and so on. The librarian would know that the roll had been divided by the notation on the magazine card "part of roll 5287", previously discussed.

When various roll widths are to be filed and file space must be carefully conserved, the rolls should be grouped by size; i.e., all 5-inch rolls should be grouped together, all 12-inch rolls should be grouped together, and so on. The finder card can carry a prefix number followed by a dash to denote the roll width; for example, a 5-prefix would indicate a 5-inch roll, a 12-, a 12-inch roll, and so forth.

The numbering system that has been outlined will appear as shown in the following example:

12–	5287.2 ←	
roll	principal	second part
width	number	of roll

QUALITY CONTROL. Adequate inspection of processed oscillograms is necessary to insure high quality production on a consistent basis. The inspection procedure may require only spot checking of finished work if personnel, equipment, and materials are properly controlled. The controls should include training of personnel, written procedures, preventive maintenance of equipment, and systems designed to provide proper coordination and communication between departments and between shifts. In addition, a priority system requiring a single authoritative approval will ease peak-load situations. All of these aspects are important to high-quality oscillograms and must not be overlooked. Poor results can only come from three causes: equipment failure, personnel failure, or material failure.[10.3] The following check list will guide the formation of systems to hold these failures to a minimum:

EQUIPMENT INSPECTION

Electrical
    1) Was there power a failure in the supply or in the equipment?
    2) Were all switches, lights, and controls functioning properly?

Mechanical
    1) Was transport system working properly?
    2) Was transport system correctly aligned?
    3) Was transport speed correct?

Plumbing
    1) Were there leaks in tanks, pumps, or lines?
    2) Was there any contamination?

Temperature
    1) Were solutions at recommended temperatures?
    2) Was drying drum within established temperature tolerences?

Light-tigthness (if daylight processor)
1) Were the machine and the magazines light-tight?

1) Did the employee receive complete training?
2) Has he had sufficient experience?
3) Is there sufficient communication between shifts?
4) Does each shift do its part in cleanup, routine maintenance, and other "chores"?
5) Does the operator check the equipment; chemical level, condition, and temperature before starting the processing operation?
6) Is the operator capable of clear-headed thinking in an emergency, and has he been trained on what to do if a jam occurs, the drier overheats, etc?
7) Does he take the line of least resistance, i.e., replenish when he should change chemicals, and use other questionable methods?
8) Does he "experiment" with valuable recordings?
9) Does he keep his work area clean, and the equipment, chemical mixing facilities, etc. free of contaminating processing (or other) chemicals?
10) Does he keep magazines and loading room dust-free and orderly?

MATERIALS

1) Are the materials recommended for this application?
2) Have these materials previously been used for this application?
3) Do the materials require any change to the equipment? any additional training of the operator?
4) Is the material a special purpose item? Can it be used in all machines? Does it require compatibility with other materials?
5) Is the proper safelight being used? at the proper illumination level? (see Safelight Tests, Chapter 9).
6) Are the materials that were received the ones that were ordered? Were they received in good condition? (Inspect for broken boxes, the possibility of storage at excessively high or low temperatures, etc.).
7) Did the sensitized material go through the correct chemicals? in the correct sequence?

In an effort to reduce failures, the above list outlines possible causes of poor results. Now let us examine common faults of finished oscillograms, and trace these faults back to their respective causes. Table 10.1 lists four main groups of deficiencies, and miscellaneous undesirable characteristics as well. From the two approaches, it is possible to determine an inspection system that will fit the individual requirements.

IMPROVING EFFICIENCY. It is a natural aim to attempt to produce oscillograms as economically as possible. Assuming that proper choices have been made in selection of recorders, that these are well matched to the task at hand, and that personnel are adept at recording and processing techniques, only minor cost reductions can be made on low-volume production. However, if production is moderately high or heavy, substantial savings can be effected by: 1) combining oscillogram processing with other photographic activities; 2) installing chemical mixing facilities for all photographic facilities; 3) reclaiming silver from photographic solutions; and 4) splicing together short ends of oscillograph paper for subsequent use. Items 1 and 2 have already been discussed in Chapter 9; silver recovery and splicing are described in the following sections.

TABLE 10.3

ERRORS AND CAUSES OF RECORD DEFICIENCIES

	Processing Error	*Recording Error*	*Material Error*
No Image	No developer Water in dev. tank Fixer in dev. tank	No illumination	Wrong material (such as print-out instead of dev.-out)
Weak Image	Exhausted developer Developer too diluted Developer temp. too low. Developer contaminated. Developer improperly mixed	Illumination not adjusted for material used. Writing speed or paper drive speed too high for material used	Wrong speed material
Stains	Exhausted developer Exhausted short-stop Exhausted fixer Excessive heat		Wrong chemicals for material used
Fog	Magazine leaked light Wrong safelight Material exposed before or after recording	Faulty magazine, improper attachment or recorder not light tight	Material too old or improperly stored Print-out material run through developer and fixer
Miscellaneous	Scorched — record-drum too hot Tears — squeegees too tight Scratches — squeegees too tight Edge tears — rollers misaligned Record stuck together — drier not hot enough, or no hardener in fixer or stabilizer Variations in record width — cycling drier thermostat; wrinkles — poor alignment, improper threading		

Silver Recovery. If you toss a penny into used fixer, its finish will change from copper to silver in a few minutes. The silver was left in solution from sensitized materials that went through the bath during the processing operation. As complex compounds in solution, they change to metallic silver when metal is put into the bath. This occurs from the low-voltage galvanic current generated. One type of silver collector works on this principle; another type uses an external source of current, and is more efficient.

The emulsion of sensitized photographic materials contains silver halide. During the processing operation, the exposed silver halide is converted to metallic silver The remaining silver halide is converted into complex silver compounds in the fixer. As additional silver is fixed out, the fixer loses its capacity to hold any more silver, and fixation is incomplete. Eventually the bath must be discarded, but this results in two losses: the loss of the fixer solution, and the loss of the silver. Both represent potential profit and economy if there is a sufficient volume. In fact, a silver recovery operation cannot be ignored in a continuous, high volume processing laboratory.

Early chemical recovery systems were bothersome, and the fixer could not be reused. However, modern electric methods are reliable,[10.4, 10.5] require very little work, and may even be obtained on a service basis. Often, the service is handled by the chemical supply firm, which will pick up the exhausted fixer when fresh developer and fixer are delivered. A percentage of the silver profit is then applied as a credit to the chemical bill. Of course, this method does not increase the life of the fixer, but it does offer some financial compensation for a waste material that otherwise would be poured down the drain.

Another service method is that supplied by firms which install electric silver-recovery units. The firm retains ownership of the recovery unit, and periodically services it. Again, a percentage of the salvaged silver is credited to the laboratory, but this time the laboratory has gained additional life from the fixer by the rejuvenation process.

Instead of using a salvage service, the laboratory may purchase its own silver recovery unit, and sell the reclaimed silver. An electromechanical silver-recovery unit can reclaim over 95 percent of the silver released into the fixer by the processed material, and can prolong the life of the fixer from 3 to 8 times its normal untreated life. Reclaimers' graphs vary, but show that it is possible to recover 1.5 to 4.7 troy ounces of silver per 1000 feet of processed 12 inch wide material. Paper, having a lower silver content will yield less recoverable silver than film.

Although the market value of silver fluctuates, the recent price of approximately 90 cents per troy ounce may be used for estimating purposes. To evaluate the worth of a silver reclaiming program, the annual volume should be used. Since the actual recovery will be based on the efficiency of the operation, it might be well to use the lowest figure given above, 1.5 troy ounces per 1000 feet of 12 inch material, for estimating purposes. Divide the annual square footage by 1000, multiply that figure by 1.5, and multiply the result by the market price of silver to obtain the annual gross return from the silver. From this gross, deduct the cost of the refiner's charges and assays, labor, and the silver reclaiming unit. The remainder may be considered profit, but to this can be added the amount saved in reduced costs for fixer.

The various types of silver collection units range from simple, electrolytic collectors, to automatic, central systems for large, continuous operation. Electrolytic devices may be obtained on rental for $ 5.00 to $ 20.00 per month, and 50 per cent of the value of the silver collected is returned from the reclaiming company.

A continuous, paper-processor electric system is available on a deposit basis for under $50.00. The supplier of this unit replaces worn parts on a guarantee-maintenance plan. Depreciation is charged at the rate of $ 10.00 per year. The unit, operating on the electroplating principle, plates out the silver on stainless steel sheets which are sent in for a 50 percent return of the salvaged silver. New stainless-steel plates are furnished at no charge.

SPLICING. It is usually impossible to predetermine how much paper will be used to record a test, so common practice is to be safe and to use a full load. The paper cost does not warrant using a short roll and running out of paper before the test is finished. A full magazine gives an adequate safety margin, but at high cost, because the short ends left over are wasted. It has been estimated that this loss may amount to 25 per cent of the total paper purchased. Since the cost of photographic paper is substantial, a tangible saving can be made if short ends can be utilized, and they can be if they are spliced together into full size rolls.[10.6] Thus, the left-over material from tests that did not use a full roll will not be wasted. The splicing can be done by present personnel after a brief training period, and the operation can be fitted into slack work periods. Other advantages are that a supply of full-length rolls can be obtained quickly when a paper order does not arrive on time; longer than normal length rolls can be made up for abnormally long recordings; and paper can be put on other size cores.

Any number of splices can be made in a roll, and if properly made by trained operators, no failures will occur in either the recording or the processing operations. The requisites for a successful splicing operation are 1) a precision splicer; 2) a competent operator, and, 3) the correct splicing tape.

The following rules may seem obvious, but if ignored can result in failure:

1) Only the same kinds of material should be spliced. Preferably, this also should be the same emulsion number, and from the same shipment. Material should be date stamped when it is received so that its use will not be haphazard; for example, it might be used on a first-in, first-out basis.

2) Splicing must be done under proper safelight conditions.

3) Room cleanliness and operator cleanliness are essential.

It has been estimated that if 8 or more 12-inch rolls or paper are used per week, that the cost of a splicer will be amortized in one year. A splicing operation has a capability of returning $ 150 to $ 200 for each hour of operation.

Improving Communications and Knowledge. Knowledge can be developed through communication with others who are pursuing similar endeavors. Through discussions and the sharing of ideas and experience, progress can be made in any field, photo-oscillography included. It is important for personnel operating the recording equipment to comprehend the basic procedure followed by those who process his records. It is equally significant for processing laboratory personnel to understand the problems associated with recording dynamic data. Both should work toward the common goal of providing the best possible oscillogram to the department that must read and reduce the information from the record.

In addition to promoting an interchange of ideas and experience within the company, the department manager (and his operating personnel) should avail themselves of opportunities that exist outside the firm to foster knowledge in this field. A common meeting ground exists in technical societies where persons engaged in similar professions join together for mutual advancement of their respective areas. In addition to several well known electronic societies, there are a number of professional photographic and instrumentation societies that are concerned with recording, processing, and display problems. Members gain knowledge from association with others at monthly meetings and technical conferences, from lectures and field trips, and from technical journals and publications. Many companies subsidize membership in professional organizations by reimbursing their employees for dues, conference expenses, and other society fees. Members are encouraged to present technical papers at conferences and for publication in society journals.

LITERATURE REFERENCES

[10.1] HEPNER, H. W., *Perceptive Management and Supervision*, Prentice-Hall, Inc. (Englewood Cliffs, New Jersey) 1961

[10.2] *Storage and Preservation of Motion Picture Film*, Eastman Kodak Company, (Rochester, New York) 1957

[10.3] HOADLEY, H. W., "Quality Improvement of Ranger 1 Oscillograph Records," *Priv. Comm.* report for Jet Propulsion Laboratory, California Institute of Technology, (Pasadena, California) May 12, 1961

[10.4] DUISENBERG, C. E., "Silver Recovery and Prolongation of Fixing Baths," *Journal of the SMPTE*, Vol. 65, No. 8, p429, August 1956

[10.5] SCHREIBER, M. L., "Present Status of Silver Recovery in Motion-Picture Laboratories — A Tutorial Paper," *Journal of the SMPTE*, Vol. 74, No. 6, p505, June 1965

[10.6] KEIDATZ, R., *Oscillograph Paper Splicing*, Rocketdyne Division of North American Aviation, Inc., (Canoga Park, California) 1961

XI READING AND DUPLICATING OSCILLOGRAMS

READING AND REDUCING DATA. Oscillograms provide quantitative analog data. This information is presented non-numerically and shows in a graphical, descriptive way what has occurred. Usually, digital data are necessary for engineering calculations, and can be obtained by reading points of the analog record to obtain numerical values. The reduction of the analog data to digital form yields quantitative data that can be used in calculations to obtain accelerations, velocities, stresses, displacements, flow, and other information.

Reading can be accomplished in several ways, with manual or semi-automatic equipment. In manual reading, the points read may be hand written in tabular form and the figures thus obtained may be used in subsequent calculations. However, if either the volume or the time schedule demands faster reduction of the data, automatic equipment becomes desirable. In addition, it is possible to feed the digital data into computer equipment to obtain final information.

The following sections describe several types of manual and semi-automatic reading equipment available today.

Chart Viewers

SANBORN CHART VIEWER. The Sanborn Model 276 motor-driven Chart Viewer, illustrated in Fig. 11.1, is designed for the convenient analysis and editing of oscillographic recordings. Strip chart recording paper, such as Sanborn 2-to-8-channel Permapaper, up to 16 inches wide and in rolls up to 200 feet may be viewed. Other chart papers up to 16 inches wide, having a maximum diameter of 4 1/2 inches, may also be viewed if wound on the hexagonal cores supplied with the viewer.

The motor provides continuously variable speeds from 3/4 inch per second to the rewind speed, 200 feet in two minutes. One knob controls both speed and direction.

The viewing table is 18 1/2 inches by 24 inches, and is finished in black to give the best viewing surface for translucent Permapaper. A transparent plastic cursor, attached to the table, slides left or right on a straight edge which can be adjusted for accurate alignment with the chart.

Chart weave is limited to 1/16 inch by guide flanges which adjust to the width of the particular chart paper being analyzed. An automatic braking system, built into the drive unit, prevents overrun of the paper supply roll when the chart speed is suddenly reduced.

The only controls are the power switch and the speed and direction control.

Overall dimensions are 36 inches long by 22 1/2 inches wide by 7 1/4 inches high; weight is 32 pounds. The frame is finished in smooth gray and all exposed hardware

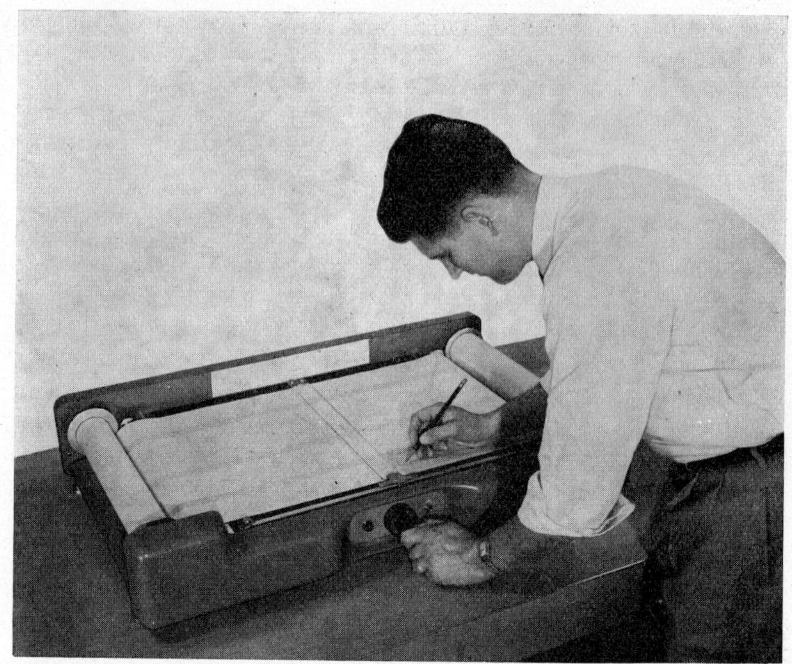

Fig. 11.1. Sanborn Model 276 chart viewer

Fig. 11.2. Houston Fearless viewing table

346

is either chrome-plated or buffed aluminum. Power requirements are 115 volts, 60 cycles, approximately 50 watts.

HOUSTON FEARLESS VIEWING TABLE. The Houston Fearless viewing table, shown in Fig. 11.2, was originally designed as a high quality interpretation device for the photogrammetrist. It will accept aerial films up to 9–1/2 inches wide. (wider film or paper models are available as custom equipment).

The top assembly, which contains the light well, motor drive and control panel, may be inclined from horizontal to a 25 degree angle for convenience of operation. The viewer has a variable-intensity light source and a mask under the panel can be adjusted to confine light to various film widths. Transport speed can be varied from continuous stall to fast forward or rewind (0–250 fpm).

Several variations of the basic viewer are available. One model features dual-path viewing for comparing two reels of film.

Accessories include a convenient film splicer and a zoom microscope.

Semi-automatic Readers

SEMI-AUTOMATIC DATA REDUCTION. In the typical semi-automatic reader-to-data link, the oscillogram, either paper or film, is threaded across a viewer, and the operator positions a set of crosshairs on each point to be read. Distances along two

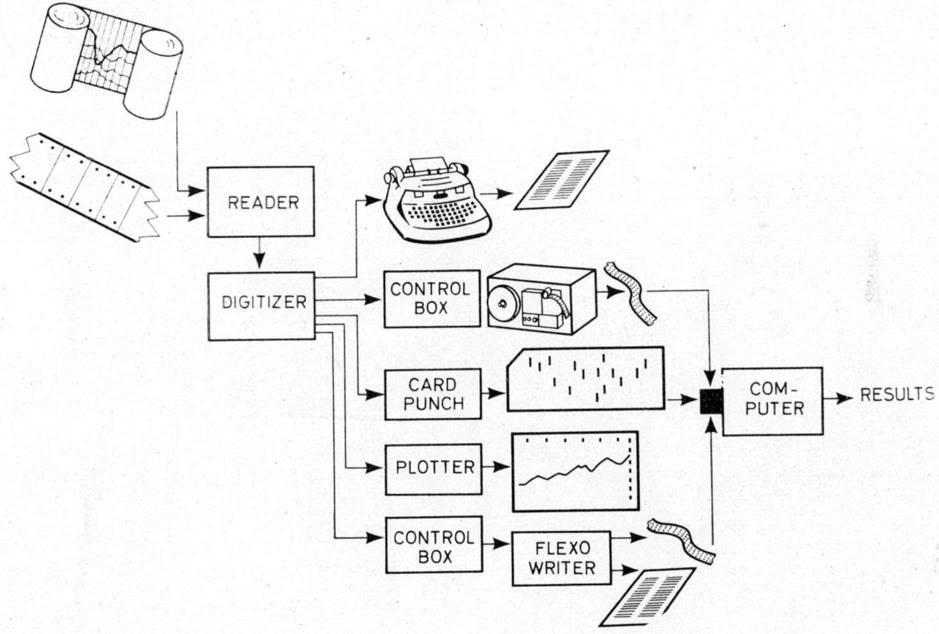

Fig. 11.3. Flow chart showing sequence of semi-automatic oscillogram reading

axes are electrically measured from fixed references, and the electrical signals are fed into a digitizer. Here, the voltage or pulses are converted into a digital output that may be read out, typewritten, punched, or otherwise stored. Fig. 11.3 illustrates symbolically and in block form, the usual sequence of the procedure.

BENSON–LEHNER READERS. Several semi-automatic models of analog-to-digital record readers are produced by the Benson–Lehner Corporation, 14761 Califa Street, Van Nuys, California. The readers, sold under the tradename "OSCAR," will accept records up to a nominal 12-inch width, apply a linear or non-linear calibration, convert data to engineering units, and automatically operate a typewriter and a keypunch. An optional film projector accessory will accommodate 16/35/70mm films.

Oscar Model S-2. This semi-automatic system shown in Fig. 11.4, is composed of the reader and the control unit containing all necessary mechanical and electronic components.

The measurement system utilizes two perpendicular cross hairs, located in front of the viewing screen, for coordinate measurement. The cross hairs have independent hand-operated controls for each axis. Other controls permit the alignment of both cross lines to the axis of the record.

A pulse-generating photoelectric encoder is directly coupled to the drive-system which controls the movement of each cross hair. The generated pulse trains are accumulated in a solid state bidirectional electronic counter. This counter contains special circuits for reversing both the sign and the direction of counting when the

Fig. 11.4. Benson-Lehner Oscar Model S-2 reader and control unit

348

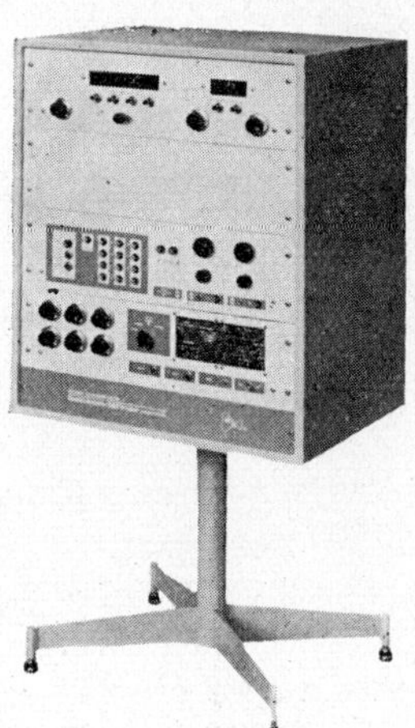

Fig. 11.5. Benson-Lehner Oscar Model F reader and decimal converter

zero point is crossed so that, under all conditions, the counter indicates the displacement of the cross hair from zero, in thousandths of inches, with appropriate sign.

Digital output is provided on the basic Oscar Model S-2 to an IBM 024/026 Keypunch and to a Benson–Lehner Electrotyper Model C. Alternate outputs are available to any IBM Summary Punch or to punched tape in any standard tape code. As an optional feature, readout to magnetic tape is also available.

The position counters and all other outputs are fully buffered so that the operator may proceed immediately with the next reading without waiting for the output punch to complete its full readout cycle.

Oscar Model F. Records up to 12–1/2 inches wide are accommodated on the reader of this two-unit machine shown in Fig. 11.5. The manually-positioned reading head furnishes a resistance output proportional to the calibrated amplitude. A typical connection to the second unit, the Model F Decimal Converter is shown in block diagram, Fig. 11.6. Its basic circuit is a ratio bridge designed to sense the position of the input potentiometer and convert it to decimal form.

Oscar Model K. This single console integrated data reading system, shown in Fig. 11.7, is composed of a record transport, viewing unit, reading head, control and display panel, and digitizing translator and output control. All of the equipment is enclosed in a console requiring less than 9.5 square feet of floor space. The machine will handle strip, folded, or rolled records transported across a lighted viewing area by a finger tip controlled motor drive.

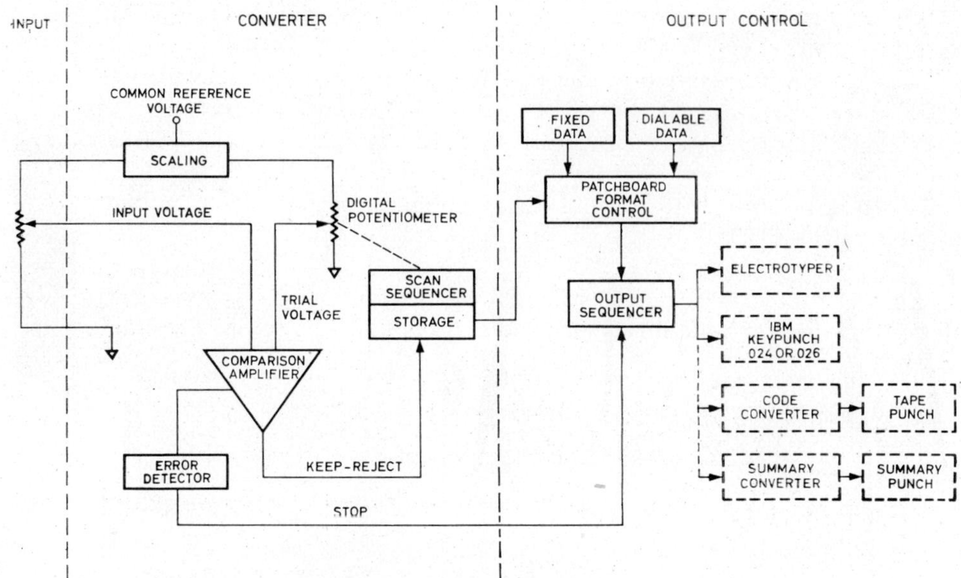

Fig. 11.6. System Diagram of Benson-Lehner Oscar Model F system

Fig. 11.7. Benson-Lehner Oscar Model K reading system with Friden tape punch

The record viewing area is illuminated by four fluorescent tubes behind the screen. Light intensity can be selected by a control.

The reading head, containing the readout pushbutton, sign selector switch and control knobs, travels on a traverse bar just below the viewing screen.

The digitizing-translator consists of a coded commutator disc and associated circuitry to convert position information to decimal digits for the output. The output format is controlled with a plugwire patchboard that permits an infinite variety of output formats.

Datareducer Type 099 Reader. The 099 oscillogram reader shown in Fig. 11.8, is used primarily for wide films and papers. It contains the record illumination and

Fig. 11.8. Benson-Lehner Datareducer Type 099 oscillogram reader

measuring system, and measures trace amplitudes on transparent, translucent or opaque records. Multi-channel records with linear or non-linear calibrations and orthogonal or curvilinear coordinates are ideally suited to this system.

The records are back-illuminated by four fluorescent tubes located directly beneath a finely-ground, 1/4 inch thick, glass screen. Controls are provided to vary the intensity of the light to suit each record and the individual operator. The width of the lighted area may be adjusted to fit the width of the record being read.

The measuring system is a group of components attached to the main frame of the reader. The principal elements in the measuring system are a measuring arm and a skewing frame. The measuring arm carries a cursor on a carriage which can be moved vertically along the length of the arm to measure translation in the Y direction. The measuring arm itself is supported on a carriage which can be moved horizontally along a way (part of the skewing frame) to measure the X direction. The arm can be raised to insert records. It can be rotated 5° from a position normally perpendicular to the horizontal way. This rotation feature is used in special reading applications, during reading operations using special templates, and also to compensate for record wander and occusionally to correct for base-line drift.

During normal reading operations, the motion of the X and Y axes is controlled by two handwheels, one for each axis, which are located in convenient, specially designed hand rests at the right and left of the screen base.

This system in its standard configuration has digitizing equipment for only one axis. However, a switch at the operator's fingertips allows switching of the digitizer

Fig. 11.9. Benson-Lehner Type 17C reader

from one axis to the other. Pressing the switch will connect the digitizer to the X axis. Simultaneous digitizing of both axes may be accomplished by adding the type 282 E Telecordex to the system (see below).

The skewing frame carries the measuring system and provides a method of aligning the measuring system to the record to be read. After the record is placed on the light table, the rotation of two controls allows the skewing frame to bring the measuring system into perfect alignment with the record.

Benson–Lehner also makes the Type 17C Reader shown in Fig. 11.9. This instrument will accommodate paper or film records from 16 mm to 12 inches wide, with or without sprocket holes. Magnifications of 2X, 4X, and 10X are provided.

Telecordex Digitizing System. The model 282E Telecordex, shown in Fig. 11.10, is an all electronic solidstate digitizing system whose output can be programmed for automatic recording in typed-list, perforated-tape, or punched-card forms.

Companion reading units for the Telecordex are equipped with magnetic reading heads. When the head is moved, its rotating shaft generates electrical pulses which

Fig. 11.10. Benson-Lehner Telecordex 282 E Digitizing System

accurately measure the amount of rotation. The number of pulses, representing the head position in digital form, is accumulated in the Telecordex by bi-directional electric counters.

The digital position conversion is displayed visually on inline (Nixie) readouts and is also stored in a load driving memory for electrical readout. Upon actuation of the electrical readout switch, the last digital entry is locked in the memory and may be applied to a wide variety of readout recorders without intermediate amplification.

Direct drive to a summary punch or parallel input printer is characteristic. By scanning, drive can be supplied for an electric typewriter, tape perforation, or keypunch.

GERBER MODEL S–2 OSCILLOGRAM SCANNER. The Gerber Oscillogram Scanner Model S–2, shown in Fig. 11.11, was designed for "quick-look" and scaling of oscillograms, films or strip charts. It is a complete, self-contained, electrically driven, variable speed transport system with a backlighted viewing area 24 inches in length.

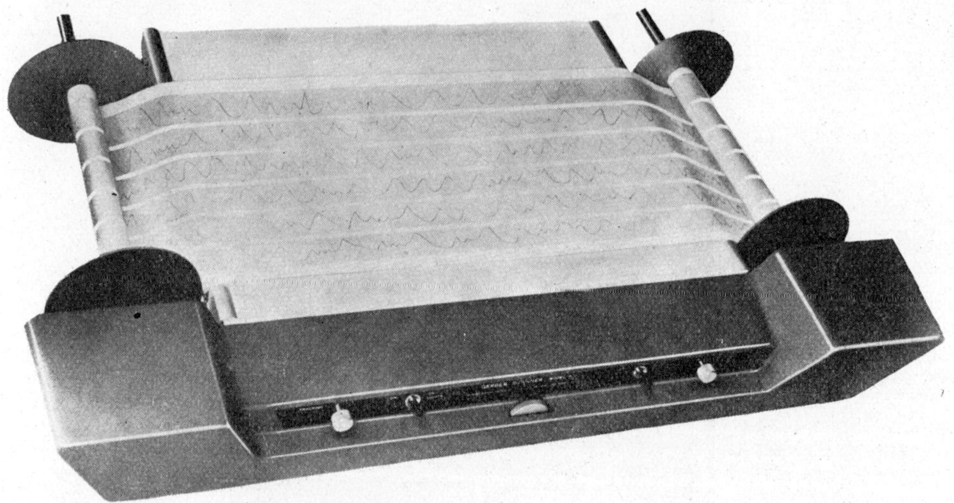

Fig. 11.11. Gerber Model S-2 oscillogram scanner

It can accommodate records up to 12 inches or 16 inches wide, depending on the model.

The drive system of this instrument consists of two independent spindles, one at either end, each containing a variable drive assembly. The spindles are cantilevered to permit rapid loading and unloading. Each spindle assembly has an adjustment knob to adjust the tracking of records. With this system, any type of record even damaged ones, can be transported and tracked very easily and efficiently.

The record speed can be varied from 0 to 500 feet per minute. Core enlargers can be obtained at an additional charge to fit any core diameter larger than the 5/8-inch standard size. A simple lever switch drives the paper in either direction and when released will stop the record instantly.

GERBER DIGITAL DATA READER. The Gerber Model GDDRS–3 Digital Data Reader, is illustrated in Fig. 11.12. This system has the capability of reducing to digital information thousands of points of data per day that can be translated to various outputs, such as typewriter, IBM key punch (024 or 026), puched paper tape, adding machine, Flexowriter, etc. The system features a back-lighted motorized paper transport, facilities for measuring on the X and Y axes, channel counter, time index counter, keyboard, and various outputs.

Two models are available: the GDDRS–3B-1 and -2. The GDDRS–3B-1 is a single encoder system, with the encoder transferable to either the X or the Y axis

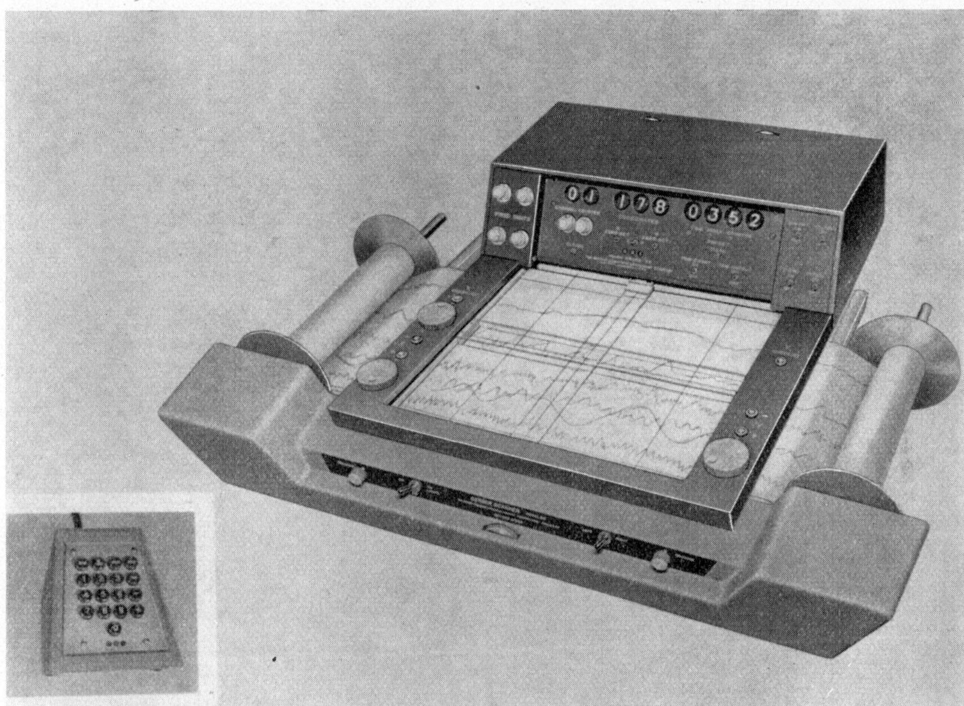

Fig. 11.12. Gerber digital data reader, Model GDDRS-3B mounted on a 16-inch S-2 scanner, with keyboard inset

by means of electric clutches. A second encoder may be added to this system at a later date without major modifications to the unit.

The GDDRS–3B-2 contains two encoders so that it is possible to read displacements simultaneously on the X and Y axes by one setting of the crosshairs.

The complete system consists of a scanner (similar to the Model S–2 already described), the reading head, inserts for scaling, time index counter, channel counter, keyboard, fixed digits, outputs to the user's **IBM** key punch and typewriter, patch board, and encoder invert switches. Numerous optional accessories are available to broaden the capability of the unit or to adapt it to specific needs.

GERBER ANALOGUE DATA REDUCTION SYSTEM. The Model S–10 Scanner, shown in Fig. 11.13, has a flow-film projector capable of handling 16–35–70 mm film. The image can be projected to a maximum of 16 by 16 inches on the reading surface.

With the X–Y reading head, program unit, and outputs, film as well as paper records can be analyzed.

The system features 15 channel capacity; variable scale factor and "0" locations; X, Y, or frequencies; program selection for each channel; continuous X reader; event counter; console; four-place analogue to digital converter; projection lamp-bank readouts; output to IBM typewriter, card punch, Flexowriter; plotter, etc.

Fig. 11.13. Gerber Model S-10 analogue data reduction system

THE VANGUARD MOTION ANALYZER. The Vanguard Motion Analyzer, is a rear-projection viewer with a crosshair assembly for measuring coordinate positions on the screen plus suitable means for advancing, frame counting and registering film. Various other design features are included to make rapid and accurate data reduction practical. Several models are available that accommodate film ranging from 8 mm to 70 mm in width.

Measurements are read off dials and are in inches, tenths, and thousandths of an inch at the screen image. The projection head may be rotated to align the image at any angle. The projected image includes the sprocket holes and film edge on the 8 mm and 16 mm models so time marks may be observed.

Means are provided for correction of the image shift due to subject shock, faulty registration, or other causes. The standard Motion Analyzer projection head does not include motion picture film advance and shutter except on the M–16C Model for 16 mm film.

The Motion Analyzer consists of a projection head, which provides the film handling and projection system; and a projection case, which provides a viewing

356

screen and a crosshair measuring system. An angle measuring screen can be supplied in place of the standard square screen, which will allow angular plus the usual coordinate measurements to be made. A tangent screw at the projection head mount can be supplied for fine positioning of the projection head (and image) rotation.

Automatic readout equipment for the Vanguard Motion Analyzer is furnished by the Coleman Engineering Company, Inc. Various models are available to feed data into a Clary Printer, IBM Summary Punch, or IBM Key Punch. In a typical application, the operator manually positions the Motion Analyzer crosshairs on the point being measured in the image on the viewing screen. The operator then actuates a switch which causes the data to be electrically read and fed to the Clary or IBM equipment.

Automatic Readers

AUTOMATIC ANALOG-TO-DIGITAL CONVERSION. With the marked increase in the use of high speed digital computers in all phases of industry has come the question of how to instrument best a project to allow the most efficient use of the computer system in the analysis of data. There has been a flurry of activity associated with the development of direct-recording digital transducers and of more sophisticated analog recording methods with later digitizing steps if desired. The fact remains that direct digital recording and analog tape recording instrumentation is far more expensive than the older paper chart or oscillographic recorders.

There is much evidence in several areas of technology that a tremendous amount of useful data could be obtained from paper charts recorded some years ago. In other areas, recent activities have resulted in a large backlog of records which are now scheduled to be analyzed by manual methods. The analysis of these data by modern computing methods requires a digitizing function to be acceptable to the computer system input channels.

UGC INSTRUMENTS ELECTROSCANNER. The Electroscanner Strip Chart Digitizer is manufactured by UGC Instruments, Inc., a subsidiary of United Gas Corporation, Shreveport, Louisiana. The instrument is designed to convert analog paper chart to digital form automatically. The use of the equipment eliminates the necessity for the operator to pick points on a curve or to perform other tasks usually associated with manual reduction of chart data. All charts are read at the same high rate of speed whether they contain one trace or fifty traces. The reading of all traces is accomplished on a single pass through the instrument. The application of the Electroscanner to chart data reduction problems results in a marked decrease in the cost per data point; a decrease which now makes practical the reading of chart data in quantities heretofore considered too monumental to be attempted with manual or semi-automatic reading devices.

The Electroscanner, shown in Fig. 11.14, measures 66 inches high, 36 inches deep and 60 inches wide including the protruding portions of the chart drive mechanism. The cabinet houses in its upper half the mechanical devices for transporting the paper chart through the reading station and for supporting the optical-scanning system and its associated electronic equipment. The lower half of the unit contains the electronic package which includes the digitizing counter and the necessary equipment to transfer the digital data to either a small computer or to an offline

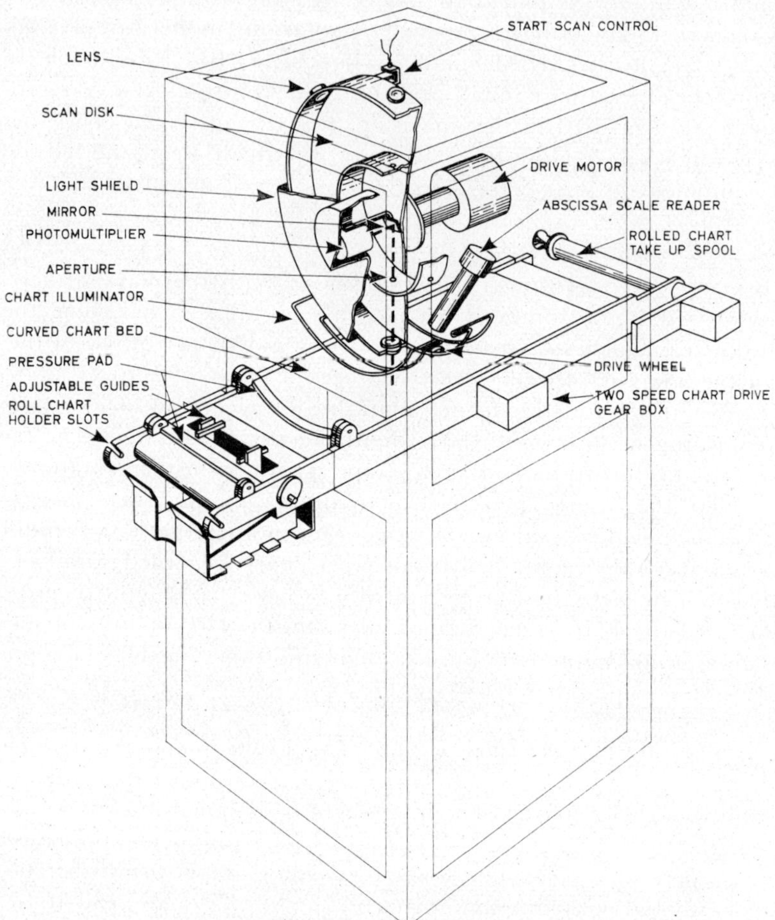

Fig. 11.14. Artist's simplified conception of electroscanner strip chart digitizer

magnetic tape recorder which writes in a computer readable format. In the event that the user's requirements call for an off-line tape write operation, a transport unit of computer-compatible specifications must be added to the system and housed in a separate cabinet. The physical size of this cabinet will be determined by the selection of the actual equipment, which is available from any of several manufacturers.

The chart, whether it be folded or rolled, is placed in the chart receiving structure, shown at the left in the illustration. A manual control lever (not shown), when actuated, lifts the two pressure pads and the pressure roller from the chart bed and drive wheel. As the chart is threaded under the first pressure pad, the chart guides are adjusted to the approximate width of the paper. The chart is then threaded under the second pressure pad and sufficiently far into the assembly so as to come between the chart drive wheel and the pressure roller when the manual control lever is actuated. The chart passes through a transition section that gradually curves the chart until the correct axis is obtained at the reading section. After being read, the curved chart is gradually flattened toward the output side. A motor-driven takeup spool is provided for rolled charts. Folded charts are received in a removable basket.

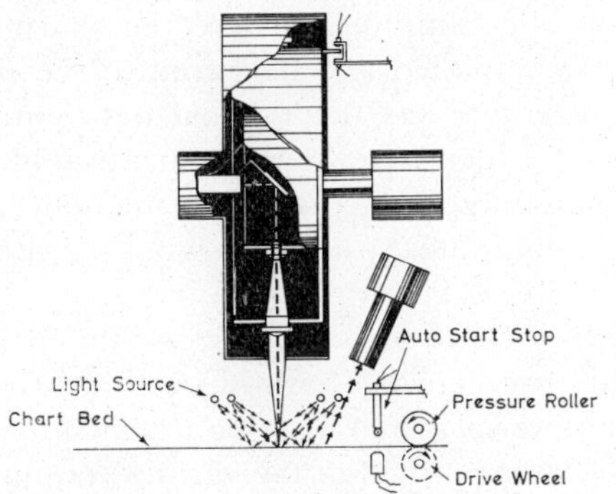

Fig. 11.15. Schematic of electroscanner scanning head

Charts to a maximum width of $12\frac{1}{2}$ inches are accepted by the drive assembly of the Electroscanner. A multiple-speed chart drive is available as an option.

In the Electroscanner, an optical device converts data from an analog form to a digital representation from previously recorded paper charts or photographic film. An optical scanning system mounted on a rotating disk, shown in Fig. 11.15, locates the position of each line on the chart by measuring the surface reflectivity of paper charts and the light transmission of negative film records. A digitizing counter tracks the position of every lens as it scans the chart. Trace signals cause the value of the digitizing counter to be transferred into a suitable buffer store from which the data may be delivered to on-line processing equipment of off-line mass storage media.

The orientation of the rotating optical scanning system with respect to the chart transport assembly is shown in Figure 11-15. The supporting structure for the optical system consists primarily of a large disk to which are attached two concentric circular rings. A suitable drive motor is provided to rotate the entire assembly at a constant speed.

Six lenses are equally spaced about the outer periphery of the large ring. The smaller ring carries an aperture and lens assembly for each lens on the outer ring. At the center of rotation of the scanning disk assembly a first-surface mirror is mounted at a forty-five degree angle with the vertical. This mirror directs the light which passes through the aperture on to the sensitive face of a photomultiplier tube mounted on the axis of rotation of the scan disk. This complete assembly is housed in a light proof shield. The chart is illuminated by a special gas-filled discharge tube located just above the chart bed.

As the lens system on the scan disk swings through its arc above the chart, the reflected light from the chart is focused by the lens on to the aperture associated with it. Thus, an image of unity magnification of the chart is presented to the aperture. The light which passes through the aperture is directed to the photomultiplier tube via the mirror. The photomultiplier tube monitors the light reflected from the chart on each scan. The small electrical signals produced by variations of the reflected light are delivered to a preamplifier. There the sharp pulses representing the scanning of the traces are amplified and delivered to the electronic package.

The inclusion of such features as the manual data entry, automatic start-stop control, and abscissa scale reader increase the utilization of the system by providing automatic methods of handling time consuming manual tasks.

Data are delivered to a computer either by an on-line connection or by way of a movable mass storage medium. The Electroscanner system is available in configurations which may be attached to a computer system for on-line operation. The general approach to this problem is to allow the Electroscanner to simulate the characteristics of an input device which may be attached to a channel of the computer. A stored program, written with all of the various timing considerations in mind, controls the transfer of data. Appropriate additions are made to the hardware described herein to provide the necessary control and buffer storage.

INFORMATION INTERNATIONAL AUTOMATIC FILM READERS. The Programmable Film Reader developed by Information International, Inc., 200 6th Street, Cambridge, Massachusetts, is a means of automatically reading and digitizing very large quantities of photographic data. Data currently recorded on media other than film, such as paper strip charts, can also be microfilmed for automatic reading and convenient permanent storage. The system is operated completely under computer control and does not require a human operator. Film is read at the rate of approximately 5000 data points per second. Data output may be recorded in digital form on IBM-compatible magnetic tape for further computer processing and analysis. Other forms of output are also available.

The film reading process involves the selective scanning of film by a rapidly moving, programmable light point on a visual display cathode ray tube. The output of this scanning operation is detected by a photo sensitive device in the film reader and relayed to a scan control and monitoring unit for further processing and analysis.

So-called "flying spot" scanning techniques scan an entire display raster and read and store the resulting data for all points on the raster. A large segment of computer memory is required to hold this raw data; in addition, extensive computer processing is then necessary to extract the significant data.

The Programmable Film Reader, on the other hand, is controlled by a stored computer program and is based on locating and tracking only the data of interest on the film. No further processing is required; the significant data are immediately available as output of the film reading system. "Noisy" data, data superimposed on grid backgrounds, and other complex types of film data may be read effectively by means of special film reading computer programs developed for use with the Programmable Film Reader.

III PFR–1 PROGRAMMABLE FILM READER

Basic Film Reader. The basic film reader contains a bi-directional transport for holding and stepping the film; and optical and electronic systems for processing film reading signals. It is composed of an optical-mechanical unit and a signal processing and logic unit.

The Optical-Mechanical Unit. Light coming from the CRT light-source enters the optical-mechanical unit of the basic film reader, Fig. 11.16. There it is divided into two beams by an optical beam splitter. One beam enters a primary optical system, passes through an enlarger and field flattener lens, and is focused on the film to be read, at a specified location corresponding to the location of the programmed light source on the CRT. The light transmitted through the film at this location passes through a set of condenser lenses, is defocused, and is sensed on the cathode of a photomultiplier tube.

The second of the two beams enters a reference system, which does not include the film being read, passes through a neutral density filter, is similarly processed, and is sensed on the cathode of a second photomultiplier tube. The signals from the photomultipliers in both systems are amplified and compared in the signal processor and logic unit of the basic film reader. Resulting digital signals are then transmitted to the scan control and monitoring unit for further processing.

A projection subsystem is also included in the optical-mechanical unit. This subsystem is not used during the film reading process itself, but rather is used as an aid in setting up film reading activities.

The optical-mechanical unit reads film with sprocket holes, either framed or continuously exposed. Film is automatically stepped under control of the Scan Control and Monitoring Unit. Three types of film transport and objective lens units are available, adapted to various sizes of film to be read: 35 mm (Type 1030), 16 mm (Type 1031), and 70 mm (Type 1032). The film step time is less than 5 milliseconds for a complete step in either direction. Transports are available with step distance as small as 1/4 of the sprocket to sprocket distance, and as much as a single frame (4 sprocket holes in the case of 35 mm film). Controls include image focus, image size, and orientation angle of the film being viewed. Film motion may be controlled automatically or manually, and modes include fast forward or rewind, and variable speed stepping or single stepping in either direction.

The Signal Processing and Logic Unit. The main function of the signal processing and logic unit is to process and convert analog signals from the optical-mechanical unit to digital signals compatible to the scan control and monitor unit (described below); and to process and transmit signals from the SCMU for control of the OMU. Specifically, the Signal Processing and Logic Unit accepts input from the Optical-Mechanical Unit, as described above; measures density; and makes the decision whether the film at a given point is "denser" or "less dense" than a given comparison standard.

Programmable CRT Light Source. The standard cathode ray tube light source has a display area of 9 $\frac{3}{8}$ inches by 9 $\frac{3}{8}$ inches containing 1024 by 1024 points. Upon designation by the SCMU of a specific X, Y coordinate pair, or sequence of selected pairs, the CRT normally illuminates each given spot at a rate of one every 50 microsconds. Special options as well as a precision CRT are also available.

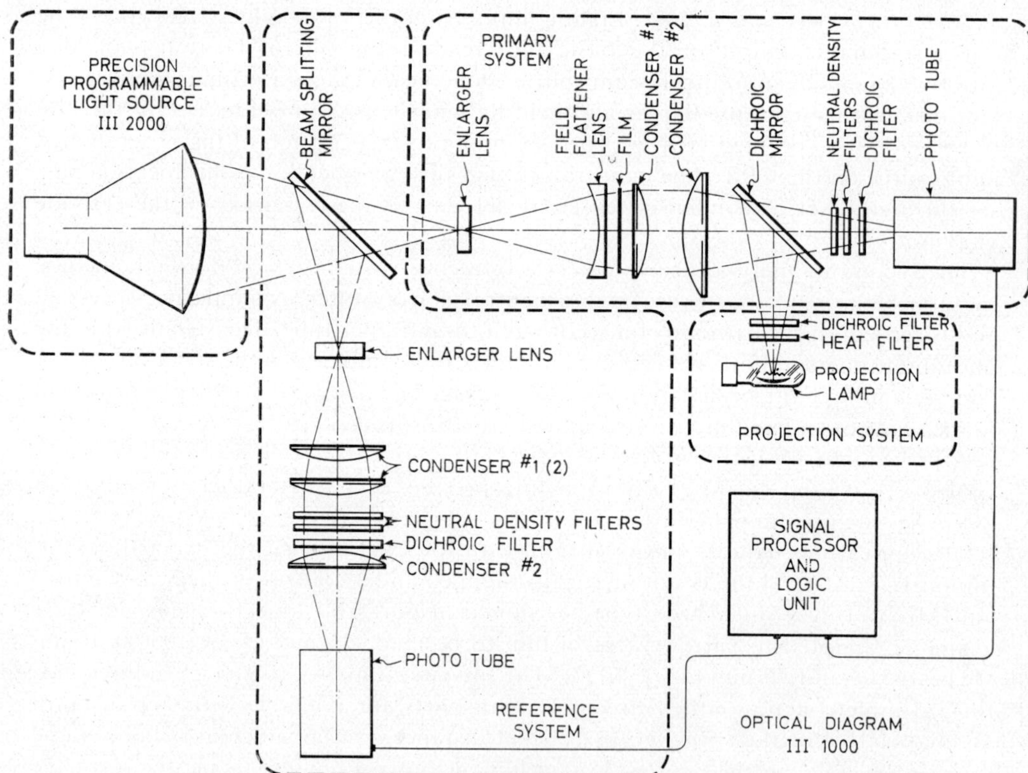

Fig. 11.16. Optical diagram of programmable film reader (Information International, Inc.)

Scan Control and Monitoring Unit. The SCMU is a high speed solid state digital device designed to operate the film reading system under the guidance of an appropriate program. It accepts information through input registers; controls the positioning of the light point; and controls the scanning actions and the film advance. Auxiliary equipment includes a magnetic tape transport and a magnetic tape control compatible with IBM tape formats.

FILM READING SERVICE. In cases where workload does not warrant purchase of a film reading system, III's film reading service is available for reading and digitizing film. Film may be mailed to the firm where it will be digitized in a suitable format, and recorded on IBM-compatible magnetic tape for further processing. Magnetic tape and film are then returned to the client.

Duplicating Oscillograms

REPRODUCING THE OSCILLOGRAM. The original oscillogram serves its prime purpose when its recorded data has been displayed and read. However, there are other functions involved in its efficient use, and nearly always, copies in one form or another become necessary for reports and reference purposes.[11.1]

The need to reproduce oscillograms exists primarily because 1) duplicate copies are needed; and 2) because reproduction in smaller size (microfilms) provides a tremendous savings in storage space. In addition, the microfilm has archival quality, which the original may or may not possess.

One solution to the problem of producing only one additional full-size copy of a record is the use of a dual-wound roll, to record two sets of equal density traces simultaneously. Widely used by offshore seismographic crews, the material in effect produced two originals: one on the extra-thin top layer, and a second on a "sandwiched" lightweight-stock bottom layer.

The obvious disadvantages of this method are that only one additional copy is immediately available, and that the recording must be made originally on special material.

DUPLICATING METHODS. Various duplication methods are shown diagrammatically in Fig. 11.17. Essentially, all reproduction methods may be divided into two

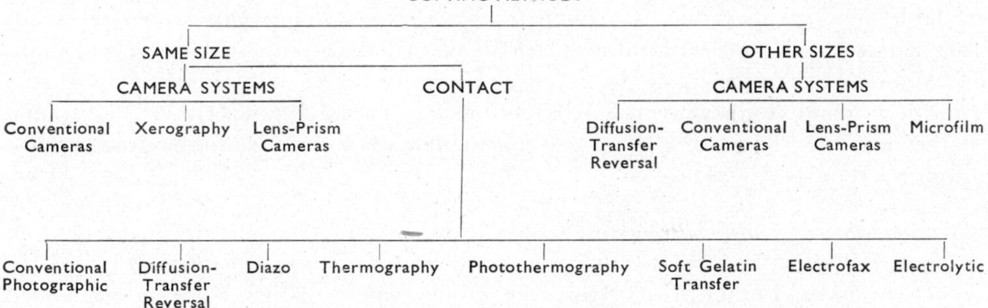

Fig. 11.17. Photoreproduction methods often used for oscillogram duplication

categories: same-size copies, and copies of a size different from the original. Same-size reproductions can be made by contact or with a lens system. However, if a size change is made, a lens system must be used.

Diazo Copies. The simplest reproduction method is a samesize duplicate made by a contact method. The one-to-one duplicating equipment, found most commonly in engineering offices, is the diazo machine. Diazo, sold under various trade names, is a dye-sensitized paper, exposed by an ultraviolet source and then dry developed in ammonia vapor. A positive image is obtained in one step; i.e., a black line yields a dark deposit on the copy. However, the duplicate print may be black, blue, brown, or some other color depending upon the type of the printing paper that is used. The process requires a high-intensity UV source, to which the original must be reasonably transparent. Therefore, if satisfactory prints are to be made, the original charts must be on film or on thin-base paper. Heavier base materials require slow running speeds, produce too much "background", and tend to obliterate lines. Unstabilized print-out records cannot be used to make duplicate diazo prints.

Generally, diazo materials are supplied in rolls much wider than oscillograms. If desired, narrow rolls of diazo paper can be ordered from the supplier, specially cut to match the width of the chart. Paper may also be cut to a standard width, 11 inches for example, to match the page height of standard bound reports and documents. Charts of smaller widths, various or odd sizes will then be standard height when bound into a report.

SILVER-HALIDE PAPER COPIES

Prism Camera. For many years, Photostat, Rectigraph, and similar cameras have been used in offices for copying all sorts of documents. Because of this widespread use, it was natural to turn to the prism camera for oscillogram reproduction. Itek Business Products, reports that its Photostat equipment has been used both for copying geological data, and for reproducing electrocardiograms. An example of a copy of a patient's record made on a Photostat machine is shown in Fig. 1.10, Chapter One.

Normal camera systems produce a negative that has a reverse (mirror image) on the emulsion side; therefore, the paper negative made by that method will not be readable. The prism camera, through its prism-lens system yields a one-step readable negative duplicate without a mirror-image negative. This saves time and money. A positive can be made by re-photographing the readable negative with the same equipment. Enlargements or reductions can be made from either the original or the negative. The size change is a particular advantage in duplicating oscillograms for reports.

Contact Silver-Halide Processes. Conventional contact photocopy processes can be used to duplicate oscillograms. In its simplest form, only a printing frame is necessary to make contact prints. However, modern methods make use of specialized machines for speed, convenience, and operation by unskilled personnel.

364

Some types of duplicators are sophisticated versions of the photographer's contact printer, some models featuring a vacuum lid. Continuous-roll printers, similar to diazo machines, but having a weaker light source, are also available.

Two types of contact prints may be made: 1) readable negatives from lightweight oscillograms; and 2) reverse (mirror image) negatives, which may be processed, dried, and printed to produce a readable positive.

To make a readable negative, the original is placed face up on the emulsion side of the silver-chloride copy paper. The original must be reasonably transparent so that traces are not obliterated by the paper texture. The sandwich is placed in the printer with the original facing the light source; the exposure is made; the sandwich is removed and separated; and the exposed, sensitized copy paper is developed in the conventional manner.

Reverse negatives are made in one of two ways. In one, the procedure outlined above is followed, except that the original and copy paper are face-to-face, i.e., emulsion-to-emulsion. In the other method, called *reflex* copying, the original and the copy paper are also face-to-face, but the sensitized paper is inserted in the printer with its back toward the light source. A shorter exposure is usually required for the reflex method.

Positives from either type of reverse negative are printed through the paper negative. The same method of printing is used for film negatives. Generally speaking, many of the photocopy papers are preferred over conventional photographic papers for duplicates, because they are lightweight, can be folded, and may cost less.

Autopositives. One-step positives can be made from original oscillograms, without any intermediate negative step, by using special autopositive materials. Such materials are made by Kodak, Anken, Ilford, and others. Both films and papers are available.

In general, papers may be handled in subdued, artificial room light, and can be exposed in the ordinary office diazo or blueprint machine. Special yellow filters, furnished by the paper manufacturer, are usually necessary with exposure-light sources of high intensity, or certain spectral-energy distribution (see manufacturers instructions). The filters are usually in the form of a yellow sheeting, interposed between the light source and the original.

Reflex autopositives can be made from opaque originals by following the manufacturer's instructions for exposure. Translucent originals can be reproduced by printing through the original. Following exposure, the material is processed by conventional photographic procedures. The finished autopositive can be used as a duplicate, or may be used as an intermediate to produce final prints by diazo or similar methods.

Incorporated Developers. Several silver-halide copy materials are available that have developers incorporated in the emulsions. Following exposure, such material can be processed in a simple one or two-step process. Table-top roller processors are available, which require little working space and a small volume of chemicals.

Soft-Gelatin Transfer Process. Generally known by its trade name "Verifax", this process utilizes a soft-gelatin printing matrix to transfer physically soft dyed gelatin to ordinary paper.

Verifax matrix material has a special emulsion containing silver halide, a tanning developing agent, and dye materials. Contact printing, and processing in an activator solution, causes the exposed areas to be tanned and hardened. Simultaneously, the unexposed areas (the dark traces of an original oscillogram) become soft, dyed-gelatin images. A readable positive copy is obtained by rolling the matrix into contact with ordinary paper; a thin layer of the dyed gelatin is transferred from the matrix to the paper receiving sheet at this point of the process. One to several additional copies can be made by reinserting the matrix in the activator, and then pressing it into contact with another receiving sheet. However, each successive copy will be lighter than the preceeding print.

Diffusion Transfer Reversal. A score of manufacturers throughout the world produce a wide variety of DTR document copy equipment. Today, even with electrophotocopying devices in common use some form of DTR equipment is found in almost every office, and because of this availability, is likely to be used to copy oscillograms.

The DTR principle has already been described in Chapter 5 as it applies to the Polaroid process. The early work of Rott, and independently, of Land, and Weyde, each with different objectives, laid the foundation for workers in their separate countries to explore many applications of the process. This led to the numerous types of equipment available. The chief differences between the Polaroid process and office copy system are that office machines use solutions, rather than the viscous chemicals in pods that are attached to the Polaroid film. Polaroid also has continuous-tone capability.

Thermography — Heat, rather than light, is used in the exposure step of the Minnesota Mining and Manufacturing Company Thermofax process. The original to be reproduced is placed in face-to-face contact with a colored wax-coated reproduction paper. The sandwich is inserted in the copy machine with the base side of the wax paper facing an infrared-radiation source. The infrared, freely transmitted by the wax, is absorbed by dark traces on the original. The dark areas of the original absorb the infrared energy creating a heat-pattern image which is conducted by contact to the wax paper. The wax is melted by the heat, and diffuses into the paper base, forming a reproduction of the original.

ANKEN POSITIVE-TO-POSITIVE DUPLICATING EQUIPMENT. The Anken Chemical & Film Corporation, 1 Hix Avenue, Newton, New Jersey, has developed the ANN-2 Printer and NN-2 Processor, shown in Fig. 11.18, to update methods used in making positive copies or diazo intermediates of all types of continuous roll documents.

Positive-to-positive prints completely eliminate intermediate negative-making steps. Ankopositive wash-off process on paper, film or Polyester base is used as the print-making medium. The base stock (there are six, including paper, triacetate and Polyester) is double-coated. The upper layer of the standard paper is a high contrast orthochromatic emulsion. After exposure, development makes the negative layer water soluble, while the unexposed and undeveloped silver (the image area) diffuses to the lower receiving layer, forming a positive image. A simple water spray in the NN-2 Processor automatically and completely removes the negative layer, leaving the finished positive print.

The completely automatic system consists of a continuous flow printer and a separate processing unit.

Anken ANN-2 Printer. The Anken ANN-2 Printer is a high speed contact printer designed to handle continuous roll originals. Exposure is made automatically on internally-stored rolls of Ankopositive which are available in roll sizes up to 500 feet by 12–1/2 inches, depending upon the type of original to be reproduced. A variable exposure drive system allows printing speed of from zero to 32 feet-per-minute, according to exposure setting used.

The dimensions of the printer are 38 by 23 by 12–1/2 inches. All rollers, side frames and base are heavy duty aluminum-alloy construction. The unit weighs 75 pounds. Power requirements are 115 v, 60 cycle ac, single phase.

Anken NN-2 Processor. The Anken NN-2 Processor is an automatic processing unit designed to handle exposed Ankopositive in continuous rolls of any size up to 500 feet long by 12–1/2 inches wide. Fast developing feature of Ankopositive (less than 60 seconds) allows processing speeds of up to 15 feet-per-minute. A one-step diffusion-transfer type developer is used for the processing cycle, eliminating the need and time required for a separate fixing operation. The NN-2 Processor automatically processes, dries and rewinds a roll of exposed print paper in one continuous operation.

Automatic Processing. The NN-2 processing cycle is divided into developing, pre-wetting-and spray-rinsing phases in sequential operation. The water spray removes the upper (negative) emulsion layer which was made water soluble in the developing stage. Remaining is a positive print. No further washing or fixation is necessary.

Because the unit is entirely self-contained, processing operations may be carried out under normal lighting conditions.

The dimensions of the processor are 60 by 23 by 60 inches, and the weight is 180 pounds. The machine requires 220 v, 60 cycle ac, single phase power.

Xerography. This dry, electro-photographic process is based on physical and electrical principles, rather than on chemical methods. Early work was done by a number of researchers in the mid- and late 1930's,[11,2] and was perfected in the 1940's by the Battelle Memorial Institute, Columbus, Ohio. The process is now marketed by Xerox in the United States, and by Rank-Xerox, Ltd., in England. Addi-

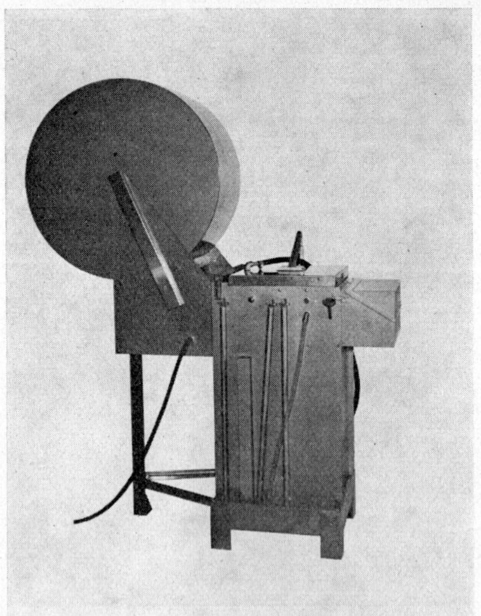

Fig. 11.18. Anken duplicating equipment. *Top*: ANN-2 printer; *Bottom*: NN-2 processor

tional development of the process has been conducted by the armed services of the United States and by the Ministry of Supply in England.

In the xerographic process, a selenium photoconductive insulating plate is evenly sensitized with an electrostatic charge. After charging, the plate is exposed in a camera system; the light action increases the electrical conductivity, reducing the charge density in the illuminated areas, and leaves an electrostatic latent image

Fig. 11.19. Xerox Model 914 Office Copier

corresponding to the original. The latent image is developed with fine-powder particles, which carry polarity charges opposite to the charge on the plate. The powder image is transferred to the final support, usually paper. The powder is fused by heat which fixes the image to the paper support. The small amount of powder remaining on the surface of the plate is brushed off, after which the plate may be used for the next copy.

Xerox 914 Office Copier. This machine, shown in Fig. 11.19, is an advanced automated device that accomplishes the copy functions described above in automatic sequence. Copies are reproduced in a 1 : 1 size, to a maximum format of 10 by 15–1/2 inches.

Electrofax. This electrophotographic process, developed by the Radio Corporation of America, differs from Xerography in one significant respect: the final

Electrofax print initially has a light-sensitive coating. The coating is photoconductive zinc-oxide, held to the surface of ordinary paper by an insulating binder, usually a synthetic resin. In general, the steps for sensitizing, exposure, and development follow the principles outlined for other electrophotographic processes.

The sensitivity of Electrofax is in the projection-printing range. The process has been used for enlargements from microfilm and for trace-recording equipment. A thin-window cathode-ray tube has been used as a high-speed printer for computer write-out. Electrofax lends itself to positive-to-positive or positive-to-negative prints, transparencies, and even color printing with pigmented developer materials.

CONVENTIONAL CAMERA COPIES. Copies made on the various office-copying machines described are adequate for many uses. Where the need exists for duplicates having quality superior to that obtainable from office equipment, copies should be made by camera-copying systems, using recommended materials and procedures. Firms having photographic departments generally have in-house copy-camera capability, or can send the chart to a vendor who can make the copy negative and duplicate prints. Large copy cameras are available so that 1 : 1 negatives and duplicate contact-prints can be made, in some cases, to a maximum length of several feet. If reproductions of longer sections of chart are required, individual portions can be photographed with some overlap, and the resulting negatives can be stripped (spliced) together. Continuous film positives or paper prints can then be made from the stripped negative. Printing is described in a separate section.

Continuous camera-reproduction machines are also available that can duplicate oscillograph charts in long lengths. The Neo-flo and Copyflo are typical machines of this type. The Neo-flo uses conventional photographic sensitized materials. Although a few 1:1 size Neo-flo cameras have been made, the standard unit produces a 50 percent size reduction. The Copyflo is a xerographic process, that produces enlarged duplicates from 4–1/2 inches to 12 inches wide (wider on some models).

If neither a large flat-top contact printer nor a continuous printer is available, the positive can be made by the step-and-repeat method on a smaller size contact printer. This requires extreme care, precise registration on the Y axis and accurate indexing on the X axis. In addition, the printer should have good contact across its printing surface, and the printing area should be sharply masked with exactly parallel strips of opaque paper, taped to the glass. A cardboard guide strip, at right angles to the masks, is attached to the glass with double-coated masking tape. The stripped negative and the unexposed photographic material, taped together at the head end, are aligned against the guide strip, the mask line marked, the platen lowered, and the exposure made. The platen is raised, the marked, exposed sandwich moved across the glass to the other mask, and the cycle is repeated. After all exposures are made, the exposed material is separated from the negative, and is processed.

EVALUATING THE OSCILLOGRAM IMAGE. The oscillogram trace ideal is considered to be a dense black line on a pure white background. In photographic parlance, such material is called "line copy" and is photographed on a very high-contrast emulsion, such as process or litho-type films. The films are then processed in high-contrast developers to yield a negative having a transparent line and a dense background.

Unfortunately, many oscillograms do not measure up to the typical black-line, white-background conception. Therefore, it is necessary to evaluate the record before deciding how to obtain the best quality reproduction from it. An understanding of the evaluation procedure can be obtained by studying Fig. 11.20. This illustration shows that because of varying frequency, trace lines are sometimes so close together that some high-contrast copy methods would render them as a single blob, obliterating the individual trace line. In other cases, at high writing speeds and amplitudes, or because of insufficient exposure from other causes, the trace will be so weak that it will be lost when copied. Other determining factors include the background, and whether it is white, gray, fogged, or stained. Print-out records present another variation: the trace is not black but blue, and the background is tan or pink in color.

Each record should be evaluated on its own merits; in fact, individual sections of a roll may require different treatment. Compensation can then be made for variations in trace density (due to change in record speed), frequency, or other differences.

Because of the extensive variety of oscillograms likely to be encountered, it is not possible to present precise copying rules. However, the following guide lines will assist in selecting the correct materials and procedure for copying.

Contact Printing: 1) The best diazo (or similar prints) are made from film positives. The second highest-quality duplicates are made from thin-base paper or vellum prints.

Regardless of the original, the greatest detail is preserved when the emulsion of the original faces the sensitized coating of the diazo paper. This condition can be achieved in the following manner: The original oscillogram should be copied on thin-base film. The resulting negative can be turned upside down on the contact printer, and a positive (either film or light-weight paper) can be obtained that is sharper than a print made in a similar manner from a heavy-base negative.

The finest detail can be preserved if a point light source is used for the printing exposure (even from heavy-base film negatives). The final positive can be run through the diazo printer, emulsion to emulsion, thus insuring the reproduction of fine detail. Film positives have excellent light transmission which allows very rapid printing.

2) If negatives are on heavy-base film, and/or a point light source is not available, the negative should *not* be "flopped"; instead use thin-base film for the film positive. Make the film positive emulsion to emulsion on diffusion contact printers, or emulsion to negative back if a point light source is used. The thin-base positive will reduce diffusion during diazo duplication, and thus produce crisp, sharp prints.

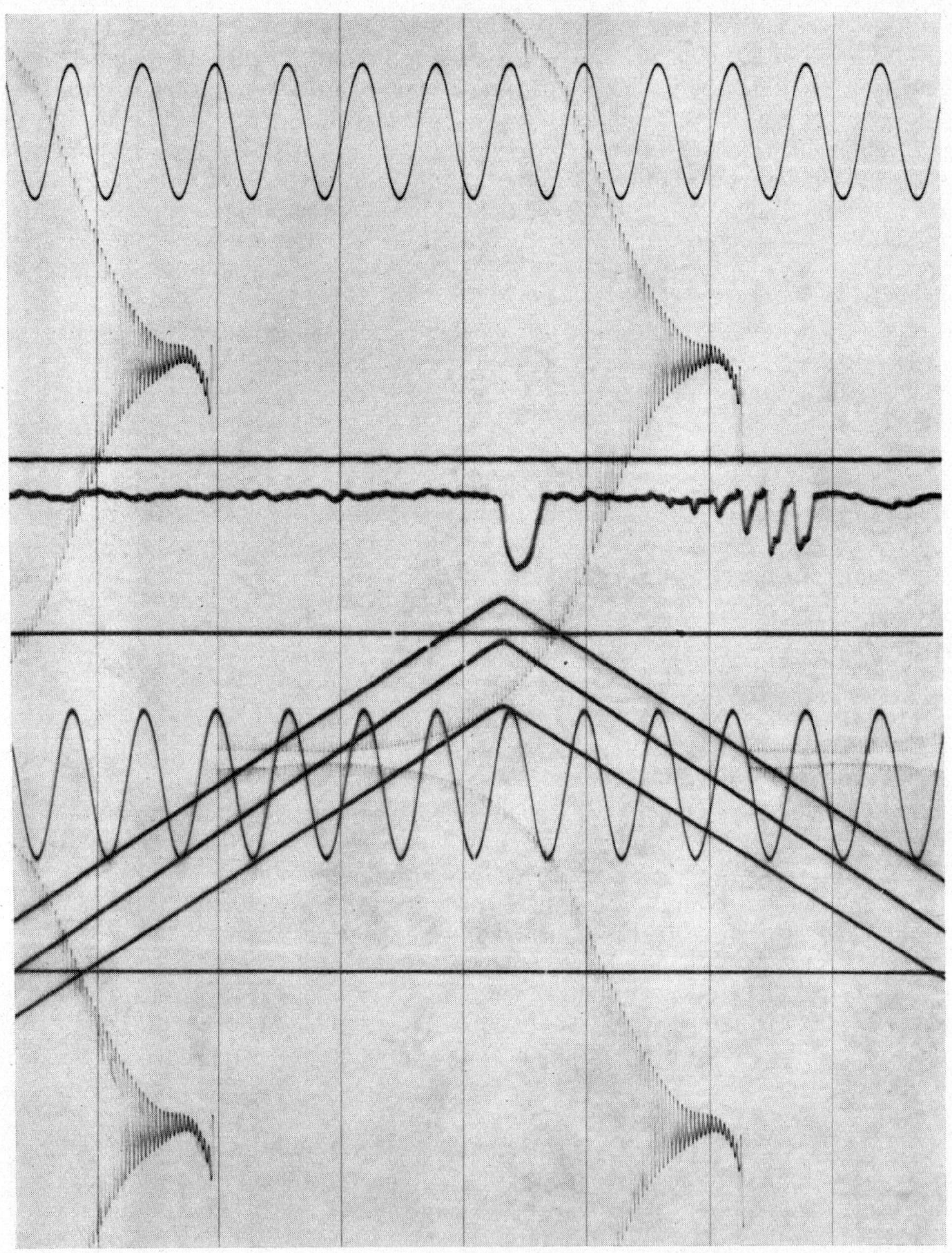

Fig. 11.20. Density differences caused by amplitude and frequency variables.

Copying with a Camera

1) Use recommended procedures for setting up and copying the original. Illumination should be an even intensity and should not cause specular reflections from the surface of the oscillogram.

2) Latensified, unstabilized print-out records may be copied under tungsten illumination under certain conditions. Illumination should not be excessive and exposure for both setup and copying should be held at a minimum. Arc lights, ultraviolet sources, and similar lamps of high actinic output should only be used for copying *stabilized* print-out records (and processed develop-out records, of course).

3) The blue-trace image from print-out charts may be enhanced through the use of Wratten K2, G and similar filters with orthochromatic film. Wratten A filters will produce similar or better results with panchromatic films. In general, if a filter is used that has a color complementary to a given color in the original, it will render that original color darker in the reproduction.

When a filter is used, the exposure must take into account the filter factor for the film being used.

4) Most oscillograms can be copied satisfactorily using a litho type film processed in a relatively low-contrast developer such as DK-50.

Microfilm Duplication

MICROFILMING OSCILLOGRAMS. The storage space required for oscillograms can be greatly reduced by microfilming, thus permitting the destruction of the original, bulky record as soon as it has been copied and verified.

Commercially available microfilm cameras which copy at a nominal 20X reduction, can reduce storage volume by a factor of about 400. At the 20X reduction, a 16 mm film record 100 feet long will contain 2000 square feet of data.

The microfilm can be viewed or a "hard copy" print can be produced with commercially available readers and reader-printers. Readers enlarge the microsized image back to its original size on a screen. Reader-printers enable a certain area of the record to be selected and printed at full size for desk use. Several comprehensive lists of equipment and manufacturers are given in references[11.3, 11.4, 11.5, 11.6]. Because of the large variety of equipment it has been necessary to limit the descriptions to a few items.

3M MICROFILM EQUIPMENT. The Minnesota Mining and Manufacturing Company makes several models of microfilm equipment found in many offices and industries. While these units have not been designed specifically for oscillogram reproduction, they may have some application for this use because they are often available in-house, and are suitable for reproducing short sections of recorder charts. A brief summary of the 3M microfilm product line follows:

1) "Filmsort" 1000d Processor-Camera. This unit is designed to record an 18 by 24 inch document field on 35 mm microfilm at a fixed reduction of 16 diameters. The film is exposed, developed, fixed, washed, dried, and delivered mounted in a Filmsort aperture card in a time cycle of the order of fifty seconds.

2) "Filmac" Reader-Printers. These units produce enlarged prints from microfilm or transparent-film negatives. Developed prints are produced in 8 to 15 seconds.

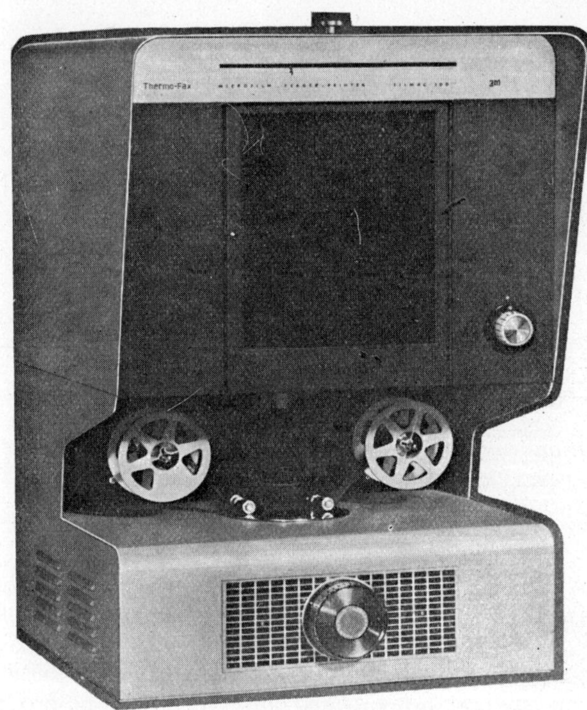

Fig. 11.21. Minnesota Mining and Manufacturing Company's Filmac reader printer

Three models are available, the Filmac "100", "200", and "300". (See Fig. 11.21). The Filmac "100" produces an 8-1/2 by 11 inch print from 16 or 35 mm film and is available with 7, 9, 13, and 26 nominal diameters of magnification.

The Filmac "300" produces prints from 4 by 11 inches continuously variable to 14 by 11 inches from an 11-inch roll stock and 4 by $8\frac{1}{2}$ to $8\frac{1}{2}$ by 14 inches from an 8 $\frac{1}{2}$-inch roll stock. Magnification is continuously variable over two ranges: 7–11.5 and 13–20 diameters.

The Filmac "200" produces either 11 by 18 or 18 by 24 inch prints at a fixed nominal enlargement of 15 diameters.

The 3M printers use an electrolytic process to make a print in about 10 seconds at a cost of approximately seven cents for an $8\frac{1}{2}\times11$.

374

After the desired frame is located and manually positioned on the viewing screen, the viewing mirror is pivoted out of the light path and an automatically timed exposure is made at the touch of a button. The exposure is made on a roll of special coated paper, consisting of a paper base with a thin conductive layer, overcoated with a white photoconductive surface. The conductive layer is electrically

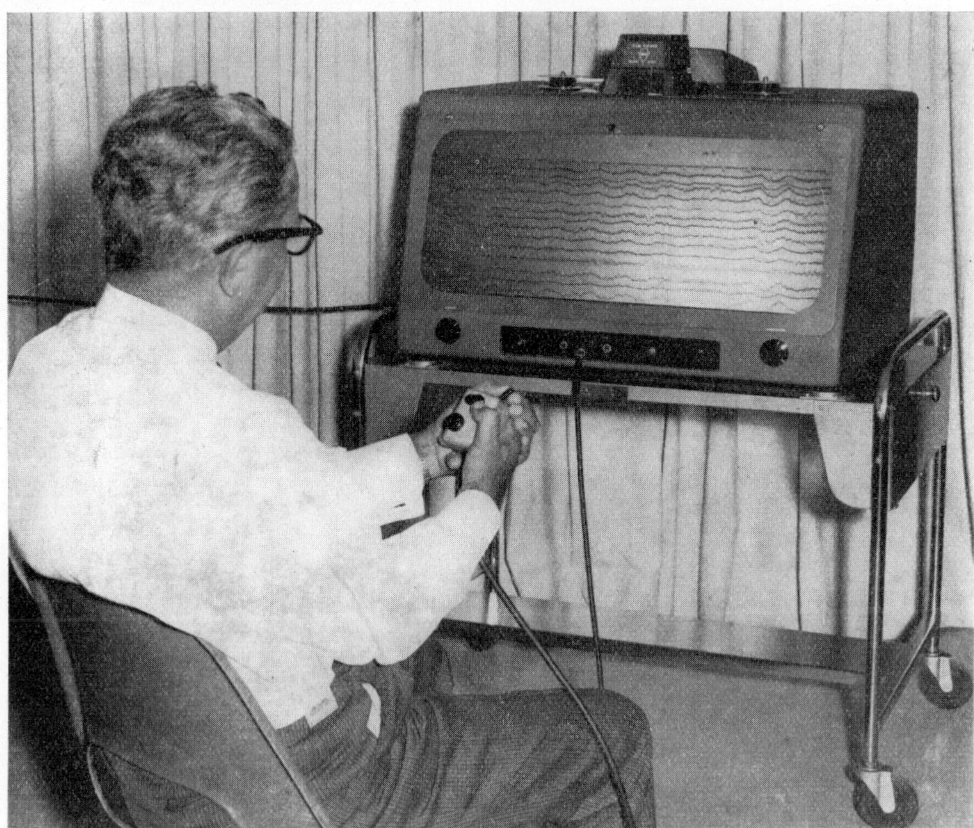

Fig. 11.22. Geotech Model 6585 microfilm viewer

grounded in the machine to a variable, low-voltage dc circuit. Following exposure, the material is transported over a single-solution process station where a slightly acid, electrolytic solution is applied by either a sponge or roller to the paper surface. The processed paper emerges from a delivery slot with the surface slightly moistened.

The principle of the process is that the coated surface becomes conductive when it is exposed to light in the projected image pattern. Thus, when the paper surface is moistened with the electrolytic solution in the activator station, a low-amperage current flows, resulting in the electrochemical deposition of metal from the plating solution. A positive image is produced from a negative, since exposure to light renders the coating conductive and promotes the electrolysis.

GEOTECH MICROFILM VIEWER. The Geotechnical Model 6585 Film Viewer, shown in Fig. 11.22, provides rapid access to microfilm copies of oscillograph records, operating logs, maps, etc. For example, any portion of a microfilm reproduction of a 12-inch wide, 2000 feet oscillograph record, can be displayed at original size in less than 45 seconds. Vernier adjustment of magnification permits an exact setting of original time scale. Edge guiding of microfilm obviates film wear and scratching even with repeated use. Operation is simple, and film reels can be changed quickly.

The viewer features a remote control, wide viewscreen, and motorized drive.

Films are projected at 20 times magnifications on the 27.5 inch by 11.5 inch screen with low distortion and high resolution.

Front panel controls are duplicated on the optional Remote Control. Model 7108 has a 10-foot extension cord for flexibility of operation.

The instrument features a high-speed search mode to enable data of interest to be located quickly. Low-speed traversing permits the detailed study of data.

Magnetic clutches permit instant stopping and reversing of the film at high speeds without damaging the film or drive system.

Separate focus and magnification controls permit a sharp focus at all times. A light-intensity control is also provided.

A cart for the film viewer is available on special order.

BELL AND HOWELL TAB-TRONIC 575 MICROFILM RECORDER. The Tab-Tronic 575 recorder, illustrated in Fig. 11.23, is designed to make high-speed microfilm copies of roll or folded records and other forms. The Model 575 automatically feeds, microfilms, and takes up the original copy at a speed of 70 feet per minute. Daylight-loading 16 mm film in 100 foot rolls copies the record at a 26:1 reduction ratio. Separate supply and take-up compartments permit partially exposed rolls of film to be unloaded. The proper amount of leader and trailer are automatically cycled by a film protection control. Film footage indicator, film compartment door lock, and exposure control are provided. A signal lamp and an audible signal indicate failure of the exposing lamp. End of film or improper film threading are indicated by a separate signal lamp. Paper feed guides are adjustable to accommodate paper widths from 6–1/2 to 15 inches.

The Tab-Tronic 575 measures 42 inches high, 22 inches wide, and 11 inches deep. It weighs 75 pounds and is movable to any location on its non-scarring casters. The unit operates on 115 volt, 60 cycle alternating current.

BELL & HOWELL MICROFILM READER-PRINTERS. The Model 530 D and Model-530 H Reader-Printers make black-and-white prints from microfilm sections view$\frac{1}{2}$ed and selected on the screen. The Model 530 D produces prints measuring 8r by 11 inches, and the Model 530 H produces prints $5\frac{1}{2}$ by $8\frac{1}{2}$ inches. Frame or location selection is speeded by a three-digit indexing meter which aids the operator in finding a specific location. After locating the desired record section to be reproduced, the PRINT button is pressed, and a squeegee-dry print is delivered automatically in 25 seconds.

Five accessory lenses may be interchanged with the lens provided with the reader. The lenses cover magnification ratios from 10.5X to 37X. The condenser can be adjusted with a simple control knob for use with different lenses.

The reader head may be rotated a full 360 degrees, and is equipped with stops every 90 degrees for easy positioning. The spindles will accommodate 100 foot rolls of 8, 16, or 35 mm film. The image is projected on the 11 by 11 inch screen for viewing.

The Bell & Howell 530 Reader-Printers measure 26 by 13 by 30 inches and weigh 68 pounds.

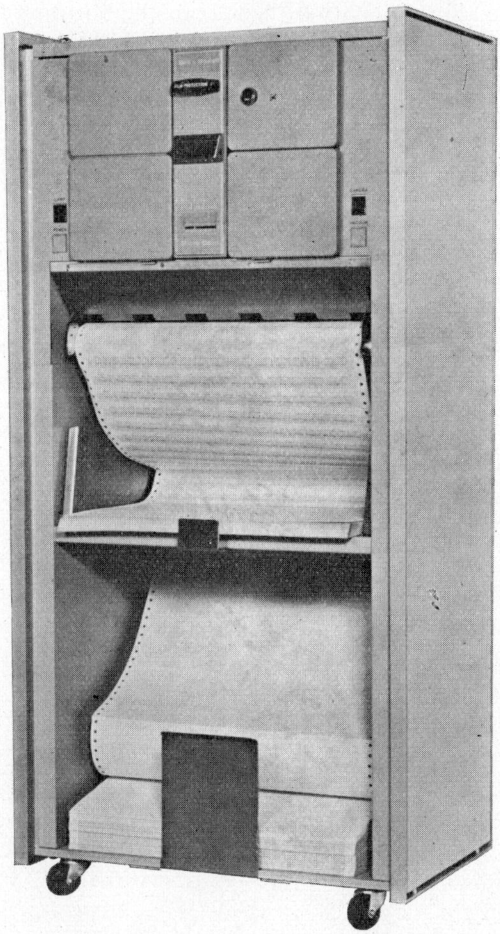

Fig. 11-23. Bell and Howell Tab-Tronic 575 microfilm recorder

BELL & HOWELL RAPID AUTOLOAD MICROFILM READER. The Bell & Howell Rapid Autoload microfilm reader utilizes an autoload cartridge which will accommodate 16 mm unperforated transparent 0.003 inch thickness film in lengths to 170 feet. The cartridge is snapped into position on two spindles over a light source mirror, and can be removed at any time regardless of the amount of film on either core. The film never leaves the cartridge.

The viewer features a two-direction, variable-speed motorized drive controlled by a single knob on the front panel. A zoom lens system provides variable magnification between 20X and 40X, also controlled by a single knob at the front of the reader. The image may be rotated 360 degrees, with click stops every 90 degrees.

A variable intensity control provides three levels of brightness on the translucent, green tinted, 14- by 14-inch viewing screen. Individual frames seen on the screen can be exposed on sensitized paper to provide a reference print.

Index scales can be affixed on either the side or the bottom of the screen for rapid retrieval by index-striped film. An index scale can also be placed directly on the Autoload cartridge, for roll diameter indexing.

The desk-top model measures 26 by 28 by $22\frac{1}{2}$ inches, weighs 95 pounds, and operates from 120 vac power.

LITERATURE REFERENCES

[11.1] HAWKEN, W. R., *The State of the Library Art*, Volume 5, Part 3, Graduate School of Library Service, Rutgers University (New Brunswick, New Jersey) 1960

[11.2] DINSDALE, A., *Photographic Science and Engineering*, 7, 1, page 1, 1963

[11.3] BALLOU, H. W., *Guide to Microreproduction Equipment*, National Microfilm Association (Annapolis, Maryland) 1965

[11.4] PERRY, W. E., *Administrative Management*, 23, 11, page 40, November, 1962

[11.5] VERRY, H. R., *Microcopying Methods*, Focal Press (London, England) 1963

[11.6] *Modern Office Procedures*, Vol. 11, No. 8, p. 27–31, (August 1966)

378

GLOSSARY

Accelerometer — An instrument or device mounted in or on the test specimen to sense accelerative forces and convert them into corresponding electrical quantities, usually for indicating or recording purposes.

Amplification — Increase in voltage or power of a signal. May be expressed as a ratio or, by extension of the term, in decibels.

Antihalation Backing — A coating on the back surface of films that absorbs the image light which reaches the rear surface and prevents its reflection back into the emulsion.

Blue-sensitive — A term applied to photographic materials which are sensitive principally to blue and ultraviolet radiation and have little or no sensitivity to other wavelengths.

Calibrate — To determine (or to correct the variations of) readings of an instrument by measurement or comparison with a standard.

Characteristic Curve — A curve formed by plotting density against log exposure to show the response of a photographic material to varying amounts of light. Sometimes called a D-log E curve or an H and D curve.

Contrast — A term used to describe the separation of tones in a negative or print. The difference between high and low densities, and the ratio of density increments.

Damping — An electromagnetic, fluid, or air retarding force applied against the rotation of a galvanometer coil to improve the fidelity of the recorded data.

Damping, Critical — A value of damping that allows the moving element of a galvanometer to come to the equilibrium position but not pass through it in the shortest time after deflection begins. Also called *aperiodic* or *dead beat* motion.

Damping Resistance — That value of resistance added across the galvanometer so that the apparent source resistance is lowered to the required value.

Densitometer — An instrument used for measuring the density of processed photographic materials.

Density — A numerical measure of the blackening, or light-stopping ability, of a photographic image. It is determined mathematically by: $D = \log \dfrac{1}{T}$ in which T = transmission or reflection expressed as a fraction of incident light.

Density Range — The difference between the measured maximum and minimum densities of a photographic image.

D_{max} — The highest density which can be obtained with a particular photographic material and its processing. When referring to a particular negative or positive, the higheset density recorded.

D_{min} — The lowest density on a positive or negative.

Emulsion — The light-sensitive coating of a photographic material, mainly silver salts suspended in gelatin.

Emulsion Speed — The rate of response of a photographic emulsion to light, determined under standard conditions of exposure and subsequent development. See *exposure index*.

Exposure — The quantity of light, the product of the intensity and its duration which is allowed to act on a photographic material.

Exposure Index — A number assigned to a photographic material that represents its speed and which is used to calculate exposure.

Film — A photographic emulsion coated on a flexible translucent or transparent plastic base.

Filter — A piece of colored transparent glass, gelatin or plastic which absorbs light of certain wavelengths.

Fixing — The process of removing the unexposed and undeveloped silver halide from a negative or print.

Fog — 1. (general), a density caused by the action of chemicals or light, that tends to veil, obscure, or obliterate the image. 2. (sensitometric) any density, not due to light, inherent in the material and its processing.

Frequency Response — The maximum and minimum number of cycles per second an oscillograph or a galvanometer is capable of recording.

Gage — (or gauge) An instrument or means for measuring or testing. By extension, the term is often used synonymously with transducer.

Galvanometer — An instrument for measuring small electric currents by the movements of a coil in a magnetic field. In an oscillograph, the galvanometer is coupled to a mirror which moves with it, reflects light through a lens, and continually exposes a trace on a roll of photorecording paper or film.

Galvanometer Unbalance — The maximum trace deflection at a specified optical arm in inches when the galvanometer is subjected to a 1g. acceleration in any direction.

Gamma (γ) — A numerical designation for the contrast of a photographic material as represented by the slope of the straightline portion of the characteristic curve. The gamma is numerically equal to the tangent of the angle which the straightline portion makes with the base line.

$$\gamma = \frac{\Delta D}{\Delta \log E}$$

Grain — Minute variations of density in a developed photographic emulsion. Caused by irregular distribution of the silver crystals.

Input — 1. The current, voltage, power, or other driving force applied to a circuit or device. 2. The terminals or other points where the electrical force may be applied to a circuit or device.

Intensification — The process of increasing the density and contrast of negatives or prints, usually accomplished by chemical treatment following the conventional development process.

Latent Image — The invisible image recorded by light action upon a photographic material and which is made visible by development.

Linearity — A galvanometer that has a spot deflection directly proportional to the driving current.

Natural Frequency — The frequency of maximum forced amplitude for an undamped vibrating system e.g., a galvanometer movement; also called resonant frequency.

Negative — The image obtained from the original in the conventional photographic process. The tones are the reverse of those in the original subject.

Noise — 1. Any unwanted disturbance within a dynamic electrical or mechanical system. 2. Any unwanted electrical disturbance or spurious signal which modifies the final data or recording.

Nonlinearity — The maximum difference between actual spot position and the theoretical value divided by the rated full-scale deflection, expressed as a percentage.

Orthochromatic — A term applied to photographic materials which are sensitive to green light, in addition to blue light and ultraviolet radiation.

Oscillograph — An instrument which provides a permanent, continuous graphic record of electrical currents fed into it.

Panchromatic — A term applied to photographic materials which are sensitive to light of all visible wavelengths.

Paper-travel Rates — The rate of speed of photographic paper traveling through a magazine. Measured in feet per second (fps) or inches per sec (ips).

Phase Shift — 1. A time difference between the input and output signals of a system. 2. A change in the phase of a periodic quantity.

Positive — A photographic image in which the dark portions of the subject appear dark and the light portions appear light. Usually made from a negative and having tone values opposite to those of the negative.

Process (Processing) — General term for the chemical treatment following exposure of the photographic material (development, fixing, etc.). The term "process" may include the sensitive material and its processing, and often the mode of exposure.

Reciprocity-law Failure — Photographic materials do not attain the same density from an exposure by a high-intensity light source acting for a short time, as from an exposure by a low-intensity light source acting for a longer time, even though the product of time and intensity is the same in both cases. For example, halving the exposure time and doubling the intensity does not result in exactly the same density. The effect is present in all photographic material but varies from one emulsion to another.

Reduction — The lightening of the density of a photographic image by removing silver, usually by a chemical solution called reducer.

Reference Traces — The straight, horizontal lines on a photo-recording paper. These lines are made by galvanometers that are not receiving an input signal. Reference traces are used to measure the amplitude of signals received by galvanometers recording test data (also called grid lines).

Resolving Power — The ability of a photographic emulsion or lens to record fine detail, usually expressed in lines per millimeter. (l/mm).

Safe Current — That value of current that may be continuously passed through the galvanometer without permanently affecting its operating characteristics so as to exceed specification tolerances.

Safelight — A lamp, for use in the darkroom, which emits light of a wavelength to which the photographic material is insensitive for a reasonable time. Different photographic materials require different safelight filters.

Sensitivity — 1. (of galvanometer) the amount deflection per unit of current. 2. (photographic) the amount of blackenning of the photographic material per unit of light.

Sensitometry — The science of measuring the sensitivity and other photographic characteristics of photographic materials and processes.

Sharpness — In photographic materials, the ability to reproduce a sharp edge; of a line, for example.

Shoulder — The portion of a characteristic curve above its straight-line section.

Spectral Sensitivity — The sensitivity of a light-sensitive material (or instrument, such as a photo-electric cell) to radiation of various wavelengths.

Spectrogram — A diagram showing the relative sensitivity of a photographic material to different colors of light. A record showing the relative output of a light source for different colors (wavelengths).

Speed — (See *emulsion speed*)

Straight-line Portion — That section of a characteristic curve which is essentially a straight line. It represents the range of exposures in which the increase in density is proportional to the increase in the logarithm of the exposure.

Telemeter — To transmit electrical signals of measurements to a distant station and there indicating or recording the quantity measured.

Time Arrow — An arrow at the top of a photorecording which shows the direction in which time is increasing.

Time-gamma Curve — A curve which indicates the changes in gamma obtained by changes in time of development.

Timing Lines — The transverse lines on a photorecording. The distance between the lighter, thinner lines usually represents 1/100 or 1/1000 second. While the distance between the heavier lines represents 1/10 or 1/100 second.

Time-temperature Chart — A chart which indicates the development times necessary at various development temperatures to produce approximately the same degree of development that would be obtained at the recommended time and temperature.

Toe — The portion of the charactistic curve below the straight-line section of the curve. It represents the area of minimum useful exposure.

Trace interruptions — The regular breaks in each photorecording trace which identify and separate it from all others.

Transducer — A device that translates physical phenomena, such as pressure, temperature, or movement into electrical currents.

Writing speed — The number of inches per second the spot formed by a beam of light, travels over the photorecording paper.

INDEX